T0212580

Communications
in Computer and Information Science   608

Commenced Publication in 2007
Founding and Former Series Editors:
Alfredo Cuzzocrea, Dominik Ślęzak, and Xiaokang Yang

More information about this series at http://www.springer.com/series/7899

Piotr Gaj · Andrzej Kwiecień
Piotr Stera (Eds.)

Computer Networks

23rd International Conference, CN 2016
Brunów, Poland, June 14–17, 2016
Proceedings

 Springer

Editors
Piotr Gaj
Institute of Informatics
Silesian University of Technology
Gliwice
Poland

Piotr Stera
Institute of Informatics
Silesian University of Technology
Gliwice
Poland

Andrzej Kwiecień
Institute of Informatics
Silesian University of Technology
Gliwice
Poland

ISSN 1865-0929   ISSN 1865-0937 (electronic)
Communications in Computer and Information Science
ISBN 978-3-319-39206-6   ISBN 978-3-319-39207-3 (eBook)
DOI 10.1007/978-3-319-39207-3

Library of Congress Control Number: 2016939355

Printed on acid-free paper

This Springer imprint is published by Springer Nature
The registered company is Springer International Publishing AG Switzerland

Preface

Computer networks today have a significant impact on our life. In our technological civilization it is impossible to run without networks, and this applies to professional activities as well as private ones. Computer networks are the part of computer science whose development not only affects the other existing branches of technical science but also contributes to the development of completely new areas. Thus, the domain of computer networks has become one of the most important fields of research.

Computer networks as well as the entire field computer science are the subject of constant changes caused by the general development of technologies and by the need for innovation in their applications. This results in a very creative and interdisciplinary interaction between computer science technologies and other technical activities, and directly leads to perfect solutions. New methods, together with tools for designing and modeling computer networks, are regularly extended. Above all, the essential issue is that the scope of computer network applications is increased thanks to the results of new research and to new application proposals appearing regularly. Such solutions were not even taken into consideration in the past decades. Whereas recent applications stimulate the progress of scientific research, the extensive use of new solutions leads to numerous problems, both practical and theoretical.

23rd International Science Conference *Computer Networks*

This book collates the research work of scientists from numerous notable research centers. The chapters refer to the wide spectrum of important issues regarding the computer networks and communication domain. It is a collection of topics presented at the 23rd edition of the International Conference on Computer Networks. The conference was held in Brunów Palace, located in Brunów – a small village near Lwówek Śląski, Poland, during June 14–17, 2016. The conference, organized annually since 1994 by the Institute of Informatics of Silesian University of Technology together with the Institute of Theoretical and Applied Informatics in Gliwice, is the oldest event of its kind in Poland. The current edition was the 23rd such event, and the international status of the conference was attained eight years ago, with the ninth international edition taking place in 2016. Just like previous events in the series, the conference took place under the auspices of the Polish section of IEEE (technical co-sponsor). Moreover, the conference partner was iNEER (International Network for Engineering Education and Research).

In 2016 the total number of submitted conference papers was 72. The presented papers were accepted after careful reviews made by at least three independent reviewers in a double-blind way. The acceptance level was 50 %, and thus the proceedings contain 36 papers, including 32 full papers and four short ones. The chapters are organized thematically into several areas within the following tracks:

- Computer Networks
  This group of papers is the largest one. General issues of networks architecture, analyzing, modeling, and programming are covered. Moreover, topics on wireless systems and wireless sensor networks, security concerns, Internet technologies, SDN, WSN, CPN, and industrial networks modeling and analysis, among others, are included.
- Teleinformatics and Communications
  This section contains topics on load balancing in LTE technology, security of speaker verification, and analysis of USB 3.1 delays.
- New Technologies
  New technologies in computer networks refers to topics on quantum network protocols, quantum direct communication, and multilevel virtualization in cloud computing.
- Queueing Theory
  The domain of queueing theory is usually one of the most strongly represented areas at the Computer Network conference. Several papers are included, e.g., a paper on developing a confidence estimation of the stationary measures in high-performance multiservers, an article on new a multidimensional Erlang's ideal grading model with queues, a contribution on queueing models with a contingent additional server while considering two-server queueing systems with a finite buffer, a paper on the usage of Markov chains in modeling the queues inside IP routers, and an article on a dual tandem queue consisting of two multiserver stations without buffers.
- Innovative Applications
  The papers in this section refer to research in the area of innovative applications of computer networks theory and facilities. There are contributions on innovative usage of cloud computing systems, frameworks for integration of IT structures and for various home devices, social network analysis, alleviation of network uncertainty in networked control systems, localization of a radio wave source, and a method of creating a signal classifier.

Each group includes highly stimulating studies that may interest a wide readership.

In conclusion, on behalf of the Program Committee, we would like to express our gratitude to all authors for sharing their research results as well for their assistance in developing this volume, which we believe is a reliable reference in the computer networks domain.

We also want to thank the members of the Technical Program Committee for their participation in the reviewing process.

April 2016   Piotr Gaj
                                                 Andrzej Kwiecień

Organization

CN 2016 was organized by the Institute of Informatics from the Faculty of Automatic Control, Electronics and Computer Science, Silesian University of Technology (SUT) and supported by the Committee of Informatics of the Polish Academy of Sciences (PAN), Section of Computer Network and Distributed Systems in technical co-operation with the IEEE and consulting support of the iNEER organization.

Executive Committee

All members of the Executive Committee are from the Silesian University of Technology, Poland.

Honorary Member:	Halina Węgrzyn
Organizing Chair:	Piotr Gaj
Technical Volume Editor:	Piotr Stera
Technical Support:	Aleksander Cisek
	Jacek Stój
Office:	Małgorzata Gładysz
Web Support:	Piotr Kuźniacki

Co-ordinators

PAN Co-ordinator:	Tadeusz Czachórski
IEEE PS Co-ordinator:	Jacek Izydorczyk
iNEER Co-ordinator:	Win Aung

Program Committee

Program Chair

Andrzej Kwiecień	Silesian University of Technology, Poland

Honorary Members

Win Aung	iNEER, USA
Adam Czornik	Silesian University of Technology, Poland
Andrzej Karbownik	Silesian University of Technology, Poland
Bogdan M. Wilamowski	Auburn University, USA

Technical Program Committee

Omer H. Abdelrahman	Imperial College London, UK
Anoosh Abdy	Realm Information Technologies, USA
Iosif Androulidakis	University of Ioannina, Greece
Tülin Atmaca	Institut National de Télécommunication, France

Rajiv Bagai	Wichita State University, USA
Zbigniew Banaszak	Warsaw University of Technology, Poland
Robert Bestak	Czech Technical University in Prague, Czech Republic
Leszek Borzemski	Wrocław University of Technology, Poland
Markus Bregulla	University of Applied Sciences Ingolstadt, Germany
Ray-Guang Cheng	National University of Science and Technology, Taiwan
Andrzej Chydziński	Silesian University of Technology, Poland
Tadeusz Czachórski	Silesian University of Technology, Poland
Andrzej Duda	INP Grenoble, France
Alexander N. Dudin	Belarusian State University, Belarus
Peppino Fazio	University of Calabria, Italy
Max Felser	Bern University of Applied Sciences, Switzerland
Holger Flatt	Fraunhofer IOSB-INA, Germany
Jean-Michel Fourneau	Versailles University, France
Janusz Furtak	Military University of Technology, Poland
Rosario G. Garroppo	University of Pisa, Italy
Natalia Gaviria	Universidad de Antioquia, Colombia
Erol Gelenbe	Imperial College London, UK
Roman Gielerak	University of Zielona Góra, Poland
Mariusz Głąbowski	Poznan University of Technology, Poland
Adam Grzech	Wrocław University of Technology, Poland
Edward Hrynkiewicz	Silesian University of Technology, Poland
Zbigniew Huzar	Wrocław University of Technology, Poland
Jacek Izydorczyk	Silesian University of Technology, Poland
Jürgen Jasperneite	Ostwestfalen-Lippe University of Applied Sciences, Germany
Jerzy Klamka	IITiS Polish Academy of Sciences, Gliwice, Poland
Zbigniew Kotulski	Warsaw University of Technology, Poland
Demetres D. Kouvatsos	University of Bradford, UK
Stanisław Kozielski	Silesian University of Technology, Poland
Henryk Krawczyk	Gdańsk University of Technology, Poland
Wolfgang Mahnke	ABB, Germany
Francesco Malandrino	Politecnico di Torino, Italy
Aleksander Malinowski	Bradley University, USA
Kevin M. McNeil	BAE Systems, USA
Jarosław Miszczak	IITiS Polish Academy of Sciences, Poland
Vladimir Mityushev	Pedagogical University of Cracow, Poland
Diep N. Nguyen	Macquarie University, Australia
Sema F. Oktug	Istanbul Technical University, Turkey
Michele Pagano	University of Pisa, Italy
Nihal Pekergin	Université de Paris, France
Piotr Pikiewicz	College of Business in Dąbrowa Górnicza, Poland
Jacek Piskorowski	West Pomeranian University of Technology, Poland
Bolesław Pochopień	Silesian University of Technology, Poland

Oksana Pomorova	Khmelnitsky National University, Ukraine
Silvana Rodrigues	Integrated Device Technology, Canada
Vladimir Rykov	Russian State Oil and Gas University, Russia
Alexander Schill	Technische Universität Dresden, Germany
Akash Singh	IBM Corp., USA
Mirosław Skrzewski	Silesian University of Technology, Poland
Tomas Sochor	University of Ostrava, Czech Republic
Maciej Stasiak	Poznan University of Technology, Poland
Janusz Stokłosa	Poznan University of Technology, Poland
Zbigniew Suski	Military University of Technology, Poland
Bin Tang	California State University, USA
Kerry-Lynn Thomson	Nelson Mandela Metropolitan University, South Africa
Oleg Tikhonenko	Częstochowa University of Technology, Poland
Arnaud Tisserand	IRISA, France
Homero Toral Cruz	University of Quintana Roo, Mexico
Leszek Trybus	Rzeszów University of Technology, Poland
Adriano Valenzano	National Research Council of Italy, Italy
Bane Vasic	University of Arizona, USA
Peter van de Ven	Eindhoven University of Technology, The Netherlands
Miroslaw Voznak	VSB-Technical University of Ostrava, Czech Republic
Krzysztof Walkowiak	Wrocław University of Technology, Poland
Sylwester Warecki	Intel, USA
Jan Werewka	AGH University of Science and Technology, Poland
Tadeusz Wieczorek	Silesian University of Technology, Poland
Józef Woźniak	Gdańsk University of Technology, Poland
Hao Yu	Auburn University, USA
Grzegorz Zaręba	University of Arizona, USA
Zbigniew Zieliński	Military University of Technology, Poland
Piotr Zwierzykowski	Poznan University of Technology, Poland

Referees

Omer H. Abdelrahman	Holger Flatt	Jerzy Klamka
Iosif Androulidakis	Jean-Michel Fourneau	Zbigniew Kotulski
Tülin Atmaca	Janusz Furtak	Stanisław Kozielski
Rajiv Bagai	Rosario G. Garroppo	Henryk Krawczyk
Zbigniew Banaszak	Natalia Gaviria	Andrzej Kwiecień
Robert Bestak	Erol Gelenbe	Wolfgang Mahnke
Ray-Guang Cheng	Roman Gielerak	Aleksander Malinowski
Andrzej Chydziński	Mariusz Głąbowski	Jarosław Miszczak
Tadeusz Czachórski	Adam Grzech	Vladimir Mityushev
Andrzej Duda	Edward Hrynkiewicz	Diep N. Nguyen
Alexander N. Dudin	Zbigniew Huzar	Sema F. Oktug
Peppino Fazio	Jacek Izydorczyk	Michele Pagano
Max Felser	Jürgen Jasperneite	Nihal Pekergin

Piotr Pikiewicz   Zbigniew Suski   Krzysztof Walkowiak
Jacek Piskorowski   Bin Tang   Sylwester Warecki
Oksana Pomorova   Kerry-Lynn Thomson   Jan Werewka
Vladimir Rykov   Oleg Tikhonenko   Tadeusz Wieczorek
Alexander Schill   Arnaud Tisserand   Józef Woźniak
Akash Singh   Homero Toral Cruz   Hao Yu
Mirosław Skrzewski   Leszek Trybus   Grzegorz Zaręba
Tomas Sochor   Adriano Valenzano   Zbigniew Zieliński
Maciej Stasiak   Peter van de Ven   Piotr Zwierzykowski
Janusz Stokłosa   Miroslaw Voznak

Sponsoring Institutions

Organizer

Institute of Informatics, Faculty of Automatic Control, Electronics and Computer Science, Silesian University of Technology

Co-organizer

Committee of Informatics of the Polish Academy of Sciences, Section of Computer Network and Distributed Systems

Technical Co-sponsor

IEEE Poland Section

Technical Partner

iNEER

Contents

Queueing Theory

Innovative Applications

Computer Networks Architectures
and Protocols

Real-Time Traffic over the Cognitive Packet Network

Lan Wang[✉] and Erol Gelenbe[✉]

Intelligent Systems and Networks Group, Department of Electrical and Electronic
Engineering, Imperial College London, London, UK
{lan.wang12,e.gelenbe}@imperial.ac.uk

Abstract. Real-Time services over IP (RTIP) have been increasingly
significant due to the convergence of data networks worldwide around
the IP standard, and the popularisation of the Internet. Real-Time applications have strict Quality of Service (QoS) constraint, which poses a
major challenge to IP networks. The Cognitive Packet Network (CPN)
has been designed as a QoS-driven protocol that addresses user-oriented
QoS demands by adaptively routing packets based on online sensing and
measurement, and in this paper we design and experimentally evaluate the "Real-Time (RT) over CPN" protocol which uses QoS goals that
match the needs of real-time packet delivery in the presence of other
background traffic under varied traffic conditions. The resulting design
is evaluated via measurements of packet delay, delay variation (jitter)
and packet loss ratio.

Keywords: Cognitive Packet Network · Real-Time over CPN

1 Introduction

Communication services such as telephone, broadband and TV are increasingly
migrating into the IP network with the worldwide deployment and operation
of Next-Generation Networks which consolidate mobile and data networks into
common IP infrastructures [1,2]. Applications such as remote security [3] and the
needs of cyber-physical systems [4] are also pushing the bounds in this direction.
Although the convergence of network services enables the transport of real-time,
video, voice and data traffic over IP networks which were originally designed
to offer best-effort services for non-real-time data traffic, our IP networks are
not well designed to guarantee Quality of Service (QoS) for such diverse communications [5], including real-time (RTIP), voice (VOIP) [6] and web based
search, video and multimedia [7–10]. Real-time constraints [11] could be well
assured in dedicated circuit-switched connections, but are difficult to cover with
IP networks [12].

Furthermore, industrial distributed systems and industrial networks [13–16]
are pushing the curve with respect to real-time communication needs. Real-time
needs for enhanced reality and telepresence [17] are also creating specific needs in

© Springer International Publishing Switzerland 2016
P. Gaj et al. (Eds.): CN 2016, CCIS 608, pp. 3–21, 2016.
DOI: 10.1007/978-3-319-39207-3_1

on-line QoS management for multimedia systems [18]. Thus in the past decade, many QoS approaches have been proposed including IntServ &RSVP, DiffServ and MPLS, which specify a coarse-grained mechanism for providing QoS, but fully satisfactory results are not yet available in this area and there is space for further investigation and research [19–21].

Thus this paper investigates the use of the Cognitive Packet Network (CPN) routing protocol for real-time traffic. CPN is a QoS-driven routing protocol based on a smart network where users (applications) can declare their desired QoS requirements, while CPN adaptively routes their traffic by means of online sensing and measurement so as to provide the best possible QoS requested by the users [22,23], as a means to support and enhance the users' QoS needs. To address the needs of real-time traffic, this paper introduces a variant of CPN with a "goal" function that addresses real-time needs with regard to "Jitter" according to RFC3393 [24], and implements new functions in CPN to support multiple QoS classes for multiple traffic flows [25]. Furthermore, since adaptive routing takes place when CPN is used in order to meet some of the requirements of this QoS goal, path switching can occur and we evaluate the correlation of end-to-end packet loss with path switching and with the delay that occurs at the end-user's re-sequencing buffer that is needed to insure that packets are received in time-stamp order. The main experimental result of this paper is that using Jitter Minimisation as the QoS goal for routing all of the traffic flows in the network, we can minimize jitter but also minimize delay and loss for real-time traffic. However we also see that adaptive schemes that switch paths to minimize delay, loss or jitter, may also cause buffer overflow and hence losses in the output re-sequencing buffers that deliver packets in time-stamp order for real-time traffic.

1.1   The Cognitive Packet Network Protocol

In CPN, QoS requirements specified by users, such as *Delay, Loss, Energy Consumption*, or a combination of the above, are incorporated in the "goal" function which is used for the CPN routing algorithm. Three types of packets are used by CPN: smart packets (SPs), dumb packets (DPs) and acknowledgments (ACKs). SPs explore the route for DPs and collect measurements; DPs carry payload and also conduct measurements; ACKs bring back the information that has been discovered by the SPs and DPs including route information and measurement results. SPs discover the route using random neural network (RNN)-based reinforcement learning (RL) [26,27] which resides in each node. Each RNN in a node corresponds to a QoS class and destination pair and each neuron of a RNN represents the choice to forward a given packet over a specific outgoing link from the node. The arrival of an SP for a specific QoS class at a node triggers the execution of the RNN-based algorithm.

During this process, the weights of the corresponding RNN are updated by Reinforcement Learning (RL) using QoS goal-based measurements that are collected by SPs and brought back by ACKs and stored in a mailbox at the node.

Following that, the output link corresponding to the most excited neuron is chosen as the routing decision. More detail on CPN routing is available in [22]. The paths explored by the SPs for the desired QoS goal are brought back to the source node by an ACK packet and stored in a list of possible paths. Among these, DPs will select one which is recent and has the best QoS goal; as DPs are sent forward, they will also create returning ACKs which can be used to update the performance (e.g. end-to-end delay or jitter) that is needed by the application or user that is forwarding the packets.

1.2   CPN for Real-Time Traffic

Previous research has suggested the ability of CPN to provide improved QoS for real-time traffic compared with the conventional Internet protocol such as OSPF [28]. This paper presents an experimental study specifically for Real-Time service over CPN, the principal building block of which is shown in Fig. 1.

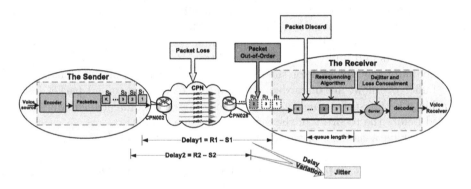

Fig. 1. RTIP system structure [29]

In Real-Time services that are important in a variety of applications, including industrial control, media transmission and real-time, the traffic sent from a source must not only arrive with a small delay, but it must also arrive with small variance not to disrupt the needs of the control application or the media receiver. However, very importantly, the end receiver must receive the sender's packets *in the order in which they are sent.*

Indeed in most control applications the order in which the control messages are received has great importance for industrial control [30,31]. Similarly, any measurements that are sent to a controller from different parts of a distributed system, have to be received in time-stamp sending order. This is similar for media where video must arrive in the order that was prescribed by the sender, and for real-time conversations where the meaning of a conversation cannot be preserved if the packets arrive in some other than time-stamp order. Note that for data packets in file transfers, this aspect is of no great importance since the packets can be reassembled in a receiving file system. Because of the human delays at the terminal, email also does not have such stringent constraints on the order of packet arrival. Thus, at the sender which resides in a RTIP application installed at a CPN node, the original analog measurement signals are sampled with a fixed frequency, which is commonly 8000 Hz if real-time standards are being used, and then each sample is encoded by using different standards (e.g. G.711, G.729, G.729a, G.723.1, G.726, G.722, G.728, for audio) which differ in the compression algorithms and the resulting bitstream bandwidth. The encoded bitstream is packetised into IP packets by adding RTP header, UDP header, and IP header.

These packets can then be transmitted across the IP network employing the CPN protocol, where IP-CPN conversion is performed at the source node by encapsulating IP packets into CPN packets which are routed based on their QoS requirements. Due to the shared nature of the IP network, real-time traffic may undergo transmission impairments including delay, jitter, packet de-sequencing and packet loss. CPN can alleviate these impairments by smartly selecting the path that provides the best possible QoS required by the user (or an application) [32] and thus the packets in a real-time traffic flow may traverse the CPN network successively along several different paths. At the receiver, packets are queued in a buffer called the re-sequencing buffer. The purpose of the buffer is to reorder packets and buffer them to reduce jitter. Packets that arrive later than the maximum time allowed for the real-time signal recovery, or those that pro- voke buffer overflow, are discarded, contributing to the end-to-end packet loss. To achieve better speech quality, several packet loss concealment techniques are applied before the recovery of the original real-time signals.

To provide a satisfactory level of QoS we have developed a "goal" func- tion which aims at "Jitter" according to its definition in the real-time pro- tocol RFC3393. We implemented new functions for CPN to support multiple QoS classes for multiple traffic flows, whereby different traffic flows (users) can declare their desired QoS goals before initiating a communication session at the same source node and then each flow is routed based on its own QoS criteria. Then, experiments were conducted to examine which QoS goal is better for RT packet delivery in a multiple traffic environment under varied traffic conditions via measurements of delay, delay variation (jitter) and loss.

Furthermore, we also consider packet end-to-end loss which consists of loss occurring within the network, plus the packet discards happening inside the RTIP receiver which are affected by path switching induced by the adaptive nature of CPN. The correlation of packet end-to-end loss and path switching were carefully studied by an off-line analysis of packet traces from the CPN testbed network, together with a discrete event simulation of the behaviour of the re-sequencing buffer that resides in the RTIP receiver.

The main results of this paper are based on the measurements we conduct on the use of CPN to support RTIP. In particular, our work results in interesting insights which are counterintuitive:

- Our measurements compare the use of Delay and Jitter Minimisation as a means to offer QoS to RTIP connections. These measurements clearly show that using Jitter Minimisation as the QoS goal for routing all of the traffic flows in the network, namely the real-time traffic and the background (other) traffic, will not only minimise jitter but also delay and loss for real-time traffic.

– Furthermore we see that packet loss in the network is clearly correlated with path switching that is induced by the adaptive scheme that is inherent to CPN. However we also see that when path switching does not occur sufficiently often, peaks in packet loss can occur because *all the traffic* including the background non-RTIP traffic, will head for paths which are viewed to be good, resulting in unwanted congestion and hence loss, which in turn can only be mitigated by switching to less loaded paths and parts of the network.

– Finally, packet loss that can be provoked by path switching and congestion in a network, also results in further buffer overflows and hence further losses in the output re-sequencing buffers of the RTIP codecs which comes on top of the losses in the earlier stages of the network.

2   Real-Time Traffic over CPN for Multiple QoS Classes

As real-time needs are sensitive to the time-based QoS metrics "delay" on the one hand, and "delay variation" on the other, our experiments will consider both of these QoS classes. In fact we will allow the foreground or primary traffic class to have one or the other of these two QoS objectives, and similarly the background traffic will have one or the other of them as well. Thus our experiments on the CPN test-bed were conducted with real-time traffic and one of the two QoS requirements between arbitrary source-destination pair within the CPN testbed network, in the simultaneous presence of several background traffic flows with the same or the other QoS goal. The measurements we conducted were then used to see which of the QoS goals in effect provided better QoS for the real-time traffic and under which conditions of background traffic this was actually happening. Thus this section presents the implementation of the goal function for the QoS criterion of "Jitter", as well as the implementation of CPN that supports multiple QoS classes for multiple traffic flows simultaneously. We note that in previous experiments reported with CPN, all flows used the same QoS goal so that this change has required some significant modifications to the CPN software that is installed at each node.

2.1   Real-Time Packets over CPN with QoS Class Jitter

In RFC3393 and RFC5481, "Packet Delay Variation" is used to refer to "Jitter". One of the specific formulations of delay variation implemented in the industry is called Instantaneous packet delay variation ($IPDV$) which refers to the difference in packet delay between successive packets, where the reference is the previous packet in the stream's sending sequence so that the reference changes for each packet in the stream.

The measurement of $IPDV$ for packets consecutively numbered $i = 1, 2, 3, \ldots$ is as follows. If S_i denotes the departure time of the i-th packet from the source node, and R_i denotes the arrival time of the i-th packet at the destination node, then the one-way delay of the i-th packet $D_i = R_i - S_i$, and $IPDV$ is

$$IPDV_i = |D_i - D_{i-1}| = |(R_i - S_i) - (R_{i-1} - S_{i-1})| \, . \tag{1}$$

To fulfill the QoS goal of minimising jitter, online measurement collects the jitter experienced by each DP. Since in CPN each DP carries the time stamp of its arrival instant at each node along its path, so when a DP say DP_i arrives at the destination, an ACK is generated with the arrival time-stamp provided by the DP, and as ACK_i heads back along the inverse path of the DP, and at each node the forward delay $Delay_i$ is estimated from this node to the destination by taking the difference between the current arrival time at the node and the time at which the DP_i reached the same node [32], divided by two. This quantity is deposited in the mailbox at the node. The instantaneous packet delay variation is computed as the difference between the value of $Delay_i$ and $Delay_{i-1}$ of the previous packet in the same traffic flow as in (1), and jitter is approximated by the smoothed exponential average of $IPDV$ with factor a smoothing factor 0.5:

$$\overline{J_i} = \frac{J_{i-1}}{2} + \frac{J_i}{2} \, . \tag{2}$$

Then, the corresponding value of jitter in the node's mailbox is also updated. When a subsequent SP for the QoS class of Jitter and the same destination enters the node, it uses data from the mailbox to compute the reward $Reward_i$ and then trigger the execution of the RNN which corresponds to the QoS class of Jitter and the destination to decide the outgoing link [32].

$$Reward_i = \frac{1}{\overline{J_i} + \epsilon} \tag{3}$$

where ϵ is used to ensure the denominator is non-zero.

2.2   Implementation of CPN Supporting Multiple QoS Classes

We enable CPN to support multiple QoS classes simultaneously, for multiple flows that originate at any node and each flow is routed based on its specific QoS criteria, we use the following components:

- The Traffic Differentiation can rely on source MAC (or IP), destination MAC (or IP) and the TCP/UDP port of the applications. For instance, the RTIP application "Linphone" has its dedicated SIP port (5060) and audio port (7080), which resides in the fields of the UDP header so that real-time traffic can be differentiated.
- The QoS Class Assignment can be defined according to the QoS requirements of different users or applications, and is stored in a configuration file which is loaded into the memory while CPN is being initiated.

– We can then treat each traffic flow according to its QoS requirement using multiple RNNs at each node, where each RNN corresponds to a QoS class and a source-destination pair. For instance, a real-time traffic flow with a specified QoS class (eg. forwarding delay or jitter) traveling across CPN corresponds to an RNN at each node along its selected path. ACKs coming back from the DPs of the real-time flow are used to collect the measurements of the specific QoS metric of the real-time flow itself and deposit the measurement results in the corresponding mailbox at each node on their way back to the source. SPs decide the outgoing link at each node they arrive by selecting the most excited neuron in the RNN based on the measurement result in the mailbox.

3 Measurement Methodology for Real-Time Packet Path Switching, Reordering and End-to-End Loss

CPN adaptively selects the path that provides best possible QoS requested for traffic transmission, leading to possible path switches so that traffic may suffer packet de-sequencing and loss. Various techniques for mitigating this effect have been demonstrated, such as the use of a switching probability that avoids simultaneous conflicting path switches among distinct flows that share some common routers, or the use of a minimum "improvement threshold" so that path switches occur only if the expected improvement in QoS is sufficiently large [28]. Accordingly, we are interested in examining the correlation between undesirable effects such as packet de-sequencing and end-to-end loss, and path switching. In the following sections, we described methods to carry out measurements and statistics for the three metrics. The measurements based on the off-line analysis of packet data which were captured by running "TCPDUMP" at the source and destination node so as to obtain the most detailed information of each packet.

3.1 Packet Path Switching

For a given flow, the path traversed by packets may change from time to time so as to satisfy the specific QoS requirement. This routing information provided by the corresponding SP is encapsulated into the cognitive map field of the DP while as it originates from the source node. Thus, the path used by each DP can be detected by extracting its routing information field after being captured at the source node. The metric we are interested in is the path switching ratio, which is defined as:

$$Ratio_{\text{path}} = \frac{Q_{\text{path}}}{N} \qquad (4)$$

where Q_{path} is the number of path switches in a given flow during the time interval being considered, and N is the total number of packets forwarded in that time interval. We can also define the path switching rate as:

$$Rate_{\text{path}} = \frac{Q_{\text{path}}}{T} \qquad (5)$$

where T is the length of the time interval.

3.2 Packet Reordering

Packet reordering or re-sequencing is an important metric for real-time because packets have to be forwarded to the end user sequentially at the receiver in the same order that they have been sent, and those that arrive later than the required maximum will have to be discarded. When packets travel over multiple paths, packets sent later may actually arrive at the receiver before their predecessors, so that to obtain a fully correct playback, packets have to be stored until all their predecessors have arrived. However the reordering buffer at the receiver is necessarily of finite length, so that packets arriving to a buffer that is full will be discarded, and packets will have to be forwarded after a given time-out even when their predecessors have not arrived in order to avoid excessive time gaps with their predecessors that have already been played back. Packet reordering is done according to the recommendation from [20], which is based on the monotonic ordering of sequence numbers. Specifically, we add a 4-byte field in the CPN header to store the unique sequence identifier which is incremented by 1 each time a new packet is sent into a traffic flow. In addition, the sequence number residing in the Real-Time Protocol (RTP) header can also be used as the identifier, which is strictly monotonically increasing with increments of "320". To detect packet reordering, at the receiver we reproduce the sender's identifier function. The Next Expected Identifier is determined by the most recently received in-order packet plus the increment (denoted by Seq_{inc}) and denoted by $NextExp$. Thus for CPN the increment is "1", while for the RTP sequence number its "320". If S is the identifer of the currently arrived packet at the receiver, if $S < NextExp$, the packet is reordered and the number of reordered packet $Q_{\mathrm{reordered}}$ is incremented by 1, else the packet is in-order and $NextExp$ is updated to $S + Seq_{\mathrm{inc}}$. For example, if the packets arrive with the identifiers $1, 2, 5, 3, 4, 6$, *packet* 3 and *packet* 4 will be reordered.

To quantify the degree of de-sequencing, we also defined the "Packet Reordering Ratio/Rate" similar to (4) and (5), and the "Packet Reordering Density" denoted by $Density_r$, so that we may differentiate between isolated and bursty packet reordering as well as to measure the degree of burstiness of packet reordering, which may affect the packet drop rate of the re-sequencing buffer at the receiver. $Density_r$ is calculated as:

$$Density_r + = \begin{cases} Cout_r^2 & \text{for} \quad \text{bursty packet reordering} \\ Cout_r & \text{for} \quad \text{isolated packet reordering} \end{cases} \tag{6}$$

where $Cout_r$ is the number of successively reordered packets; it resets to zero when the in-order packets arrive and is incremented when reordering occurs.

3.3 Packet End-to-End Loss

The recommendation in [19] states that packet loss should be reported "separately on packets lost in a network, and those that have been received but then discarded by the jitter buffer" at the receiver for real-time packet delivery, because both have an equal effect on the quality of real-time services. The combination of packet loss and packet discard is usually called packet end-to-end loss. In this section, we present the approaches we use to study these two metrics.

Here, packet loss will only refer to the packets lost in the network, as detected for a packet that is sent out but not received by its destination node. Since the sending packets are consecutively numbered by monotonically increasing identifiers at the source while entering CPN Sect. 3.2, for sent packets falling within the test interval, each packet received at the destination node will be matched with the combination of packet identifier, the source and destination IP address. Non-matched sent packets are identified as the lost packets, so that the timestamp of each lost packet and the number of lost packets are obtained. The packet loss Ratio/Rate is a commonly-used metric to quantify the degree of loss as in (4) and (5).

At the receiver in the RTIP application at CPN node, the received packets are stored in the "jitter buffer" which reorders packets and buffers them to reduce jitter. Packets that arrive later than the expected time which is required for the real-time signal recovery, and those that cause buffer overflow, will be discarded, contributing to the end-to-end loss. To achieve better speech quality, RTIP applications provide several packet loss concealment approaches to compensate packet loss before real-time playback. Thus, packet discard cannot be measured inside a RTIP application, and we cannot directly access the run-time version of the RTIP application. Accordingly, we have had to simulate the operation of a jitter buffer which employs resequecing so as to study packet discards and the buffer queue length, and their correlation with packet reordering and packet loss.

The discrete event simulation approach we use simulates the Arrival Events which correspond to the arrival of packets at the receiver, the Departure Events which indicate the departure of packets after being processed by the server, and the Wait Event. Due to the real-time demand of real-time traffic, the waiting time of packets that are delayed for re-sequencing in the buffer will have an upper bound. Each of these events are stored in the Future Event List (FEL) and are executed in time sequence. The queue used in the simulation consists of a waiting queue $LQ_{\text{reordered}}(t)$ for packets waiting for re-sequencing, and a processing queue $LQ(t)$, for packets waiting to be delivered to the end user. The event scheduling process which is integrated with the Re-sequencing Algorithm is as follows.

To capture a packet trace, we collect the packets in a real-time stream received at the CPN destination node during a test interval. The arrival of each packet corresponds to an arrival event scheduled in the FEL with its arrival time. It is assumed that the service time of the server follows an exponential distribution with average value 1 ms. We define $WT_{\text{threshold}}$ as the maximum wait time for a packet and assign it the value of 200 ms which is substantially larger than the typical inter-arrival time of real-time packets is 20 ms. As has been mentioned in Sect. 3.2, $NextExp$, a key variable to identify the next expected packet consistent with the sending sequence for reordering detection, is initialised as the identifier of the first received in-order packet and updated by reproducing the sender's identifier function.

For each arrival event, If $Current\_Event\_ID > NextExp$, the packet should be delayed in the waiting queue (increament $LQ_{\text{reordered}}(t)$ by 1) to wait for the expected in-order packets. Moreover, if the early arrived packet waits for more than one packet, the sequence discontinuity size (indicated by d_s) is required to indicate the difference between the identifier of the current arrived packet and the $NextExp$. Therefore, the waiting time is calculated as:

$$tw = d_s * WT_{\text{threshold}} . \tag{7}$$

This produces a wait event which is scheduled to be executed at the time $t + tw$, where t is the arrival time of the current packet. The wait event executes when the waiting time of the packet in the waiting queue expires. The packet is then ordered at the head of the queue and the $NextExp$ is assigned the identifier of this packet and incremented by Seq_{inc} which is a constant equal to the increment between the identifiers of the two successive packets at the sender. Subsequently, the checking loop starts in the waiting queue. Each remaining packet in the waiting queue $LQ_{\text{reordered}}(t)$ will be checked to find the next expected packet. If the next expected packet is found, it is ordered and moved in the processing queue ($LQ = LQ + 1$) and $NextExp$ is udated as described above again. In addition, each time the expected packet arrives or is found in the waiting queue, the waiting time of each waiting packet is reduced by the length of the waiting threshold $tw \leftarrow tw - WT_{\text{threshold}}$. The checking runs until there none of the next expected packets are in the waiting queue.

If $Current\_Event\_ID == NextExp$, the packet will be buffered in the processing queue ($LQ \leftarrow LQ + 1$) if the server is busy; otherwise, the packet will be processed by the server, which produces the departure event with the occurrence time at $t + ts$ (ts is the simulated service time of the server). $NextExp$ is incremented by Seq_{inc}. Subsequently, the same checking loop executes in the waiting queue as described above to reorder the packets. On the other hand if $Current\_Event\_ID < NextExp$, the packet will be discarded because it arrives too late for real-time playback, and we increment $Q_{\text{discarded}}$ by 1.

During the simulation, we count the number of discarded packets $Q_{\text{discarded}}$ due to excessively late arrivals, as well as the time of the occurrence of the discard. The summation of $LQ(t)$ and $LQ_{\text{reordered}}(t)$ accounts for the total queue length in the re-sequencing buffer observed at time t and packet end-to-end loss is measured. To record the difference between isolated and bursty packet loss, we have defined the Packet End-to-End Loss Density and measure it similarly to $Packet Reordering Density$.

4   Experimental Results

Our experiments were carried out on a wired test-bed network consisting of 8 nodes with the topology shown in Fig. 2, whereby multiple paths are available for packet delivery between source-destination pairs. CPN was installed as a loadable kernel module [33] running under Linux 2.6.32 at each node. Adjacent nodes are connected with 100 Mbps Ethernet links. The purpose of experiments is to examine two issues:

1. Network performance for real-time traffic with respect to the packet delay, jitter and packet loss, under varied background and obstructing traffic loads that use different QoS goals in their routing algorithm.
2. The correlation of packet end-to-end loss and path switching.

Note that we use 20 % of SPs in the total traffic and the real-time traffic rate is at 50 pps.

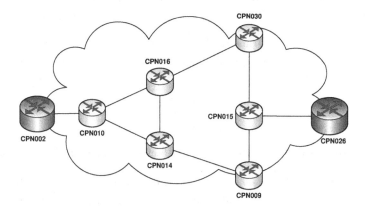

Fig. 2. CPN network testbed topology used in the experiments

To generate actual real-time traffic, "Linphone", a RTIP phone, was installed at each node in the network testbed, whereby real-time traffic can be originated everywhere within the network. Its SIP port is 5060 and audio port is 0780. Thus, real-time traffic can be differentiated according to the UDP port number exclusively used by Linphone. We used three flows of real-time traffic as shown in Table 1.

Table 1. Real-time Traffic Distribution used in the experiments

Source node	Destination node	Traffic rate
cpn002	cpn026	50 pps
cpn010	cpn015	50 pps
cpn016	cpn009	50 pps

Due to the constant rate of the real-time traffic produced by "Linphone", background traffic flows with a range of data rates and constant packet sizes of

Table 2. Background Traffic Distribution used in the experiments

Source node	Destination node
cpn002	cpn026
cpn010	cpn015
cpn016	cpn009
cpn030	cpn010
cpn014	cpn002
cpn015	cpn016

Table 3. Distinct QoS goal combinations used in our measurements for the real-time and the background UDP traffic flows

Real-time traffic	Background traffic
QoS:Delay	QoS:Delay
QoS:Jitter	QoS:Delay
QoS:Jitter	QoS:Jitter

1024 bytes were introduced into CPN and allowed to travel between any source-destination pair to provide varied traffic conditions. For the sake of simplicity, we used six UDP traffic flows as the background traffic, which were distributed as shown in the Table 2. We repeated each experiment with data rates of 1 Mbps, 2 Mbps, 3.2 Mbps, 6.4 Mbps, 10 Mbps, 15 Mbps, 20 Mbps, 25 Mbps, and 30 Mbps.

Furthermore, CPN routing was implemented both for the real-time and the background traffic, and we set a distinct QoS goal setting for the real-time traffic and the background UDP traffic. To this effect, we used three scenarios as shown in Table 3. Thus in one case we allowed both the real-time and the background traffic to use Delay Minimisation as their QoS goal, in two other cases we used Jitter and Delay Minimisation (and vice-versa) for the two types of traffic, and finally we used Jitter Minimisation for both the real-time and the background traffic, and we report measurements for each of these cases.

For each test, three flows of real-time packets and six flows of background packets with a specified rate were originated using one of the three QoS goal setting scenarios for a duration of ten minutes. The real-time traffic flow from CPN002 to CPN026 was selected as the object of our measurement, since it had the greatest number of intermediate nodes between this source and destination pair in the testbed. The average delay and delay variation, and the packet loss ratio for this flow were measured so as to examine which QoS goal is better for real-time packets delivery in the multiple QoS goal environment we considered under widely varying traffic conditions.

From the results shown in the Fig. 3, we can find the same trend for the three performance metrics in the three QoS goal setting scenarios under varying traffic conditions. With low background traffic load, the two different QoS goals

(Delay and Jitter) that are used for either the real-time or the background traffic have little effect, as may be expected, since overall delay, jitter and loss are very low.

We see quite clearly, that using Jitter as the QoS goal for *both* the real-time itself and the background traffic (BT) provides the *best results* for the real-time traffic at medium to high loads. Surprisingly enough, using Jitter for real-time and Delay for the background traffic (BT) yields the worst results. While, quite surprisingly, if Delay is used *both* for real-time and the background traffic, then the results in all three metrics (delay, jitter and loss) are not as good as when Jitter is used for both, but provides an intermediate result.

To explain this results, we have to refer to the definition of *Jitter* which means delay variation. Adaptive routing will necessarily cause some path switching, as well as some resulting end-to-end packet loss, and switching causes significant delay variation. Thus the QoS requirement of Jitter Minimisation applied to both real-time and the background traffic is likely to result in the least amount of path switching, even though at heavy traffic some route oscillations will still occur and will potentially degrade network performance. Therefore overall, we see that routing based on the QoS goal of *Jitter* is effective in alleviating route oscillations and losses, although the fact that it seems to reduce end-to-end delay itself (top figure) was definitely an unexpected result of these experiments.

5   Correlation Between Real-Time Packet Path Switching, Reordering and End-to-End Loss

While the previous result show that Jitter Minimisation is overall useful in the network to better satisfy the QoS needs of real-time traffic, not just when it is applied directly to real-time traffic but also when it is used for the background traffic, we were also interested in better understanding the detailed behaviour of the real-time traffic during these experiments. Thus we looked more carefully at the data collected in the experiments of Sect. 4 and measured the real-time traffic flow between CPN002 and CPN026 during the test interval being considered.

As summarised in Fig. 3, there is negligible packet loss under low traffic load. As we increase the background traffic rate, Fig. 4 shows us that as the six background traffic flows reach 20 Mbps, looking at the timestamp at which packets are send (the *x-axis*), the path switching rate of real-time traffic is around 10 packets per second. Though most of the time hardly any packets are lost, *we were indeed surprised to observe that in several time intervals there was was indeed a burst of packet loss.* During three time intervals (800–900 s, 900–1000 s, 1300–1400 s), *packet loss occurred in bursts while the path switching rates became very low.* This is the contrary of what we would have expected. However we feel that this does have an intuitive explanation.

The explanation is that when a given path for real-time traffic satisfies the Jitter Minimisation QoS criterion for a relatively long time, and hence the path switching rate is close to "0", *this path becomes attractive for background traffic and then becomes saturated*, resulting in bursty packet loss with the loss rate

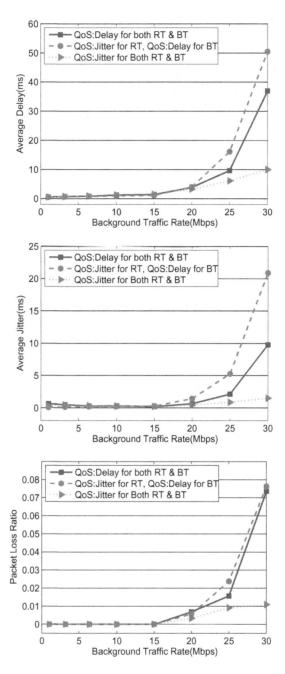

Fig. 3. The performance for real-time Traffic under varied background traffic conditions. We show the average delay (top), the delay variation or jitter (middle), and the packet loss ratio (bottom) for the real-time flow from CPN002 to CPN026 for different values of the background traffic load.

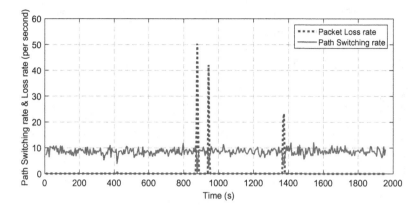

Fig. 4. The correlation of Packet Loss and Path Switching under medium traffic conditions: an apparently good path attracts more background traffic (BT), resulting in spurious bursty performance degradation and packet loss, followed by switching to better paths thanks to CPN's ability to adapt

increasing sharply and the loss ratio reaching "1". However, shortly after that, CPN reacts as it should to this QoS degradation: the SPs detect the performance degradation and another path is selected. Subsequently, loss rate decreases to "0" and the path switching rate increases.

As the rate of the six background traffic flows increased to 30 Mbps respectively, loss occurred more frequently and a large amount of packet de-sequencing was observed at the destination node. Figure 5 describes the correlation of Packet End-to-End Loss, Reordering and Path Switching under heavy traffic conditions in a test run of duration of 700 s.

We can see that packet reordering rate (caused by de-sequencing) varies in proportion to path switching. This strong linear correlation between the two metrics is confirmed by regression analysis, showing that packet reordering in CPN is mainly due to path switching. Furthermore, it was found that packets were lost more frequently when the loss rates/ratio were not high.

It is not easy to observe the correlation of packet path switching and packet loss from the figure. By applying regression analysis, we also obtained a very weak correlation of the two metrics. It is possibly because under heavy traffic conditions, packet loss is not only due to link saturated, route oscillation induced by heavy traffic loads also leads to the occurrence of loss. We can found that the proper path switching rate (or ratio) is beneficial to loss reduction. If path switching rate is increased excessively, it is converted to route oscillation which also lead to packet loss.

We have observed that packet discards in the re-sequencing buffer contributes to packet end-to-end loss for real-time traffic, and heavy traffic load does provoke packet reordering and loss. To illustrate this, we used the real-time packets traveling between CPN002 at CPN026 in an experiment where six background traffic flows were simultaneously forwarded at a rate of 30 Mbps, under the same

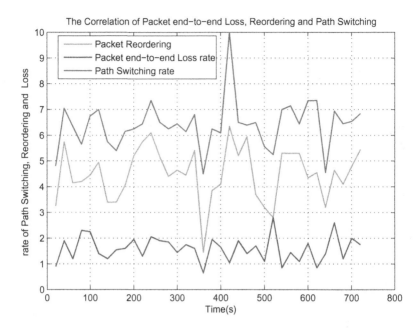

Fig. 5. The correlation of Packet end-to-end Loss, reordering and Path Switching under heavy traffic conditions: the data that is collected in the experiment, is processed using regression analysis

conditions as those discussed in the simulation of Sect. 3.3. It was found that the successive occurrence of bursty packet loss and reordering, are together the main reason for the significant increase in queue length at the re-sequencing buffer, causing buffer overflow. Note that packet loss in the network will cause packets to wait for their predecessors (who do not arrive) in the re-sequencing buffer; thus paradoxically buffer length increases when packets are lost. The higher the degree of burstiness of packet reordering and loss, the longer the waiting queue in the re-sequencing buffer and hence the higher the probability of buffer overflow.

Thus, the reordering density and loss density defined in Sects. 3.2 and 3.3, together with the queue length translated into the length of time spent in the buffer, are shown in Fig. 6. We can see that during the measurements queue length was relatively small when the reordering and loss density were low. At the time of around 520 s, many packets were lost or reordered successively, and subsequently the queue length in the buffer increased sharply. This data confirms our previous statement that packet reordering provokes large buffer queue lengths and actually accentuates (or creates a positive correlation) to the overall end-to-end packet loss, which results from the initial packet losses plus the buffer overflows.

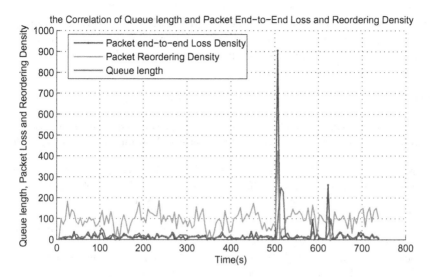

Fig. 6. The Correlation of the Queue Length in the re-sequencing Buffer, with Packet Loss and Reordering

6 Conclusions

This paper has presented an extension of the CPN routing algorithm so that multiple QoS classes can inhabit all routers of the network. Then these scheme has been applied to networks which carry real-time as well as other background traffic. Detailed experiments on CPN were conducted for real-time traffic, as well as other background traffic, with two distinct QoS requirements: *Delay* and *Jitter* Minimisation for any source-destination pair within the CPN testbed.

Surprisingly enough, measurement results for real-time traffic regarding average delay, jitter and loss have shown that when "Jitter" is specified as the QoS goal for both real-time and background traffic, better performance is achieved for all QoS metrics and all network load conditions. This seems to result from the useful effect of "Jitter" as a means to reduce route oscillations, also generally giving rise to less packet loss. However, we notice that long periods of path stability can also result in congestion with all traffic tending to use the same paths, when these paths do not oscillate and hence have low jitter: at that point, bursty effects of loss occur and CPN can mitigate for them by switching paths. Thus novel sensible routing schemes which distribute traffic over many paths in the network [34] may be useful even though they may result in the other disadvantages of multi-path packet forwarding. Furthermore we observe that packet loss in some parts of the network will provoke delays for other packets in the re-sequencing output RTIP codec buffers, which in turn can provoke buffer overflow and further losses. Thus, through a detailed and careful measurement study in a well instrumented network test-bed, this paper presents a complex series of cause and effects that allow us to better understand and better design networks that need to support real-time traffic. Future work in this area is planned to consider a specific application in distributed industrial control of manufacturing systems, so as to apply this research in a specific practical context.

References

1. FCC: 2013 measuring broadband america. In: Office of Engineering and Technology and Consumer and Governmental Affairs Bureau, Washington, DC, USA, Federal Communications Commission, February 2013
2. Baldi, M., Martin, J.D., Masala, E., Vesco, A.: Quality-oriented video transmission with pipeline forwarding. IEEE Trans. Broadcast. **54**(3), 542–556 (2008)
3. Gelenbe, E., Cao, Y.: Autonomous search for mines. Eur. J. Oper. Res. **108**(2), 319–333 (1998)
4. Gelenbe, E., Wu, F.J.: Large scale simulation for human evacuation and rescue. Comput. Math. Appl. **64**(12), 3869–3880 (2012)
5. Soucek, S., Sauter, T.: Quality of service concerns in IP-based control systems. IEEE Trans. Ind. Electron. **51**(6), 1249–1258 (2004)
6. Soldatos, J., Vayias, E., Kormentzas, G.: On the building blocks of quality of service in heterogeneous IP networks. Commun. Surv. Tutorials **7**(1), 69–88 (2005)
7. Dong, H., Hussain, F.K.: Focused crawling for automatic service discovery, annotation, and classification in industrial digital ecosystems. IEEE Trans. Ind. Electron. **58**(6), 2106–2116 (2011)
8. Dong, H., Hussain, F.K., Chang, E.: A service search engine for the industrial digital ecosystems. IEEE Trans. Ind. Electron. **58**(6), 2183–2196 (2011)
9. Santos, R., Pedreiras, P., Almeida, L.: Demonstrating an enhanced ethernet switch supporting video sensing with dynamic QoS. In: DCOSS, pp. 293–294 (2012)
10. Borzemski, L., Kaminska-Chuchmala, A.: Distributed web systems performance forecasting using turning bands method. IEEE Trans. Ind. Inform. **9**(1), 254–261 (2013)
11. Toral-Cruz, H., Argaez-Xool, J., Estrada-Vargas, L., Torres-Roman, D.: An introduction to VoIP: End-to-end elements and QoS parameters. In: InTech, pp. 79–94 (2011)
12. Roychoudhuri, L., Al-Shaer, E.S.: Real-time packet loss prediction based on end-to-end delay variation. IEEE Trans. Netw. Serv. Manage. **2**(1), 29–38 (2005)
13. Canovas, S.R.M., Cugnasca, C.E.: Implementation of a control loop experiment in a network-based control system with lonworks technology and IP networks. IEEE Trans. Ind. Electron. **57**(11), 3857–3867 (2010)
14. Cucinotta, T., Mancina, A., Anastasi, G., Lipari, G., Mangeruca, L., Checcozzo, R., Rusina, F.: A real-time service-oriented architecture for industrial automation. IEEE Trans. Ind. Inform. **5**(3), 267–277 (2009)
15. Felser, M., Jasperneite, J., Gaj, P.: Guest editorial special section on distributed computer systems in industry. IEEE Trans. Ind. Inform. **9**(1), 181 (2013)
16. Gaj, P., Jasperneite, J., Felser, M.: Computer communication within industrial distributed environment - a survey. IEEE Trans. Ind. Inform. **9**(1), 182–189 (2013)
17. Gelenbe, E., Hussain, K., Kaptan, V.: Simulating autonomous agents in augmented reality. J. Syst. Softw. **74**(3), 255–268 (2005)
18. Silvestre-Blanes, J., Almeida, L., Marau, R., Pedreiras, P.: Online qos management for multimedia real-time transmission in industrial networks. IEEE Trans. Ind. Electron. **58**(3), 1061–1071 (2011)

19. Friedman, T., Caceres, R., Clark, A.: RFC3611: RTP control protocol extended reports (RTCP XR). Request for Comments, November 2003
20. Morton, A., Ciavattone, L., Ramachandran, G., Perser, J.: RFC4737: packet reordering metrics. Request for Comments, November 2006
21. Larzon, L.Å., Degermark, M., Pink, S.: Requirements on the TCP/IP protocol stack for real-time communication in wireless environments. In: Ajmone Marsan, M., Bianco, A. (eds.) QoS-IP 2001. LNCS, vol. 1989, pp. 273–283. Springer, Heidelberg (2001)
22. Gelenbe, E., Lent, R., Xu, Z.: Design and performance of cognitive packet networks. Perform. Eval. **46**((2,3)), 155–176 (2001)
23. Gelenbe, E.: Cognitive packet network. In: U.S. Patent 6,804,201, October 2004
24. Demichelis, C., Chimento, P.: IP packet delay variation metric for IP performance metrics (IPPM). The Internet Society. https://www.ietf.org/rfc/rfc3393.txt
25. Gelenbe, E., Kazhmaganbetova, Z.: Cognitive packet network for bilateral asymmetric connections. IEEE Trans. Ind. Inform. **10**(3), 1717–1725 (2014)
26. Gelenbe, E.: The first decade of g-networks. Europ. J. Oper. Res. **126**(2), 231–232 (2000)
27. Gelenbe, E., Lent, R., Nunez, A.: Self-aware networks and QoS. Proc. IEEE **92**(9), 1478–1489 (2004)
28. Gellman, M.: QoS routing for real-time traffic. Ph.D. thesis, Imperial College London (2007)
29. Baccelli, F., Gelenbe, E., Plateau, B.: An end-to-end approach to the resequencing problem. J. ACM **31**(3), 474–485 (1984)
30. Masala, E., Quaglia, D., de Martin, J.C.: Variable time scale multimedia streaming over IP networks. IEEE Trans. Multimedia **10**(8), 1657–1670 (2008)
31. Baldi, M., Marchetto, G.: Time-driven priority router implementation: analysis and experiments. IEEE Trans. Comput. **62**(5), 1017–1030 (2013)
32. Gelenbe, E., Lent, R., Montuori, A., Xu, Z.: Cognitive packet networks: QoS and performance. In: Proceedings of the IEEE MASCOTS Conference, Ft. Worth, TX, pp. 3–12. Opening Keynote Paper, October 2002
33. Gelenbe, E.: Steps toward self-aware networks. Commun. ACM **52**(7), 66–75 (2009)
34. Gelenbe, E.: Sensible decisions based on QoS. Comput. Manage. Sci. **1**(1), 1–14 (2003)

Expanding the Ns-2 Emulation Environment with the Use of Flexible Mapping

Robert R. Chodorek[1]($\boxtimes$) and Agnieszka Chodorek[2]

[1] Department of Telecommunications, The AGH University of Science
and Technology, Al. Mickiewicza 30, 30-059 Krakow, Poland
chodorek@agh.edu.pl
[2] Department of Information Technology, Kielce University of Technology,
Al. Tysiaclecia Panstwa Polskiego 7, 25-314 Kielce, Poland
a.chodorek@tu.kielce.pl

Abstract. The Berkeley's ns-2 simulator was, for a long time, one of the most popular open-source simulation tools. Although the new tool in the ns family, the ns-3, replaced it in the above ranking, the simplicity of the ns-2, with its flexibility and ability to operate at higher levels of abstraction caused the simulator to remain in use. This paper presents our enhancements to the mapping of incoming and outgoing traffic in the ns-2 simulator when it works in emulation mode. Our enhancements expand the build-in 1-to-1 MAC address mapping to 1-to-many address/port mapping, which allows the emulator to connect to more end-systems or subnetworks than the number of interfaces of the emulation server.

Keywords: ns-2 simulator · Performance evaluation · Emulation of computer networks · Elastic traffic · Streaming

1 Introduction

The name ns is an acronym for "network simulator" and refers to the family of open-source simulation tools, beginning with the discrete event LBNL[1] Network Simulator (later known as ns-1). Now the ns family consists of two versions, ns-2 [1] and ns-3 [2]. Both are discrete event-driven environments for the simulation of computer networks, and both are publicly available for research, education and development under the GNU license [3].

The ns-2 simulator is focused mainly on simulations of computer networks with relatively weak possibilities of network emulation. The simulator is a flexible tool for computer network analysis, able to provide large numbers of network types, topologies, technologies and protocols. It allows the user to carry out simulation experiments at different levels of abstraction. From low levels, demanding detailed description and exact simulation models (e.g. simulation

---

[1] LBNL stands for Lawrence Berkeley National Laboratory, the ns-1 place of development.

© Springer International Publishing Switzerland 2016
P. Gaj et al. (Eds.): CN 2016, CCIS 608, pp. 22–31, 2016.
DOI: 10.1007/978-3-319-39207-3_2

of TCP transmission with full protocol processing and functionality, or simulation of Wi-Fi LAN [4] with RTS/CTS frame exchange process) to high levels, demanding only a rough description of the problem and general network situation (ftp applications modelled as an infinitive source of data packets, long-distance links modelled as twin simplex links with defined throughput and propagation delay, etc.).

The ns-3 has a reputation of being more emulator than simulator. This programming tool achieves high functionality and high accuracy by using Linux mechanisms as accurate models of network functions. Full code and full processing of base network functions (e.g. protocols: TCP, UDP, IP and base routing) in the ns-3 simulator is mainly borrowed from the Linux operating system. Such an approach to network modelling allows the simulator to obtain a natural ability to network emulation at the level of protocol processing.

In the lower layers of the OSI/ISO Reference Model, the ns-3 simulator implements detailed models of many network solutions (including detailed and accurate propagation models for Wi-Fi [5], LTE [6], etc.). It also allows the running of user's code in applications and services. Both those advantages improves accuracy of simulations and supports network emulation. However, in event-driven environments, this accuracy and support is done at the cost of performance.

Generally, execution of certain operations of the ns-3 simulator requires more processor time than the execution of the corresponding operations of the ns-2. Simulations in the ns-3 are carried out using more precise models and the obtained results are more accurate, but the experiment lasts longer. In the context of real-time emulation, it means that on the same machine we'll be able to emulate a less complicated network environment than if the ns-2 is used. Moreover, for detail-oriented approaches to simulation based on very precise models, the testing of new ideas at early stages of deployment where higher levels of abstraction are usually needed is inhibited. Last but not least, there is no possibility of an easy transfer between these tools because of different programming user interfaces[2]) and lack of backward compatibility between ns-3 and ns-2 [7]. As a result, after initial enthusiasm with ns-3 which caused users to desert the ns-2, ns-2 has come back into favor. Nowadays, both simulation tools are under development and active maintenance. A new development version of the ns-2 tool (version 2.36) is planned for release in February 2016.

The aim of the paper is to present our proposition for extensions to the ns-2 simulator, which enable the ns-2 emulation platform to flexibly map incoming and outgoing traffic. The paper is organized as follows. Section 2 describes emulation possibilities of the ns-2. Section 3 presents proposed extensions. Section 4 summarizes the paper.

2   The Network Emulator Embedded in the Ns-2

A network emulator is a device, program or software-hardware environment that imitates network behavior to deceive real-world networks or real-world

---

[2] TCL-based user interface in the ns-2 and the Python-based interface in the ns-3.

end-systems connected to the emulator into behaving as if they are connected to a real (usually novel or very complex) network. The ns-2 simulation environment has several mechanisms to support network emulation. Moreover, although protocol agents included in the ns-2 simulation platform operate at rather higher levels of abstraction (for example, the TCP protocol is modelled as two separate parts, a sending one and a receiving one – Agent/TCP and Agent/TCPSink, respectively), some of them (for example, Agent/TCP/FullTCP) are a close enough match to their real-world equivalents to be used for emulation purposes.

2.1   Operating Modes

The assumption is that the cooperation of the ns-2 simulation platform with the real-world network environment allows one to build more or less complex network environments inside the simulation platform shared with the real-world networks or network equipment. Connecting real and virtual environments facilitates the implementation of new algorithms defining the work of network nodes (for instance, new queuing), as well as the implementation of new protocols or protocol mechanisms, and tests of selected elements of a system.

Depending on the interpretation of live data, introduced from a real network to the emulated one, designers of the ns-2 emulator have defined two operating modes of the designed platform [1]:

- opaque mode, where live data are treated as opaque data packets,
- protocol mode, where live data may be interpreted, processed and even generated by the simulator.

In the first, opaque operating mode, packets from real fragments of the network are captured and then introduced into the ns-2 simulator while the content of captured packets (headers, data segments) will not be processed. Thus, captured packets can be transmitted inside the ns-2 with a given delay. They can be buffered, dropped in network nodes (due to congestions, or according to established order or assumed error rate) and transmitted out of sequence. Then, if the captured packet is not dropped or damaged, such packet will leave the simulator and will be injected back into the real fragment of the network.

The opaque mode shows partial real-virtual interference, so both intermediate and end systems do not have to be fully implemented. The second, protocol operating mode, is a mode of full real-virtual interference. The ns-2 emulation platform (emulated network devices, especially endpoints) interferes with the content of captured packets and, also generates its own packets which interfere with captured ones. Interference with captured packets can be seen when headers and if necessary data segments are processed and modified. Such interference takes place mainly in intermediate nodes. Generation of new packets typically takes place in endpoints (e.g. the TCP endpoints send packets according to settings of the connected traffic generator). In practice, in the ns-2 emulation platform, implementation of the protocol operating mode is very limited. So limited that the manual reports that only the opaque mode was implemented. Existing functionality of the protocol mode can be the basis for further development.

In both operating modes, input and output modules[3] and the real-time scheduler are used to enable the ns-2 to work as an emulator and, as a result, to cooperate with real-world networks.

2.2   Real-Time Scheduler

The ns-2 is an event-driven simulator. Events are stored in the event calendar (the calendar queue), where they are queued in non-descending time order. The process of the extraction of events from the calendar, according to their chronology of execution, is served by a scheduler. During simulation, events in the ns-2 platform are executed in real-time, and order of execution is set according to the simulator's virtual time. It means that in reality, events are executed as soon as possible, and usually much faster than the virtual passage of time. Only in the case of simulations of very complex networks is it possible that simulated time will be longer than simulation time (time that has passed in the real world).

Cooperation of simulated networks with real-world networks demands that all events in the simulation platform must be executed in real-time. Because of this demand, in the ns-2 emulator the default system scheduler was replaced with a real-time scheduler able to execute events in real-time. If the testbed computer, on which the ns-2 emulator is running, is too inefficient (CPU "horsepower" is insufficient), there is a danger that delays in the execution of some events might happen. If such delay exceeds the threshold value, stored in the slop factor parameter, the ns-2 will produce a warning.

The default value of the slop factor is 10 ms, and it is only three or four times less than the video frame period (frame display time). In the case of the analysis of video streams, transmitted in real-time, such and even smaller delays will cause substantial falsification of research results. An answer to this problem were extensions to the ns-2 simulator, developed at the University of Magdeburg (Germany) [8,9]. Extensions included improvements to the real-time scheduler module, modification to modules enabling the emulator to cooperate with the real-world network (network objects and tap agents) and trace extensions.

Improvements reported in [8,9] were intended for the emulation of mobile networks. In the case of the emulation of video transmission in a wired network, despite the usage of all above mentioned extensions, limitations to the emulated execution of real-time transmission still significantly falsified the test results. Therefore, further enhancements to the ns-2 emulator were needed. The first attempt at these enhancements was briefly mentioned in the paper [10]. In the second attempt, current extensions to the ns-2 simulator working in emulation mode were developed. Full extensions included network interface handling, address/port mapping, initial and final packet processing, improvements to the real-time scheduler and extensions for the emulator's protocol operating mode. The first two will be described in the next section.

---

[3] Input and output modules, respectively, captures traffic and (after processing) inject it into a real-world network.

3   Flexible Mapping of Incoming and Outgoing Traffic

When creating the ns-2 network emulator, it was assumed that it must cooperate not only with directly connected computers (which entails the usage of a homogeneous network, made with the use of one technology), but also with many computer systems connected via a complex, heterogeneous IP network. This assumption makes it possible to test the cooperation of applications executed on computer systems geographically located in different places and in different fragments of the Internet (for instance, in Kielce and Cracow). It also allows one to apply emulation only in chosen fragments of the end-to-end network path and only there, where we want to introduce test modifications. In other fragments of the end-to-end path, packet routing and processing will be carried out using existing, real-world network infrastructure, and test traffic will be subjected to a natural interaction with the real-world Internet traffic.

Practical implementation of the above assumption requires the development of new programming modules to serve the network interface, which allows the ns-2 emulator to flexibly capture packets received from the real network, and inject packets into the real-world network after processing. Improvements to modules capable of cooperation with the internal routing of the emulator also are needed. All modification and improvements are supposed to be able to serve all (unicast, broadcast and multicast) IPv4 addressing, and unicast and multicast IPv6 addressing (anycast addressing is omitted).

In order to achieve the assumptions, two programming modules were implemented: `Network/newIP` and `Agent/Tap/newIP`. They are derived classes of, respectively, `Network` and `Agent/Tap` superclasses. The `Network/newIP` module[4] realizes cooperation with real-world IP networks. The `Agent/Tap/newIP` module is associated with a given network node and performs the conversion of packets derived from the real-world network of packets used by the ns-2 emulator. The `Agent/Tap/newIP` module also performs the reverse conversion, where ns-2 packets are mapped as packets that will be injected into the real network. This module assures proper cooperation between the `Network/newIP` module and the given node of the emulated network.

The relationship between elements of the ns-2 emulator are illustrated with an example of a simple, five-node emulated network, shown in the Fig. 1. Nodes of the emulated network – R1, R4 and R5 – receive (or send) live data (to) the real-world network. This live traffic is transmitted through Tap agents ($a1$, $a2$, and $a3$), connected to emulated nodes R1, R4 and R5. The Tap agents are instances of class `Agent/Tap/newIP`. Agents which cooperate with real-world networks are seen by emulated nodes as typical agents of protocols or services (i.e. as other instances of the `Agent` base class).

The Tap agents also cooperate with instances of the `Network/newIP` class – *net*1 and *net*2. As we can see from the Fig. 1, both 1-to-1 and 1-to-many

---

[4] The identifier of the module (`newIP`) denotes, that the `newIP` is a newly written module for cooperation with the IP protocol. This name was given, because `Network/IP` class already exists in the ns-2.

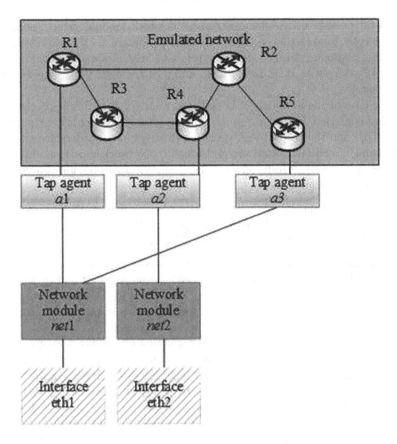

Fig. 1. Emulation system

cooperation is possible. In the picture, one `Network/newIP` module cooperates with one `Agent/Tap/newIP` agent (*a2* and *net2*), and one `Network/newIP` module (*net1*) cooperates with many (precisely, two: *a1* and *a3*) instances of the `Agent/Tap/newIP` class.

It's worth remarking that the original ns-2 mapping, available for emulation, permits only 1-to-1 mapping. In this typical solution, traffic introduced via a given interface from a real network to the emulated one is sending to a node of the emulated network using the typical ns-2 flat addressation (unlike hierarchical IP addressation), obtained by the agent. For example, IP address: 2, port: 3. If a packet is injected from the emulated network to the real one, it is injected via the output interface "as is" – it's just copied from the input buffer without any changes (including content of the TTL field).

In the case of the described extensions, the `Network/newIP` network module classifies packets according to rules defined by the user and sends the packet to a corresponding agent or agents. The module also retrieves parameters from the agent(s). If traffic is injected from the emulator to the real-world network, Tap agents send packets to the `Network/newIP` network module, which, according to stored rules, makes the necessary changes in content to the packet, including address mapping. Address mapping refers to MAC addresses (in link layer's frames), IP addresses (in IP datagrams), and port numbers (in transport protocol's packets).

In the case of the mapping of IP addresses and/or port numbers, checksum processing is needed. Checksum processing is performed for transport protocols (TCP, UDP) and the IP protocol, version 4 (in IPv4 only headers are protected by checksum). Such operations are performed by Network/newIP modules.

Mapping of traffic introduced to the ns-2 emulator can be done on the basis of typical IP flow identifiers:

- quintuple (IPv4 and IPv6): source IP address, destination IP address, source port, destination port and type of transport protocol,
- triple (IPv6 only): flow label, source IP address, destination IP address.

Flow identification allows for the implementation more sophisticated traffic engineering. For instance, traffic introduced via interface eth1 (Fig. 1) can be transmitted through the emulated network to the eth2 interface according to assumed rules. For example, the TCP traffic is transmitted via routers R1, R3 and R4 and the UDP traffic is transmitted via routers R5, R2 and R4.

1-to-many mapping plays a crucial role in the emulator, because of hardware limitations. If we want to carry out emulation experiments in which we want to perform empirical analysis of interactions between different traffic sources (applications) executed on many servers, then we would connect N computers (application servers) to N interfaces with high performance emulation servers. However, in the case of laboratory tests, we usually have at our disposal many PC computers, each equipped with only one network card. Those computers are ideal to work as end systems and live traffic generators, but if we try to use one of them as an emulation server, in the case of 1-to-1 mapping we will only be able to connect one generator of live data to the ns-2 emulator. And we have no possibility of connecting separate, physical receivers of the transmitted data. In the case of tests of video transmission, where visual quality check (analysis of quality of experience) is important, the possibilities offered by 1-to-1 mapping are woefully inadequate.

Server computers are usually equipped with two, or four, different communication interfaces. However, further expansion to more interfaces can be difficult (for example, in the case of the smallest rack-mount servers, sized 1U). Additionally, a large number of interfaces can complicate the organization of experiments and time-consuming patching.

The proposed solution allows us to skip this problem. It assumes that the end system computers can create a group of machines connected through an efficient network to a physical switch, and the switch is, in turn, connected to the server interface. Such topology of the test network was used during experiments described in [11], where 1 server interface was able to serve a group of multicast receivers. A situation where a group of senders was connected to one interface of emulation servers also were investigated.

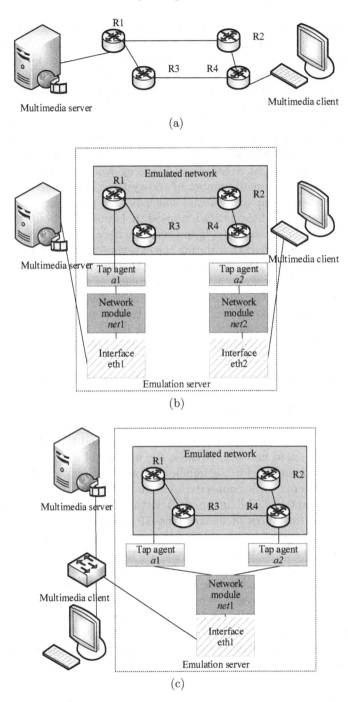

Fig. 2. Network topologies: (a) schematic diagram, (b) wiring diagram (emulation server equipped with two interfaces), (c) wiring diagram (emulation server equipped with one interface)

To validate the solution, performance tests were carried out. These tests were to transmit the TCP and UDP traffic between two endpoints (the endpoints were real computers outside an emulation environment). The proposed extension was verified by a long-term video traffic transmission in the presence of foreign traffic. It was checked to see whether the Ethernet frame loss reached zero and the emulated bandwidth was equal to the predetermined one.

On the basis of the results of a broad range of performance tests (conducted by the Authors during their experiments on SD and HD video transmission [11]) it can be stated that for many experiments 2 or 4 interfaces of an emulation server should be enough, if a 1-to-many mapping will be used. In the case of simple networks, with a small number of end-systems, 2 interfaces are enough. In the case of more complex systems, which need a partition, e.g. according to traffic sources, 4 interfaces should be used. It is possible that if the testbed network will be complex enough 4 is too small number of interfaces.

It should be remembered that during the experiments the critical infrastructure must be controlled. At a minimum, an Ethernet frame loss measurement at the switch must be performed. If the experiment is well prepared, Ethernet frame loss should be zero or, at the very least, negligible.

A simple emulation experiment, where the testbed network was partly realized in reality and partly emulated, is shown in Fig. 2. The network consists of one sender (multimedia server), one receiver (multimedia client) and four routers (R1 ... R4). A schematic diagram of the testbed network is depicted in Fig. 2a. Traffic, generated by the sender, enters the interface of router R1. Then, via router R2 or R3 (according to internal routing rules), it is directed to router R4, which sends the traffic to the receiver.

Figure 2b shows a diagram of cable connections in the case of an emulation server equipped with two network interfaces. The R1 and the R4 are boundary routers of the emulated network. Because the number of available server interfaces is greater than or equal to the number of connected real-world end systems, senders and receivers can be connected to separate physical interfaces.

Packets are captured and mapped to virtual ns-2 data structures in the network modules (instances of `Network/newIP` class). One network module is associated with one network interface of the emulation server. ns-2 packets are transferred to Tap agents (instances of `Agent/Tap/newIP` class). Tap agents are associated with data streams or flows and are attached to nodes of the emulated network. The 1-to-1 relation between network module and Tap agents is created.

The emulation server, presented in Fig. 2c, has only one network interface. As is shown in the wiring diagram, an auxiliary equipment (switch) was introduced to the real-world network to enable sharing of the interface. Because of the coexistence of input and output data streams in the same server's interface, the 1-to-many relation between network module and Tap agents is created.

4   Conclusions

In the paper our enhancements to the network simulator ns-2, working as an emulator, were presented. Improvements refer to the mapping of incoming and outgoing live packets onto internal ns-2 packets. Improvements were designed to exceed a hardware limitation that restricts the number of end-systems connected to the emulator to the number of network interfaces of the emulation server. The idea of the enhancement is that a high-speed switch, connected to the interface via a high-speed network, serves as a hardware expander, and the proposed flexible 1-to-many mapping (instead of the build-in 1-to-1 one) based on both MAC and IP addresses, as well as port numbers, serves as a multiplexer/demultiplexer of live traffic.

The proposed solution was successfully tested for HD video transmission.

Acknowledgment. The work was supported by the contract 11.11.230.018.

References

1. Fall, K., Varadhan, K.: The ns Manual (2014). http://ftp.isi.edu/nsnam/dist/release/rc1/doc/
2. Riley, G.F., Henderson, T.R.: The ns-3 network simulator. In: Wehrle, K., Güneş, M., Gross, J. (eds.) Modeling and Tools for Network Simulation, pp. 15–34. Springer, Berlin (2010)
3. Henderson, T.R., et al.: Network simulations with the ns-3 simulator. SIGCOMM demonstration (2008)
4. Bhaskar, D., Mallick, B.: Performance Evaluation Of MAC Protocol For IEEE 802.11, 802.11Ext. WLAN And IEEE 802.15. 4 WPAN Using NS-2. International Journal of Computer Applications 119.16 (2015)
5. Pei, G., Henderson, T.: Validation of ns-3 802.11b PHY model (2009). http://www.nsnam.org/~pei/80211b.pdf
6. Piro, G., Baldo, N., Miozzo, M.: An LTE module for the NS-3 network simulator. In: Proceedings of the 4th International ICST Conference on Simulation Tools and Techniques. ICST (Institute for Computer Sciences, Social-Informatics and Telecommunications Engineering) (2011)
7. Rene, S., et al.: Vespa: Emulating infotainment applications in vehicular networks. IEEE Pervasive Comput. **13**(3), 58–66 (2014)
8. Mahrenholz, D., Svilen, I.: Real-time network emulation with NS-2. In: Proceedings of the 8-th IEEE International Symposium on Distributed Simulation and Real Time Applications. Budapest Hungary (2004)
9. Mahrenholz, D., Svilen, I.: Adjusting the ns-2 Emulation Mode to a Live Network. In: Proceedings of KiVS'05. Kaiserslautern, Germany (2005)
10. Chodorek, R., Chodorek, A.: An analysis of QoS provisioning for high definition video distribution in heterogeneous network. In: Proceedings of the 14th International Symposium on Consumer Electronics (ISCE 2010), Braunschweig, Germany (2010)
11. Chodorek, R.R., Chodorek, A.: Providing QoS for high definition video transmission using IP Traffic Flow Description option. In: Proceedings of the IEEE Conference on Human System Interaction, pp. 102–107, Warsaw, Poland (2015)

Monitoring and Analysis of Measured and Modeled Traffic of TCP/IP Networks

Olga Fedevych$^{(\boxtimes)}$, Ivanna Droniuk, and Maria Nazarkevych

Lviv Polytechnic National University, 12 Bandera Street, Lviv, Ukraine
olhafedevych@gmail.com, ivanna.droniuk@gmail.com,
mar.nazarkevych@gmail.com
http://www.lp.edu.ua/ikni

Abstract. The software system for simulation of traffic flow for TCP/IP networks based on differential equations of oscillating motion with one degree of freedom was described. The interface of the software system, and the algorithm of its work were presented. Topology of TCP/IP network in ACS department was shown. Developed software that outputs the results in graphical and tabular forms was represented. The relationship between real and simulated traffic was considered. To verificate the results of the experiment the maximum correlation and coefficient that shows the ratio of standard deviation to the maximum were chosen. They show that the proposed equations allow predicting the behavior of traffic flow in TCP/IP networks.

Keywords: Ateb-functions · Traffic flow · Differential equations · Traffic simulation · Maximum correlation

1 Introduction

According to the Cisco forecasts, until 2020 Internet traffic volume will increase to 2 zettabytes [1]. With such an increase of load on the network equipment, it is an important task to ensure the effectiveness of its use. One of the methods of solving this problem is to simulate the behavior of traffic flow in computer network that will allow to predict certain trends and enable the creation of tools for efficient management of network equipment. The scientific literature mentions the computer network modeling methods based on Markov processes [2], as well as modeling of Internet traffic based on properties of self-similarity and Markov Modulated Poisson Process (self-similarity of Internet Traffic and Markov Modulated Poisson Process) [3]. Also, an approach to modeling traffic based on Diffusion and Fluid-Flow Approximations [4] is known. In this article the authors develop the approach, proposed by them before, for modeling traffic based on differential equations of oscillating movement [5]. In order to achieve this goal based on the equations proposed by the authors, a software package that monitors the network and simultaneously calculates modeled traffic with proposed equations was developed. The software also allows to compare real and modeled traffic.

© Springer International Publishing Switzerland 2016
P. Gaj et al. (Eds.): CN 2016, CCIS 608, pp. 32–41, 2016.
DOI: 10.1007/978-3-319-39207-3_3

2   Mathematical Model Predicting Traffic Flow

Ateb-functions mathematical apparatus made it possible to solve analytical system of differential equations that are describing essentially nonlinear oscillating processes in the systems with one degree of freedom [6]. To predict the traffic flow in a segment of computer network, differential equation that describes oscillating movement with a small perturbation was used in the form

$$\ddot{x} + a^2 x^n = \varepsilon f(x, \dot{x}, t), \tag{1}$$

where $x(t)$ – the number of packets in the network at time t; a – constant, which determines the amount of traffic fluctuations period, $f(x, \dot{x}, t)$ – arbitrary analytic function used to simulate small traffic deviations from the main component of fluctuations, n – a number that determines the degree of nonlinearity of the equation, which exudes on the period of a main component of oscillation.

While executing the following conditions on a and n $a \neq 0$, $n = \frac{2k_1+1}{2k_2+1}$, $k_1, k_2 = 0, 1, 2, \ldots$ it was proved that the analytical solution (ξ, ζ) of Eq. (1) is shown in the form of Ateb-functions [6]

$$\begin{cases} \xi = aCa(n, 1, \phi) - \varepsilon\widetilde{f}, \\ \zeta = a^{\frac{1+n}{2}} hSa(1, n, \phi) - \varepsilon\widetilde{f}, \end{cases} \tag{2}$$

where $a = \max_t |\zeta| \vee \min_t |\xi|$ – oscillation amplitude, $Ca(n, 1, \phi)$, $Sa(1, n, \phi)$, Ateb-cosine and Ateb-sine accordingly, $h^2 = \frac{2a^2}{1+n}$. Variable ϕ is related with time t by special ratio

$$\phi = \frac{a^{\frac{n-1}{2}}}{L} t + \phi_0, \tag{3}$$

where $L = \frac{2B\left(\frac{1}{2}, \frac{1}{1+n}\right)}{\pi(1+n)h}$, B – Beta function, ϕ_0 – initial phase of fluctuations.
Function f is represented as:

$$f(t) = \sum_{i=1}^{N} \alpha_i \delta_i(t_i), \tag{4}$$

where α_i – perturbation amplitude, $-A \leq \alpha_i \leq A$, A – maximum perturbation amplitude (based on observation data), δ_i – Dirac function, t_i – time, in which there is the i-th disturbance generated randomly, N – number of randomly generated perturbations.

3   The Algorithm of the Software Work

This section describes the software package designed for traffic simulation based on a mathematical model represented by Eqs. (2)–(4). The software interface is shown in Fig. 1.

As shown in Fig. 1 the menu provides the following options: *Show* – to show traffic data across time marks To and From; *Log* – logarithmic scale of Scale Y;

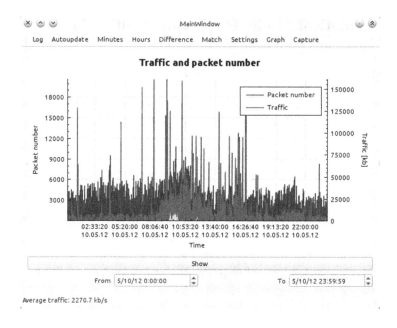

Fig. 1. The interface of the developed software system

Autoupdate – to show data for the last 10 min with automatic updates; *Minutes* – to show data on the scale of minutes; *Hours* – to show data on the scale of hours; *Difference* – to show the difference between a given Ateb-function and traffic; *Match* – to show graphs of traffic and Ateb-function simultaneously; *Graph* – to open an interface to work with graphs; *Capture* – to open an interface to capture traffic; *Settings* – see sidebar; *Ateb-settings* – Ateb-function settings, its type, range, and step for computing in Difference mode. In addition, the developed software makes it possible to predict the traffic flow parameters for more efficient use of nodal equipment in the segment of computer network. The algorithm of work of the software is described as follows. Initially, the observation time T for the flow of traffic and time t_0-tick-interval are set. Observation time T is limited by the capabilities of the hardware and can be chosen arbitrarily. Time t_0-tick-interval is set within the range specified in the data capture and may be in the range of 0.001 s to 10 mins. The developed software allows to analyze and store data traffic. There is a search for the minimum and maximum values of traffic in a corresponding moment of time during a given tick-interval. The next step is to find the average value (on the axis OY, along the axis OX preset interval is displayed on the interval $[-\Pi(n,1),\Pi(n,1)]$, where $\Pi(n,1)$ – Ateb-function period) between the minimum and the maximum, to make a comparison with an Ateb-function selected in program. The average value should be zero, and the maximum and minimum 1 and -1 respectively, as a limit of Ateb-functions value. All intermediate values are normalized by dividing by half of the difference between the maximum and minimum values of traffic.

Table 1. Description of customizable features to forecast the traffic flow in the segment of computer network

№	Function	Interval [min]	Step of prediction [min]	Observation time [min]
1	Sa(1/3, 1)	60	1	1440

Further arrays $P = \{P_i | i = 1, \ldots, 100\}$ and $Q = \{Q_i | i = 1, \ldots, 100\}$ containing values of traffic flow and calculated values of selected Ateb-function during tick-interval t_0 are considered. The next step is calculating the difference $|P_i - Q_i|$ between the stored values P_i and the calculated values of the selected Ateb-function Q_i at the selected interval with step $t_0/100$, i.e. the number of points of calculation $J = 100$. Then it searches the mean deviation in the array of the created differences. The mean deviation is calculated as follows

$$\mu = \frac{\sum_{i=1}^{J} |P_i - Q_i|}{J}. \tag{5}$$

Then according to the equation

$$M = \frac{2 - \mu}{2} \tag{6}$$

ordinate value of point for plotting the graph of similarity of Ateb-function to real traffic flow values during the observation time T is recalculated. A flow chart of the algorithm of work of the developed software system is shown in Fig. 2.

The software implements calculation of Ateb-function depending on given parameters and for predicting traffic flow provides selection of Ateb-function based on the value of time between peaks of traffic flow, which corresponds to the Ateb-function period. At time t_0 the J periods of Ateb-function (see Eq. (5)) were realized. The developed program contains 2246 strings. The C++ software program was developed in Qt Creator environment that implements selection of Ateb-functions (represented in Table 1), performs scaling by the Eqs. (5), (6) and describes conducted experiments.

4   Application and Experiments

In this section the real network of ACS department in Lviv Polytechnic National University (LPNU) is shown, for which the verification of mathematical model Eqs. (1)–(4) was conducted. The following Fig. 3 presents the topology of computer network of ACS department in LPNU. Switches used in the network of LPNU ACS department are the following:

1. D-Link DES-1024 R;
2. 3Com Swich 3300 xM;
3. 3Com Swich 4226 T;
4. Allied Telesyn AT8026T.

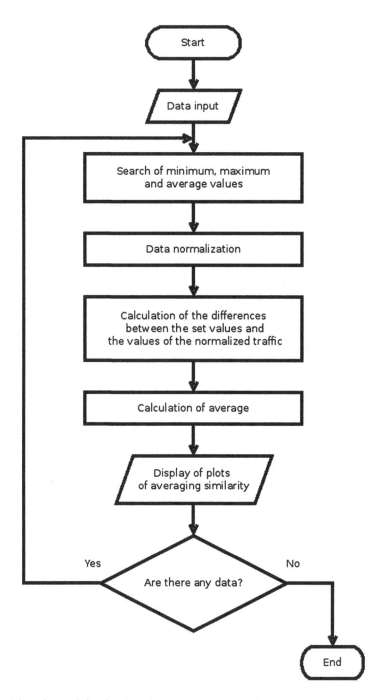

Fig. 2. Flow chart of the developed program, responsible for forecasting the traffic flow in the segment of computer network

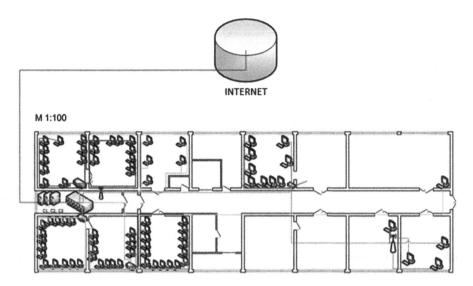

Fig. 3. The topology of computer network of LPNU ACS department

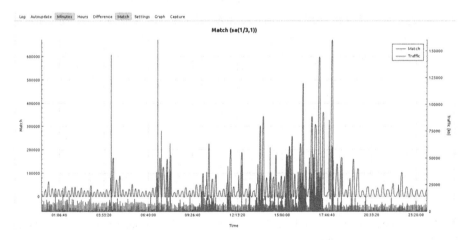

Fig. 4. Computer network traffic flow of LPNU ACS department (May 3, 2015) and appropriate Ateb-function values

The method of analysis and forecasting of traffic flow in the segment of computer network is based on differential equations of oscillating motion. To achieve this goal, a software package for analysis and forecasting of traffic flow in the segment of computer network was created. The choice of source data in Figs. 4, 5 and 6 corresponds to the values of parameters of the experiment № 1 from Table 1. In these figures lower graph corresponds to the real value of the traffic flow, and the top graph – to calculated values of predicted traffic flow. Observations were conducted during 24 h.

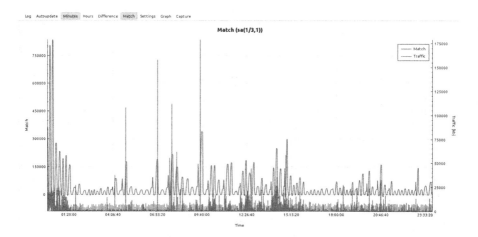

Fig. 5. Computer network traffic flow of LPNU ACS department (May 6, 2015) and appropriate Ateb-function values

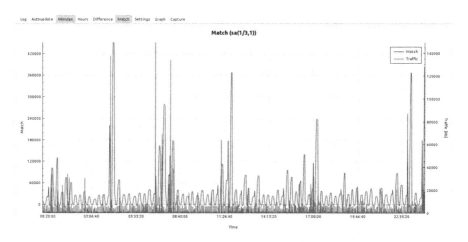

Fig. 6. Computer network traffic flow of LPNU ACS department (May 5, 2015) and appropriate Ateb-function values

The observations were made within the May 2015 at the LPNU ACS department in the computer network shown in Fig. 3, but observation only for 3 days are shown in Figs. 4, 5 and 6. Described software has been tested not only on TCP/IP traffic of computer network of ACS department, but also on traffic samples of IITIS PAN [7], which were obtained using Wireshark network protocol analyzer. To confirm the representativeness of TCP/IP traffic [8] and applicability of described software, analysis and comparison of packets in traffic samples of IITIS PAN and ACS department were implemented. The results of this analysis are presented in Fig. 7. As shown in Fig. 7, traffic structure is the same in both cases. The analysis shows that the TCP/IPv4 packets in traffic samples of ACS

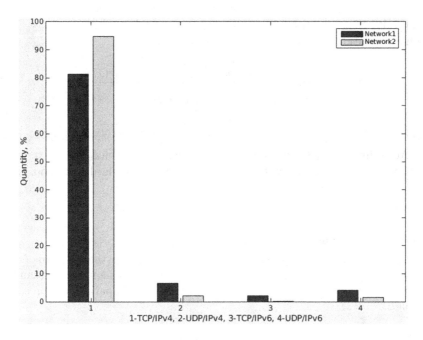

Fig. 7. Comparison of TCP/IP traffic of IITIS PAN (Network1) and ACS department (Network2)

Table 2. Comparison of traffic simulation results (Sa (1/3,1)) with real data network traffic of LPNU ACS department

Date	03.05.2015	06.05.2015	05.05.2015
K	66.5 %	67.5 %	65.9 %
ρ	0.44	0.44	0.15

department are prevailing over the same type of packets in traffic samples of IITIS PAN, but UDP/IPv4, TCP/IPv6 and UDP/IPv6 packets in traffic samples of IITIS PAN are prevailing over the same type of packets in traffic samples of ACS department accordingly.

5   The Research Results

To calculate the results of the experiment, the following quantities describing the maximum correlation and coefficient that shows the ratio of standard deviation to the maximum were chosen. The original data for calculation of the values in Table 2 are the data presented in Figs. 4, 5 and 6.

Formulas that were used for computing:
The correlation r_{xy}:

$$r_{xy} = \frac{\sum_{i=1}^{J}(x_i - \overline{x})(y_i - \overline{y})}{\sqrt{\sum_{i=1}^{J}(x_i - \overline{x})^2 \sum_{i=1}^{J}(y_i - \overline{y})^2}} = \frac{\text{cov}(x,y)}{\sqrt{s_x^2 s_y^2}}, \tag{7}$$

where $s_x = \sqrt{\frac{1}{J-1}\sum_{i=1}^{J}(x_i - \overline{x})^2}$, $s_y = \sqrt{\frac{1}{J-1}\sum_{i=1}^{J}(y_i - \overline{y})^2}$, $\text{cov}(x,y)$ – covariance. According to Eq. (7) the maximum correlation $\rho = \max_{xy} r_{xy}$.
Coefficient K of the ratio of the maximum to the standard deviation:

$$K = \frac{s_{\max}}{s}, \tag{8}$$

where $s = \sqrt{\frac{J}{J-1}\sigma^2}$ and $s_{\max} = \max_s s$.

Comparison of simulation experiments was conducted by the criterion of maximum correlation ρ. Coefficient K of the ratio of the maximum to the standard deviation was calculated. The calculated results of the comparison are presented in Table 2.

6   Conclusion

This article describes the software system developed by the authors to simulate the traffic flow for TCP/IP networks founded on a mathematical model based on differential equations of oscillating motion with one degree of freedom. The interface of the software system and the algorithm of its work were presented and described. The software system was tested in the computer network of LPNU ACS department. The topology of TCP/IP network in this department was presented. The results of the program in operation were graphically illustrated. The relationship between the real and the simulated traffic was shown. Parameters of correlation between the real traffic and the traffic calculated by equations were computed. Comparison of real and simulated traffic shows a sufficient matches level from 65.9 % to 67.5 %. The monitoring of the daily traffic during a month was conducted. The daily traffic samples were taken for investigation and for each of these samples the correlation value and coefficient K were obtained, and some of these experiments and calculated values were shown in this article. To improve the load of nodal equipment it is planned to develop the device for adaptive control of traffic flow that will be placed inside of the switching device. To do this, into the device for adaptive control of traffic flow the described software will be included in the block of prediction of traffic flow parameters.

References

1. Cisco Visual Networking Index Predicts IP Traffic to Triple from 2014–2019. http:// newsroom.cisco.com
2. Morozov, E., Nekrasova, R., Potakhina, L., Tikhonenko, O.: Asymptotic analysis of queueing systems with finite buffer space. In: Kwiecień, A., Gaj, P., Stera, P. (eds.) CN 2014. CCIS, vol. 431, pp. 223–232. Springer, Heidelberg (2014)
3. Domańska, J., Domański, A., Czachórski, T.: Modeling packet traffic with the use of superpositions of two-state MMPPs. In: Kwiecień, A., Gaj, P., Stera, P. (eds.) CN 2014. CCIS, vol. 431, pp. 24–36. Springer, Heidelberg (2014)
4. Czachórski, T., Pekergin, F.: Diffusion approximation as a modelling tool. In: Kouvatsos, D.D. (ed.) Next Generation Internet: Performance Evaluation and Applications. LNCS, vol. 5233, pp. 447–476. Springer, Heidelberg (2011)
5. Droniuk, I.M., Nazarkevich, M.A.: Modeling nonlinear oscillatory system under disturbance by means of Ateb-functions for the internet. In: Proceedings of 6th Working International Conference Het-Nets 2010, Gliwice, pp. 325–335 (2009)
6. Dronjuk, I., Nazarkevych, M., Fedevych, O.: Asymptotic method of traffic simulations. In: Vishnevsky, V., Kozyrev, D., Larionov, A. (eds.) DCCN 2013. CCIS, vol. 279, pp. 136–144. Springer, Heidelberg (2014)
7. Vlieg, E.J.: Representativeness of Internet2 usage statistics. A comparison with an other network. http://referaat.cs.utwente.nl/conference/8/paper/6891
8. Foremski, P.: Mutrics: multilevel traffic classification. http://mutrics.iitis.pl/

Investigating Long-Range Dependence in E-Commerce Web Traffic

Grażyna Suchacka[1(✉)] and Adam Domański[2]

[1] Institute of Mathematics and Informatics,
Opole University, ul. Oleska 48, 45-052 Opole, Poland
gsuchacka@uni.opole.pl
[2] Institute of Informatics, Silesian Technical University,
Akademicka 16, 44-100 Gliwice, Poland
adamd@polsl.pl

Abstract. This paper addresses the problem of investigating long-range dependence (LRD) and self-similarity in Web traffic. Popular techniques for estimating the intensity of LRD via the Hurst parameter are presented. Using a set of traces of a popular e-commerce site, the presence and the nature of LRD in Web traffic is examined. Our results confirm the self-similar nature of traffic at a Web server input, however the resulting estimates of the Hurst parameter vary depending on the trace and the technique used.

Keywords: Long-range dependence · Self-similarity · Hurst parameter · Hurst index · H index · Web server · Web traffic · HTTP traffic

1 Introduction

Analysis and modelling of Web traffic has been a hot research issue in recent years. HTTP requests' arrival times at a Web server may be easily observed and analyzed. In reality, a request arrival process on a Web server has been proven to reveal significant variance (burstiness): peak request rates can exceed the average request rate even tenfold and surpass the server capacity, resulting in the poor quality of Web service [1,2]. When this process is bursty on a wide range of time scales, it may have a feature of *self-similarity*. As a consequence of burstiness on many time scales, the arrival process may show *long-range dependence* (LRD), which means that values at any instant are non-negligibly positively correlated with values at all future instants [3]. Although the concepts of self-similarity and long-range dependence are not equivalent, in the literature they are often used interchangeably which may be attributed to the fact that the presence of both self-similarity and LRD may be estimated with the Hurst parameter (Hurst index), denoted as H.

Self-similarity has been discovered not only in Web server workload [2–4] but also in computer network traffic [5–9] or Web query traffic [10]. The synthetic self-similar traffic can be constructed by multiplexing a large number of

© Springer International Publishing Switzerland 2016
P. Gaj et al. (Eds.): CN 2016, CCIS 608, pp. 42–51, 2016.
DOI: 10.1007/978-3-319-39207-3_4

on/off sources characterized by heavy-tailed on and off period lengths. Analysis of the Web traffic [3] showed that the self-similarity feature of such traffic can be attributed to several factors, including heavy-tailed distributions of Web document sizes and user "think times", the effect of caching, and the superimposition of many such transfers in the network.

Self-similarity may have a significant negative impact on system performance and scalability [11]. That is why taking into consideration this Web traffic feature is essential when developing a synthetic workload model used to test the server system capacity – otherwise system performance may be overestimated. A number of traffic models and synthetic traffic generators implementing self-similarity and burstiness have been proposed [12–15].

Very few studies have investigated self-similarity and LRD of the arrival process at e-commerce websites so far [2,4]. The main impediment for this fact is a difficulty in obtaining traffic traces from online retailers, mainly due to e-business profitability and e-customer privacy concerns. In this paper, we investigate LRD in traffic arriving on a popular e-commerce Web server. The additional motivation for our study was a huge increase in popularity of online marketing and Web analytics in recent years, which could induce changes in Web traffic patterns at e-commerce servers, mainly due to the increased share of bot-generated traffic.

The paper is organized as follows. Section 2 presents background information on self-similary, LRD, and some methods for investigating these phenomena in time series. Section 3 presents datasets analyzed in our study and discusses the results of LRD intensity estimation. Section 4 concludes the paper.

2   Background

In this section notions of self-similarity and long-range dependence are briefly presented and some methods for estimating these phenomena are introduced. For detailed discussion on these issues refer e.g. to [16].

2.1   Self-similarity and Long-Range Dependence

Self-similarity may be defined in terms of the process distribution as follows. A stochastic process $Y(t)$ is *self-similar* with a self-similarity parameter H if for any positive stretching factor c, the distribution of the rescaled process $c^{-H}Y(ct)$ is equivalent to that of the original process $Y(t)$ [17].

A self-similar process shows *long-range dependence* if its autocorrelation function follows a power law: $r(k) \sim k^{-\beta}$ as $k \to \infty$, where $\beta \in (0,1)$ [3] (it is worth noting that LRD can be also defined for non self-similar processes).

A presence and a degree of self-similarity and long-range dependence is expressed by the *Hurst parameter*, H. When H is in the range of 0.5 and 1, one can say that a process is self-similar [18] and the higher H is, the higher degree of self-similarity and LRD is revealed by the series [2] (although a process can be self-similar even if $H \leq 0.5$, e.g., for the special case of Fractional Brownian motions).

2.2 Selected Methods for Estimating the Hurst Parameter

We apply five popular methods for assessing self-similarity and LRD of the Web traffic [3,8,12,19]. Four of them are graphical methods: aggregate variance method, R/S plot, periodogram-based method, and wavelet-based method. The last method is Local Whittle estimator.

The aggregate variance method and the R/S plot method are in the time domain. Let us consider a time series $X = (X_t; t = 1, 2, \ldots, N)$. In the *aggregate variance method*, the m-aggregated series $X^{(m)} = (X_k^{(m)}; k = 1, 2, \ldots)$ is defined by summing the time series X over nonoverlapping blocks of length m. The variance of series $X^{(m)}$ is plotted against m on a log-log plot and the points are approximated by a straight line, e.g., by using the least squares method. Then, the slope of the line, $-\beta$, is established and the Hurst parameter is computed as $H = 1 - \beta/2$. For a self-similar series variance decays slowly so $-\beta$ is greater than -1, which gives H higher than 0.5.

In the *R/S plot method*, the rescaled range, i.e., the R/S statistic, is plotted against m (which has been traditionally denoted by d in this method) on a log-log plot. For a self-similar series, R/S grows according to a power law with exponent H as a function of d and the plot has slope which is an estimate of H.

Other three methods are in the frequency domain. In the *periodogram-based method*, a periodogram of a time series X is defined by:

$$I_N(\lambda) = \frac{1}{2\pi N} \left| \sum_{t=1}^{N} X_t e^{i\lambda t} \right|^2, \tag{1}$$

where $i = \sqrt{-1}$. Usually it is evaluated at the Fourier Frequencies $\lambda_{j,N} = \frac{2\pi j}{N}$, where $j \in [0, n/2]$. The estimation of H is based on the slope γ of a log-log plot $I_N(\lambda_{j,N})$ versus $\lambda_{j,N}$ as frequency approaches zero. The relationship between the periodogram slope and the Hurst parameter is given by the formula $\gamma = 1 - 2H$.

Local Whittle estimator is a non-graphical method based on periodograms. This method assumes that the spectral density $f(\lambda)$ of the series can be approximated by the function:

$$f_{c,H}(\lambda) = c\lambda^{1-2H} \tag{2}$$

for frequencies λ as frequency approaches zero. The Local Whittle estimator of H is defined by minimizing:

$$\sum_{j=1}^{m} \log f_{c,H}(\lambda_{j,N}) + \frac{I_N(\lambda_{j,N})}{f_{c,H}(\lambda_{j,N})} \tag{3}$$

with respect to c and H; I_N is defined in (1) and $f_{c,H}$ is defined in (2).

In the *wavelet-based estimator* of the Hurst parameter, wavelets are considered as a generalisation of Fourier transform. For the series X the wavelet coefficients are determined; based on their values a time average μ_j is performed at a given scale (for the j-th octave). The relationship between μ_j and H is given by the formula:

$$E \log_2(\mu_j) \sim (2H - 1)j + C, \tag{4}$$

where E means the average, C depends only on H. Using this relationship, H may be determined based on the slope of an appropriate weighted linear regression.

Some other methods for determining the Hurst parameter in time series have been also proposed, e.g. detrended fluctuation analysis (DFA) [20] or multifractal analysis [21]. We do not discuss them in the paper due to space limitations.

3 Estimation of the Hurst Parameter for E-Commerce Traffic

3.1 Data Collection

The main goal of our analysis was to investigate LRD in e-commerce Web traffic. The analysis was done for data recorded in Web server log files obtained from an online retailer trading car parts and accessories. HTTP description lines were converted into time series reflecting the request arrival process at the Web server during the successive 14 days. 14 one-day traces were separately analyzed (traces are named with dates of traffic collection).

To verify the results obtained for the e-commerce traces, we decided to perform an additional LRD analysis of traffic at an actual non e-commerce server. To this end, we used seven traces from a server hosting a specialized mailing list.

The number of samples (i.e., the number of HTTP requests) in each trace is presented in Table 1.

Table 1. Cardinality of the analyzed data sets

E-commerce trace		Non e-commerce trace	
Trace (date)	Number of samples	Trace (date)	Number of samples
01.12.2015	13 643	10.01.2016	12 151
02.12.2015	56 284	11.01.2016	13 832
03.12.2015	9 642	12.01.2016	13 640
04.12.2015	17 842	13.01.2016	13 552
05.12.2015	25 082	14.01.2016	14 010
06.12.2015	16 092	15.01.2016	15 438
07.12.2015	15 860	16.01.2016	1 765
08.12.2015	16 138		
09.12.2015	190 934		
10.12.2015	170 529		
11.12.2015	41 249		
12.12.2015	9 758		
13.12.2015	14 594		
14.12.2015	17 453		

Package R [22] was used to estimate the Hurst parameter for both sets of traces with the application of the five methods described in Subsect. 2.2.

3.2 Results and Discussion

Figures 1, 2, 3, 4 provide examples of the application of four graphical methods to analyze two e-commerce traces, collected on 1 and 5 December 2015. Traffic in the 01.12.2015 trace is characterized by rather low LRD intensity compared to other e-commerce traces. On the other hand, for traffic registered in the 05.12.2015 trace, the highest mean H estimate was achieved in our analysis. Thus, in Figs. 1, 2, 3, 4 one can compare plots for Web traffic characterized with a moderate and a high level of long-range dependence.

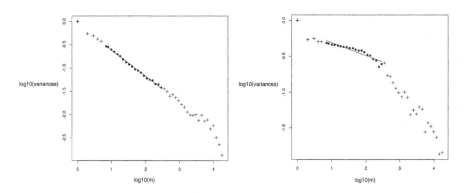

Fig. 1. Aggregate variance plot for the 01.12.2015 trace (left) and the 05.12.2015 trace (right)

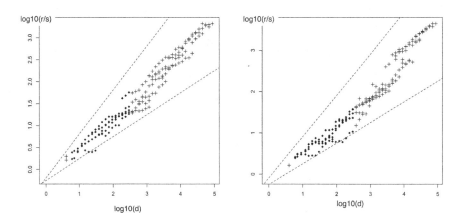

Fig. 2. R/S plot for the 01.12.2015 trace (left) and the 05.12.2015 trace (right)

Figure 1 shows the aggregate variance plots. One can observe that the linear plots are characterized by a slope clearly different from −1 which confirms the self-similarity of the analyzed time series. The slope of the plot for 01.12.2015 data (left) was estimated as −0.56, giving an estimate for the Hurst parameter of 0.72. The slope estimated for a 05.12.2015 data plot (right) is −0.18 which results in H of 0.91.

The R/S plots in Fig. 2 have an asymptotic slope between 0.5 and 1 (the corresponding lines are shown for comparison). The slope, being an estimate of H, was determined using regression as 0.65 for the 01.12.2015 trace and 0.54 for the 05.12.2015 trace.

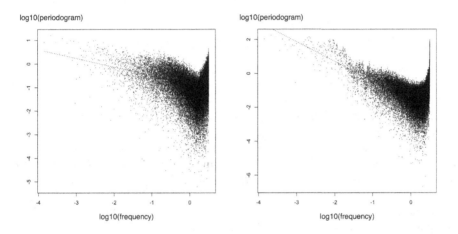

Fig. 3. Periodogram for the 01.12.2015 trace (left) and the 05.12.2015 trace (right)

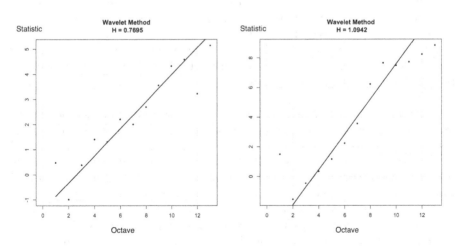

Fig. 4. R/S plot for the 01.12.2015 trace (left) and the 05.12.2015 trace (right)

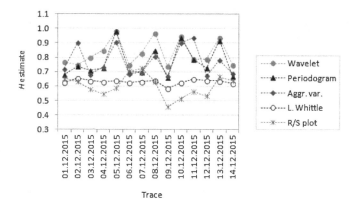

Fig. 5. Comparison of H estimates for the e-commerce traces

Figure 3 presents example results achieved using the periodogram-based method. Regression lines for periodogram plots have a slope of -0.36 and -0.5, giving the estimates of H as 0.68 and 0.75 for the 01.12.2015 and 05.12.2015 traces, correspondingly.

Figure 4 shows results of application of the wavelet-based estimator of the Hurst parameter to the two example e-commerce traces. The corresponding H of 0.77 and 1.09 were estimated. For H determined with this method confidence intervals are provided (Table 2).

Table 2 summarizes the results of our study across the different methods for all 14 e-commerce traces. In general, the H estimate exceeds 0.5 which indicates the self-similar character of the traffic. Only H estimated for the 09.12.2015 trace using the R/S plot method was 0.46. Other values of the Hurst parameter exceed 0.5 and they vary significantly, ranging from 0.51 to even 0.98.

Mean H values estimated for each e-commerce trace (the last column) show significant fluctuations in LRD intensity depending on a day, with H ranging from 0.6 for the 09.12.2015 trace to 0.8 for the 05.12.2015 trace. The last row of Table 2 shows even bigger differences in H estimates depending on the method applied.

Fluctuations in H estimates depending on the trace and the method applied are graphically presented in Fig. 5. One can observe that for the Local Whittle method, the estimate of H stays relatively consistent across all 14 analyzed datasets (with the mean value of 0.63). On the other hand, for the graphical methods it varies greatly. The wavelet-based method tends to give the highest H estimates (with the mean of 0.85) whereas H estimates for the R/S plot method are the lowest (with the mean of 0.6). We cannot give reasons for such a big variance of H estimates across various methods. However, such variance is not uncommon - it has been also obtained in some previous studies, e.g., for network traffic [8,12] and MPEG-1 encoded video sequences2 [23].

Table 3 presents estimates of the Hurst parameter for the non e-commerce traces and Fig. 6 illustrates fluctuations in these estimates depending on the

Table 2. H estimates for the e-commerce traces

Trace	Aggregate variance method	R/S plot	Periodogram-based method	Local Whittle estimator	Wavelet-based method	MEAN
01.12.2015	0.72	0.65	0.68	0.62	0.77 ± 0.05	0.67
02.12.2015	0.90	0.63	0.74	0.65	0.75 ± 0.04	0.73
03.12.2015	0.68	0.58	0.71	0.64	0.80 ± 0.06	0.65
04.12.2015	0.73	0.54	0.73	0.63	0.85 ± 0.07	0.66
05.12.2015	0.91	0.59	0.98	0.64	0.98 ± 0.03	0.80
06.12.2015	0.68	0.70	0.70	0.62	0.75 ± 0.05	0.68
07.12.2015	0.70	0.72	0.70	0.63	0.83 ± 0.06	0.69
08.12.2015	0.81	0.63	0.85	0.64	0.97 ± 0.03	0.74
09.12.2015	0.67	0.46	0.66	0.58	0.74 ± 0.03	0.60
10.12.2015	0.90	0.51	0.94	0.63	0.95 ± 0.07	0.74
11.12.2015	0.94	0.56	0.79	0.65	0.79 ± 0.05	0.73
12.12.2015	0.67	0.53	0.73	0.64	0.79 ± 0.05	0.64
13.12.2015	0.78	0.67	0.92	0.64	0.94 ± 0.03	0.75
14.12.2015	0.69	0.64	0.67	0.62	0.75 ± 0.05	0.65
Mean	0.77	0.60	0.79	0.63	0.85	

Table 3. H estimates for the non e-commerce traces

Trace	Aggregate variance method	R/S plot	Periodogram-based method	Local Whittle estimator	Wavelet-based method	MEAN
10.01.2016	0.60	0.65	0.62	0.62	0.66 ± 0.02	0.63
11.01.2016	0.61	0.62	0.64	0.62	0.66 ± 0.01	0.63
12.01.2016	0.59	0.64	0.60	0.62	0.65 ± 0.02	0.62
13.01.2016	0.68	0.60	0.74	0.63	0.75 ± 0.04	0.68
14.01.2016	0.73	0.62	0.73	0.65	0.73 ± 0.03	0.69
15.01.2016	0.61	0.61	0.61	0.62	0.65 ± 0.02	0.62
16.01.2016	0.79	0.74	0.73	0.68	0.71 ± 0.02	0.73
Mean	0.66	0.64	0.66	0.63	0.69	

trace and the method used. For this traffic the Hurst parameter (with the mean of 0.66) seems to be a little lower than the one for the e-commerce traffic (with the mean of 0.7). At the same time, H estimates for non e-commerce traffic are much more consistent across days and methods applied (c.f. Fig. 5).

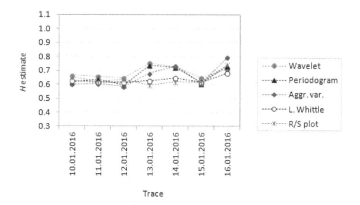

Fig. 6. Comparison of H estimates for the non e-commerce traces

4   Conclusions

Application of five popular Hurst parameter estimators to e-commerce traffic shows that this traffic reveals a significant level of long-range dependence. The mean H estimate ranges from 0.6 to 0.8 depending on a day. This result is consistent with results for request arrival process at other e-commerce sites: 0.66 in [2] and 0.73–0.8 in [4]. Furthermore, our study confirms previous findings that one cannot rely on a single method to estimate the Hurst parameter since different methods usually give different results. In our case, the mean H estimate for the e-commerce traffic ranges from 0.6 to 0.85 depending on the method. For the non e-commerce traffic, analyzed in the paper for comparative purposes, these fluctuations are much smaller and range from 0.63 to 0.69.

A coarse analysis of our results shows that the Hurst parameter determined for 24-hour intervals does not depend on the number of HTTP requests arrived on the server within these intervals. It also does not depend on the share of Web bot requests in the intervals. A deeper LRD analysis, performed for intervals shorter than 24 hours, is being planned to investigate these possible relationships. Furthermore, we plan to use traces from multiple Web servers to inspect if it is possible to use the Hurst parameter to distinguish between e-commerce and non e-commerce source.

References

1. Kant, K., Venkatachalam, M.: Transactional characterization of front-end e-commerce traffic. In: IEEE Global Telecommunications Conference (GLOBE-COM 2002), vol. 3, pp. 2523–2527. IEEE Press, New York (2002)
2. Vallamsetty, U., Kant, K., Mohapatra, P.: Characterization of e-commerce traffic. Electron. Commer. Res. **3**(1), 167–192 (2003)
3. Crovella, M., Bestavros, A.: Self-similarity in world wide web traffic: evidence and possible causes. IEEE/ACM Trans. Networking **5**(6), 835–846 (1997)
4. Xia, C.H., Liu, Z., Squillante, M.S., Zhang, L., Malouch, N.: Web traffic modeling at finer time scales and performance implications. Perform. Eval. **61**(2–3), 181–201 (2005)

5. Domańska, J., Domański, A., Czachórski, T.: A few investigations of long-range dependence in network traffic. In: 29th International Symposium on Computer and Information Sciences, Kraków, Poland. Information Sciences and Systems, part III, pp. 137–144. Springer, Heidelberg (2014)
6. Dymora, P., Mazurek, M., Strzałka, D.: Computer network traffic analysis with the use of statistical self-similarity factor. Ann. UMCS Informatica AI **13**(2), 69–81 (2013)
7. Leland, W.E., Taqqu, M.S., Willinger, W., Wilson, D.V.: On the self-similar nature of ethernet traffic. IEEE/ACM Trans. Networking **2**(1), 1–15 (1994)
8. Park, C., Hernández-Campos, F., Le, L., Marron, J.S., Park, J., Pipiras, V., Smith, F.D., Smith, R.L., Trovero, M., Zhu, Z.: Long-range dependence analysis of internet traffic. J. Appl. Stat. **38**(7), 1407–1433 (2011)
9. Olejnik, R.: Charakter Ruchu HTTP w Lokalnych Sieciach Bezprzewodowych (Title in English: The nature of HTTP traffic in wireless local area networks). Metody Informatyki Stosowanej **4**, 175–180 (2011)
10. Balakrishnan, R., Kambhampati, S.: On the self-similarity of web query traffic: evidence, cause and performance implications. Technical report, Arizona State University (2009)
11. Hernandez-Orallo, E., Vila-Carbo, J.: Analysis of self-similar workload on real-time systems. In: 16th IEEE Real-Time and Embedded Technology and Applications Symposium (RTAS 2010), pp. 343–352. IEEE Press, New York (2010)
12. Domańska, J., Domański, A., Czachórski, T.: Estimating the intensity of long-range dependence in real and synthetic traffic traces. In: Gaj, P., Kwiecień, A., Stera, P. (eds.) CN 2015. CCIS, vol. 522, pp. 11–22. Springer, Heidelberg (2015)
13. Kaur, G., Saxena, V., Gupta, J.P.: Characteristics analysis of web traffic with Hurst index. Lect. Notes Eng. Comput. Sci. **2186**(1), 234–238 (2010)
14. Lu, X., Yin, J., Chen, H., Zhao, X.: An approach for bursty and self-similar workload generation. In: Lin, X., Manolopoulos, Y., Srivastava, D., Huang, G. (eds.) WISE 2013, Part II. LNCS, vol. 8181, pp. 347–360. Springer, Heidelberg (2013)
15. Suchacka, G.: Generating bursty web traffic for a B2C web server. In: Kwiecień, A., Gaj, P., Stera, P. (eds.) CN 2011. CCIS, vol. 160, pp. 183–190. Springer, Heidelberg (2011)
16. Beran, J.: Statistics for Long-Memory Processes. Monographs on Statistics and Applied Probability. Chapman and Hall, New York, NY (1994)
17. Cavanaugh, J.E., Wang, Y., Davis, J.W.: Locally self-similar processes and their wavelet analysis. Handbook Stat. **21**, 93–135 (2003)
18. Stolojescu, C., Isar, A.: A comparison of some Hurst parameter estimators. In: 13th International Conference on Optimization of Electrical and Electronic Equipment (OPTIM 2012), pp. 1152–1157. IEEE Press, New York (2012)
19. Clegg, R.G.: A practical guide to measuring the Hurst parameter. Int. J. Simul. Syst. Sci. Technol. **7**(2), 3–14 (2006)
20. Krištoufek, L.: Rescaled range analysis and detrended fluctuation analysis: finite sample properties and confidence intervals. AUCO Czech Econ. Rev. **4**, 315–329 (2010)
21. Sanchez-Ortiz, W., Andrade-Gómez, C., Hernandez-Martinez, E., Puebla, H.: Multifractal Hurst analysis for identification of corrosion type in AISI 304 stainless steel. Int. J. Electrochem. Sci. **10**, 1054–1064 (2015)
22. The R Project for Statistical Computing. https://www.r-project.org
23. Cano, J.C., Manzoni, P.: On the use and calculation of the Hurst parameter with MPEG videos data traffic. In: 26th Euromicro Conference, vol. 1, pp. 448–455. IEEE Press, New York (2000)

More Just Measure of Fairness for Sharing Network Resources

Krzysztof Nowicki[1], Aleksander Malinowski[2(✉)], and Marcin Sikorski[3]

[1] Gdansk University of Technology, 80-233 Gdansk, Poland
know@eti.pg.gda.pl
[2] Bradley University, Peoria, IL 61625, USA
olekmali@bradley.edu
[3] CUBE.ITG, Gdansk, Poland

Abstract. A more just measure of resource distribution in computer networks is proposed. Classic functions evaluate fairness only "on average". The proposed new fairness score function ensures that no node is left without resources while on average everything looks good. It is compared with well-known and widely adopted function proposed by Jain, Chiu and Hawe and another one recently proposed fairness function by Chen and Zhang. The function proposed in this paper meets most of the properties both earlier proposed functions and at the same time is more restricted and has additional nonzero assignment property.

Keywords: Fairness · Fairness score function · Resource distribution · Network · Network performance evaluation

1 Introduction

Fairness and congestion control are popular and important topics in designing computer networks. Fairness is the main criterion for new real-time applications that are not yet implemented such as tele-stock, large-scale distributed real-time games, real-time tele-auctions, etc. [1]. These fair applications need fair network mechanisms. Fairness became one of the main requirements in designing RPR (Resilient Packet Ring) IEEE 802.17 standard for Metropolitan Area Networks, where special fairness algorithm is implemented [2]. Mayer et al. [3] attempted to generalize the solution to arbitrary topology networks. Durvy et al. [4] showed that unfairness of network media access in small networks does not have to persist when scaled up to large decentralized networks. Fairness of resource allocation in queueing has been considered by Avi-Itzhak et al. in [5]. Recently She et al. considered the problem of fairness and its measurement in the context of wireless networks [6].

First and most popular quantitative measure for fairness was proposed by Jain et al. [7]. Even though it was developed more than twenty years ago it is still widely used. A quick search in Google Scholar shows more than 2200 citations including more than 1000 since 2011 [8].

© Springer International Publishing Switzerland 2016
P. Gaj et al. (Eds.): CN 2016, CCIS 608, pp. 52–58, 2016.
DOI: 10.1007/978-3-319-39207-3_5

2   Related Work

There is a need for quantitative measures to evaluate the fairness of the network resource distribution. The quantitative measure of fairness should be a function of the amounts of resource users receive, that meets the following requirements:

1. if the distribution is completely unfair, the value of the function should be 0;
2. if the distribution is perfectly fair, the value of the function should be 1;
3. when the distribution becomes fairer, the value of the function should increase.

The mentioned earlier quantitative measure for fairness that was proposed by Jain et al. in [7] is called the Fairness Index and is shown in Eq. (1) below.

$$F(x_1, x_2, \ldots, x_n) = \frac{\left(\sum_{i=1}^{n} x_i \right)^2}{n \cdot \sum_{i=1}^{n} x_i^2} \tag{1}$$

where x_i denotes allocation of resources to the i-th user. The allocation of resources shall be in the range $0 \leqslant x_i \leqslant 1$, where 0 denotes no allocation of resources and 1 denotes allocation of all resources to the i-th user. The sum of all resource allocations shall add up to 1.

The Fairness Index possesses following properties:

1. $0 \leqslant F(x_1, x_2, \ldots, x_n) \leqslant 1$ for $x_i >= 0$, where $i = 1, 2, \ldots, n$,
2. if the distribution is completely unfair

$$F(x_1, x_2, \ldots, x_n) = \frac{1}{n}, \tag{2}$$

3. if the distribution is perfectly fair

$$F(x_1, x_2, \ldots, x_n) = 1, \tag{3}$$

4. the fairness score function does not depend on the scale,
5. the fairness score function continuously reflects changes in allocations,
6. if k out of n users share the entire resource equally when others do not receive any

$$F(x_1, x_2, \ldots, x_n) = \frac{k}{n}. \tag{4}$$

As we can see, Properties 2 and 6 do not meet the requirement (1). It means that Fairness Index does not fit the real situation well for the completely un-fair situation.

Chen and Zhang in [9] proposed fairness score function $G(x_1, x_2, \ldots, x_n)$ as follows:

$$G(x_1, x_2, \ldots, x_n) = 1 - \frac{n \cdot \sum_{i=1}^{n} a_i^2 - 1}{n - 1}, \tag{5}$$

where

$$a_i = \frac{x_i}{\sum_{j=1}^{n} x_j}, \quad i = 1, 2, \ldots, n.$$

$G(x_1, x_2, \ldots, x_n)$ possesses Properties 1, 3, 4, and 5. According to [9] it performs better than $F(x_1, x_2, \ldots, x_n)$ in completely unfair case:

2 if the distribution is completely unfair $G(x_1, x_2, \ldots, x_n) = 0$,

6 if k out of n users share the entire resource equally, when others do not receive any

$$G(x_1, x_2, \ldots, x_n) = \frac{n(k-1)}{k(n-1)}. \tag{6}$$

The idea of completely unfair distribution is subjective. Authors of [10] claim that a property of non-zero assignment to all competitors is identified as a fairness aspect of the fairness algorithm. If any number of users does not receive any of the resources, it is seen as an unfairness property. We agree with this issue. In some applications, the situation in which any number of allocations is equal to zero is unacceptable.

Nonzero allocation property was the main goal while designing new fairness score function proposed in this paper. This function satisfies all properties that have, except the Property 6.

3   A New Fairness Score Function

We define fairness function $G(x_1, x_2, \ldots, x_n)$ as follows,

$$S(x_1, x_2, \ldots, x_n) = n^n \cdot \prod_{i=1}^{n} a_i, \tag{7}$$

where

$$a_i = \frac{x_i}{\sum_{j=1}^{n} x_j}, \quad i = 1, 2, \ldots, n.$$

Theorem 1. *When the distribution becomes fairer, the value of the function increases. Here, distribution becomes fairer if becomes more equal. For $\eta > 0$ define*

$$D(\eta) = S(x_1, x_2, \ldots, x_s - \eta, \ldots, x_t + \eta, \ldots, x_n) -$$
$$S(x_1, x_2, \ldots, x_s, \ldots, x_t, \ldots, x_n) \tag{8}$$

If $x_1, x_2, \ldots, x_n \neq 0$ then

$$D(\eta) = \begin{cases} > 0 & \text{if} \quad \eta < x_s - x_t \\ = 0 & \text{if} \quad \eta = x_s - x_t \\ < 0 & \text{if} \quad \eta > x_s - x_t \end{cases}. \tag{9}$$

Proof.

$$D(\eta) =$$

$$n^n \left(\prod_{i=1}^{s-1} \sum_{j=1}^{n} \frac{x_i}{x_j} \right) \left(\prod_{i=s+1}^{t-1} \sum_{j=1}^{n} \frac{x_i}{x_j} \right) \left(\prod_{i=t+1}^{n} \sum_{j=1}^{n} \frac{x_i}{x_j} \right) ((x_s - \eta)(x_t + \eta) - x_s x_t)$$

$$= n^n \left(\prod_{i=1}^{s-1} \sum_{j=1}^{n} \frac{x_i}{x_j} \right) \left(\prod_{i=s+1}^{t-1} \sum_{j=1}^{n} \frac{x_i}{x_j} \right) \left(\prod_{i=t+1}^{n} \sum_{j=1}^{n} \frac{x_i}{x_j} \right) \eta((x_s - x_t) - \eta). \tag{10}$$

The proof then follows.

Theorem 2. *If the distribution is completely unfair*

$$S(x_1, x_2, \ldots, x_n) = 0. \tag{11}$$

Here, distribution is completely unfair if any number of users do not receive any of the resources.

Proof. If $x_1 = 0 \vee x_2 = 0 \vee \ldots \vee x_n = 0$ then $S(x_1, x_2, \ldots, x_n) = 0$ is obvious.

Theorem 3. *If the distribution is perfectly fair*

$$S(x_1, x_2, \ldots, x_n) = 1. \tag{12}$$

Here, distribution is perfectly fair if all users share the entire resource equally.

Proof. If $x_1 = x_2 = \ldots = x_n = \frac{1}{n}$ then

$$S(x_1, x_2, \ldots, x_n) = n^n \prod_{i=1}^{n} \frac{1}{n} = n^n \cdot n^{-n} = 1.$$

Theorem 4.
$$0 \leq S(x_1, x_2, \ldots, x_n) \leq 1. \tag{13}$$

Proof. According to Theorems 1 and 3, for any $n > 1$, $S(x_1, x_2, ..., x_n) \leq 1$. $S(x_1, x_2, \ldots, x_n) \geq 0$ is obvious. The proof follows.

Theorem 5. *The fairness score does not depend on scale. For $\delta > 0$,*

$$S(x_1\delta, x_2\delta, \ldots, x_n\delta) = S(x_1, x_2, \ldots, x_n). \tag{14}$$

Proof.

$$S(x_1\delta, x_2\delta, \ldots, x_n\delta) = n^n \prod_{i=1}^{n} \frac{x_i\delta}{\delta \sum_{j=1}^{n} x_j} = S(x_1, x_2, \ldots, x_n).$$

4   Comparison of Fairness Score Functions

We gathered some results of fairness functions $F(x_1, x_2, ..., x_n)$, $G(x_1, x_2, ..., x_n)$, and $S(x_1, x_2, ..., x_n)$. Our fairness function is more restricted. It has additional nonzero allocation property. It acts differently than $F(x_1, x_2, \ldots, x_n)$ and $G(x_1, x_2, \ldots, x_n)$ when disproportions between users are large. For example, when $n = 4$ and one of the users receives 1000 times less than others, the situation is 75 % fair using Jain, Chiu and Hawe function and almost 90 % fair using Chen and Zhang function. In the same situation, our fairness function returns 0.0031 which means that situation is only 0.31 % fair. In our opinion, this result better fits the real world situation than $F(x_1, x_2, \ldots, x_n)$ and $G(x_1, x_2, \ldots, x_n)$ do.

Figure 1 compares the three fairness functions for the case of $n = 2$. Fairness Index $0.5 \leqslant F(x_1, x_2) \leqslant 1$ and is less restricted than $G(x_1, x_2)$ and $S(x_1, x_2)$. We can also notice that the values of fairness score functions $G(x_1, x_2)$ and $S(x_1, x_2)$ are the same. This is not a coincidence. Simple proof can be made as follows,

$$G(x_1, x_2) = 2 - 2 \left(\frac{x_1^2 + x_2^2}{(x_1 + x_2)^2} \right) = \frac{4x_1 x_2}{(x_1 + x_2)^2} = S(x_1, x_2). \tag{15}$$

Figure 2 shows properties of the proposed fairness score function proved in Sect. 2, while Fig. 3 shows differences between the proposed fairness function and compared existing metrics.

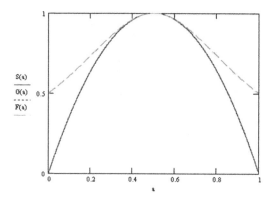

Fig. 1. Resource allocation between two users $S(a)$ – our proposed fairness score function for $n = 2$, $G(a)$ – fairness score function proposed by Chen and Zhang for $n = 2$, $F(a)$ – fairness score function proposed by Jain, Chiu and Have for $n = 2$

It can be seen that proposed fairness function is more restricted that fairness score functions proposed by Jain, Chiu and Hawe and function proposed by Chen and Zhang. In both cases, the biggest differences occur when one of the users is discriminated (zero allocation) while other two users share resources equally.

We realize that, for example, if only one out of one million users is left out for some reason and all others share resources equally, proposed fairness score function would return 0 while the value of other fairness functions would be slightly less than 1. In this specific case, for many network applications, proposed function will not be suitable metric. It does not show a different degree of fairness when no resources are distributed to any of users.

However, there are applications in which this nonzero property would be desired such as security, alarm systems or systems based on the bus, where no number of users can be left out. It would be also desired property to measure fairness for high priority traffic where Committed Information Rate (CIR) [11] is contracted.

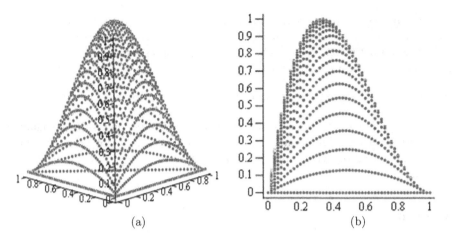

(a)   (b)

Fig. 2. Our fairness score function for $N = 3$

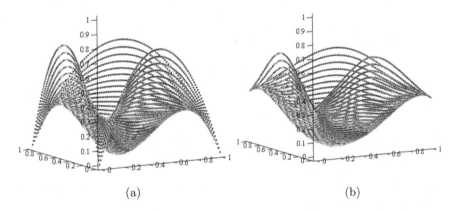

(a)   (b)

Fig. 3. The difference between Chen and Zhang fairness function and our fairness function (a) and between Jain, Chiu and Hawe fairness function and our fairness function (b) for $N = 3$

5 Conclusion

In this paper we propose new measure for fairness in computer networks, as follows,

$$S(x_1, x_2, \ldots, x_n) = n^n \cdot \prod_{i=1}^{n} a_i, \tag{16}$$

where

$$a_i = \frac{x_i}{\sum_{j=1}^{n} x_j}, \quad j = 1, 2, \ldots, n.$$

Our fairness function meets all meritorious demands proposed in [7]. The proposed fairness score function is independent of the scale of the allocation metric. It is bounded between 0 and 1 so it can be easily expressed as a percentage. The discrimination function can be defined as

$$D(x_1, x_2, \ldots, x_n) = 1 - S(x_1, x_2, \ldots, x_n).$$

It is also continuous which means that any change of allocation results in a change of fairness function value. In addition, it has nonzero assignment property which means that situation when any of users receives no resources is totally unfair distribution.

References

1. Wu, W., Cai, Y., Guizani, M.: Auction-based relay power allocation: Pareto optimality, fairness, and convergence. IEEE Trans. Commun. **62**(7), 2249–2259 (2014). doi:10.1109/TCOMM.2014.2331072
2. Yuan, P., Gambiroza, V., Knightly, E.: The IEEE 802.17 media access protocol for high-speed metropolitan-area resilient packet rings. IEEE Netw. **18**(3), 8–15 (2004). doi:10.1109/MNET.2004.1301017
3. Mayer, A., Ofek, Y., Yung, M.: Local and congestion-driven fairness algorithm in arbitrary topology networks. IEEE/ACM Trans. Networking **8**(3), 362–372 (2000). doi:10.1109/90.851982
4. Durvy, M., Dousse, O., Thiran, P.: On the fairness of large CSMA networks. IEEE J. Sel. Areas Commun. **27**(7), 1093–1104 (2009). doi:10.1109/JSAC.2009.090907
5. Avi-Itzhak, B., Levy, H., Raz, D.: A resource allocation queueing fairness measure: properties and bounds. Quing Syst. **56**(2), 65–71 (2007). doi:10.1007/s11134-007-9025-x
6. She, H., Prasad, R.V., Onur, E., Niemegeers, I.G.M.M.: Fairness in wireless networks: issues, measures and challenges. IEEE Commun. Surv. Tutorials **16**(1), 5–24 (2014). doi:10.1109/SURV.2013.050113.00015
7. Jain, R., Chiu, D., Hawe, W.: A quantitative measure of fairness and discrimination for re-source allocation in shared computer systems. DEC research report TR-301 (1984). http://www1.cse.wustl.edu/jain/papers/ftp/fairness.pdf
8. Google Scholar. Accessed 18 July 2014
9. Chen, Z., Zhang, C.: A new measurement for network sharing fairness. Comput. Math. Appl. **50**(5–6), 803–808 (2005). doi:10.1016/j.camwa.2005.03.015
10. Bharath-Kumar, K., Jaffe, J.M.: A new approach to performance - oriented flow control. IEEE Trans. Commun. **29**, 427–435 (1981). doi:10.1109/TCOM.1981.1095007
11. Su, H., Atiquzzaman, M.: ItswTCM: a new aggregate marker to improve fairness in DiffServ. In: Proceedings of IEEE Global Telecommunications Conference, vol. 3, pp. 1841–1846 (2001). doi:10.1109/GLOCOM.2001.965893

Classification of Solutions to the Minimum Energy Problem in One Dimensional Sensor Networks

Zbigniew Lipiński$^{(\boxtimes)}$

Institute of Mathematics and Informatics, Opole University, Opole, Poland
zlipinski@math.uni.opole.pl

Abstract. We classify solutions of the minimum energy problem in one dimensional wireless sensor networks for the data transmission cost matrix which is a power function of the distance between transmitter and receiver with any real exponent. We show, how these solutions can be utilized to solve the minimum energy problem for the data transmission cost matrix which is a linear combination of two power functions. We define the minimum energy problem in terms of the sensors signal power, transmission time and capacities of transmission channels. We prove, that for the point-to-point data transmission method utilized by the sensors in the physical layer, when the transmitter adjust the power of its radio signal to the distance to the receiver, the optimal transmission is without interference. We also show, that the solutions of the minimum energy problem written in terms of data transmission cost matrix and in terms of the sensor signal power coincide.

Keywords: Sensor network · Energy management · Channel capacity

1 Introduction

Characteristic feature of sensor networks is that these consist of small electronic devices with limited power and computational resources. Typical activity of sensor network nodes is collection of sensed data, performing simple computational tasks and transmission of the resulting data to a fixed set of data collectors. The sensors utilize most of their energy in the process of data transmission, this energy grows with the size of the network and the amount of data transmitted over the network. Generally in sensor networks there are two models of energy consumption. The goal of the first model is to maximize the functional lifetime of the network [1,2]. For this type of problems the data transmission in the network is modeled in such a way, that the energy consumed by each sensor is minimal. To extend the network lifetime the sensors must share their resources and cooperate in the process of data transmission. For typical solutions of such problems the consumed energy is evenly distributed over all nodes of the network. The second type of problems is to optimize the energy consumed by the whole network [3–5]. Such problems arise when the network nodes are

© Springer International Publishing Switzerland 2016
P. Gaj et al. (Eds.): CN 2016, CCIS 608, pp. 59–71, 2016.
DOI: 10.1007/978-3-319-39207-3_6

powered by the central source of energy or the node batteries can be recharged and the total energy consumed by the network is to be minimized. To solve this type of problems it is enough to find the optimal energy consumption model of each sensor and summed up their energies. In this paper we discuss solutions of the minimum energy problem in one dimensional wireless sensor network S_N when each sensor generates the amount Q_i of data and sends it, possible via other sensors, to the data collector. We assume, that the network S_N is build of N sensors and one data collector. The sensors are located at the points $x_i > 0$ of the line and the data collector at the point $x_0 = 0$. We prove, that for any exponent $a \in R$ in the data transmission cost function

$$E(x_i, x_j) = |x_i - x_j|^a, \tag{1}$$

where $|x_i - x_j|$ is a distance between transmitter and receiver, when in the interval $(x_0, \frac{1}{2}x_N)$ there are N' sensors, there are $(N' + 1)$ solutions of the problem. We show, how to determine the optimal solutions of the minimum energy problem for the data transmission cost function which is sum of two factors

$$E(x_i, x_j) = |x_i - x_j|^a + \lambda |x_i - x_j|^b, \tag{2}$$

where $a, b \in R$ and $\lambda \geq 0$. We also discuss the solutions of the minimum energy problem in wireless sensor networks when the energy utilized by the network is expressed in terms of node signal power, transmission time and there are constraint on the transmission channels due to presence of the noise and interference [6].

We represent a sensor network S_N as a directed, weighted graph $G_N = \{S_N, V, E\}$ in which S_N is a set of graph nodes s_i, $i \in [1, N]$, V is the set of edges and E set of weights. Each directed edge $T_{i,j} \in V$ defines a communication link between i-th and j-th node of the network. To each edge $T_{i,j}$ we assign a weight $E(x_i, x_j) \equiv E_{i,j}$, which is the cost of transmission of one unit of data between i-th and j-th node. The data flow matrix $q_{i,j}$ defines the amount of data transmitted along the edge $T_{i,j}$. By $U_i^{(\text{out})} \subseteq S_N$ we denote a set of the network nodes to which the i-th node can send the data, i.e., $U_i^{(\text{out})} = \{s_j \in S_N | \exists T_{i,j} \in V\}$. The set $U_i^{(\text{out})}$ defines the maximal transmission range of the i-th node. In the paper we assume, that each sensor of S_N can send the data to any other node of the network $\forall_{s_i \in S_N} U_i^{(\text{out})} = S_N$. If we assume, that each sensor of the S_N network generates the amount Q_i of data, $i \in [1, N]$, and the data have to be sent to the data collector, then the energy consumed by the i-th sensor in the process of data transmission can be written in the form $E_i(q) = \sum_{j=1, j \neq i}^{N} q_{i,j} E_{i,j}$. For the total energy consumed by the network $E_T(q) = \sum_{i=1}^{N} E_i(q)$ the minimum energy problem can be defined by the set of following formulas

$$\begin{cases} \min_q E_T(q), \\ \sum_i q_{i,j} = Q_i + \sum_j q_{j,i}, \\ E_{i,j} \geq 0, \quad q_{i,j} \geq 0, \quad Q_i > 0 \quad i, j \in [0, N], \end{cases} \tag{3}$$

where the second formula defines the feasible set of the problem. It states that the amount of data generated by the i-th node Q_i and the amount of data received from other nodes $\sum_j q_{j,i}$ must be equal to the amount of data which the node can send $\sum_i q_{i,j}$.

Because the objective function $E_T(q)$ of the problem is continuous and linear from this we can deduce simple but helpful fact, that any local minimum of $E_T(q)$ is a global one and thus it is a solution of (3).

If we assume, that we search for a solution of (3) in the integers, i.e. $q_{i,j} \in Z_+^0$ for $Q_i \in Z_+$, then we get the mixed integer linear programming problem. It is easy to see that such problem is NP-hard. To find the minimum of $E_T(q)$ first we must find an integer matrix q satisfying the set of equations given by the second relation in (3). Because this requires solution of the partition problem [7], we get the reduction of the partition problem to the minimum energy problem (3) with the requirements $q_{i,j} \in Z_+^0$, $Q_i \in Z_+$.

2   Solution of the Problem with the Monomial Cost Function

In this section we solve the minimum energy problem (3) for the data transmission cost function (1) with arbitrary real value of the exponent a. As can be seen, the monomial (1) for $a \geq 1$ and $x_i \geq x_j \geq x_k \geq 0$ satisfies the inequality

$$|x_i - x_j|^a + |x_j - x_k|^a \leq |x_i - x_k|^a, \tag{4}$$

and it is an example of a super-additive function [8]. This is because for $x_i - x_j = x$, $x_j - x_k = y$ (4) can be written in the form $|x|^a + |y|^a \leq |x + y|^a$. Solutions of the minimum energy problem in S_N with the cost function (1), where $a \geq 1$, can be easily generalized to any data transmission cost function $E(x_i, x_j)$ which satisfies the inequality

$$\forall_{x_i \geq x_j \geq x_k \geq 0} \ \ E(x_i, x_j) + E(x_j, x_k) \leq E(x_i, x_k). \tag{5}$$

From (5) it follows that the energy consumed by each sensor is minimal when it sends all of its data to the nearest neighbor in the direction of data collector. Let us assume, that the data is transmitted between two nodes located at the points x_i and x_k and (5) is satisfied, then the cost of transmission $E(x_i, x_k)$ can be reduced by transmitting the data via the j-th node located between them, i.e., via the the point x_j for which the inequality is satisfied $x_i \geq x_j \geq x_k \geq 0$. Because the total energy consumed by the network is a sum of energies consumed by its nodes, then the solution of the minimum energy problem (3) with (1) and $a \geq 1$ can be described by the transmission graph $T^{(0)} = \{T_{i,i-1}^{(0)}\}_{i=1}^N$, with the weight of each edge $T_{i,i-1}^{(0)}$ equal to $q_{i,i-1}^{(0)} = \sum_{j=i}^N Q_j$. The graph $T^{(0)}$ defines the next hop data transmission along the shortest path, where the shortest path means transmission along the distance $d(x_i, x_j)$ between the transmitter and the receiver.

For $a \leq 1$ elements of the data transmission cost function (1) are the sub-additive functions, i.e., satisfy the inequality $|x|^a + |y|^a \geq |x + y|^a$ [8]. Solutions of the minimum energy problem (3) with (1) and $a \leq 1$ can be generalized to the data transmission cost function $E(x_i, x_j)$ which satisfy the inequality

$$E(x_i, x_j) + E(x_j, x_k) \geq E(x_i, x_k). \tag{6}$$

The optimal behavior of the sensors which minimizes the total network energy $E_T(q)$ can be deduced from the inequality (6), but it does not uniquely determine the solution of (3). To get the unique solution of (3) we need a concrete form of the data transmission cost function, for example (1) or (2). From (6) it follows that the cost of transmission between two nodes located at the points x_i and x_j is minimal when the data is transmitted along the longest hops, i.e., any transmission via node which lie between x_i and x_j is less optimal. For the data transmission cost function (1) and $a \in (-\infty, 1]$ one may expect, that the optimal data transmission is given by the graph $T^{(1)} = \{T_{i,0}^{(1)}\}_{i=1}^{N}$, with the weights $q_{i,0}^{(1)} = Q_i$. This is true for sensors which lie in the interval $[x_N, \frac{1}{2}x_N]$ of S_N. When $x_i \in [x_N, \frac{1}{2}x_N]$, then the distance $d(x_i, x_0)$ between the transmitter and the data collector is maximal and the inequality (6) for x_i, $x_k = 0$ and any sensor $j \in [1, N]$, i.e., not only for $x_j < x_i$ but also for $x_j > x_i$, is satisfied. For sensors which lie in the interval $(0, \frac{1}{2}x_N)$ to find the optimal transmission it must be taken into account two data transmission paths to the data collector. The directly to the data collector transmission path $T_i^{(1)} = \{T_{i,0}^{(1)}\}$ and the two hops transmission given by the path $T_i^{(1')} = \{T_{i,N}^{(1')}, T_{N,0}^{(1')}\}$. Selection of which one depends on the value of the parameter $a \in (-\infty, 1]$ in (1). Let us assume, that in the interval $(0, \frac{1}{2}x_N)$ there are N' sensors. For each sensor k from the interval $(0, \frac{1}{2}x_N)$ we split the network S_N into two sets $V_1^{(k)}$ and $V_2^{(k)}$. To the set

$$V_1^{(k)} = \{s_i \in S_N \mid d(x_i, x_0) < d(x_k, x_0)\}, \quad k \in [1, N']$$

belong sensors which lie to the left the k-th sensor. The set $V_2^{(k)}$ is the completion of $V_1^{(k)}$, i.e., $V_2^{(k)} = S_N \setminus V_1^{(k)}$. The sensors from the interval $(0, \frac{1}{2}x_N)$ can be used to classify solutions of the minimum energy problem for the data transmission cost function (1) and any value of the exponent $a \in (-\infty, 1]$. For the k-th sensor from the interval $(0, \frac{1}{2}x_N)$, we must check whether the optimal data transmission path from the k-th sensor to the data collector is $\{T_{k,0}^{(k)}\}$ or $\{T_{k,N}^{(k)}, T_{N,0}^{(k)}\}$. In other words, we must check the values of the parameter a for which the inequality holds

$$E(x_k, x_N) + E(x_N, x_0) \geq E(x_k, x_0), \quad k \in [1, N'].$$

Instead solving these inequalities, we solve the set of equations

$$E(x_k, x_N) + E(x_N, x_0) - E(x_k, x_0) = 0, \quad k \in [1, N'],$$

which for (1) have the form

$$|x_N - x_k|^a + x_N^a - x_k^a = 0, \quad a \in (-\infty, 1], \quad k \in [1, N']. \tag{7}$$

For N' roots a_k of (7) we can form N' intervals $a \in [a_{k+1}, a_k]$, $k \in [0, N']$, where $a_0 = 1$ and $a_{N'+1} = -\infty$. For any $a \in [a_{k+1}, a_k]$ the following set of inequality holds

$$\begin{cases} |x_N - x_k|^a + x_N^a - x_k^a \leq 0, \\ |x_N - x_{k+1}|^a + x_N^a - x_{k+1}^a \geq 0, \end{cases} \tag{8}$$

which means, that for $a \in [a_{k+1}, a_k]$ the nodes $i \in [1, k]$ transmit data along the two hops path $\{T_{i,N}^{(k+1)}, T_{N,0}^{(k+1)}\}$ and the nodes $i \in [k+1, N]$ along the one hop path $\{T_{i,0}^{(k+1)}\}$. The above results summarizes the following:

Lemma 1. *The solutions of the minimum network energy problem for the data transmission cost matrix $E_{i,j} = |x_i - x_j|^a$ and $a \in R$ is given by the data transmission graphs*

$$\begin{cases} T^{(0)} = \{T_{i,i-1}^{(0)}\}_{i=1}^N & \text{for } a \in [1, \infty), \\ T^{(1)} = \{T_{i,0}^{(1)}\}_{i=1}^N & \text{for } a \in [a_1, 1], \\ T^{(k+1)} = \{T_{i,N}^{(k+1)}, T_{i',0}^{(k+1)}\}_{i=1,i'=k+1}^{k,N} & \text{for } a \in [a_{k+1}, a_k], k \in [1, N'-1], \\ T^{(N'+1)} = \{T_{i,N}^{(N'+1)}, T_{i',0}^{(N'+1)}\}_{i=1,i'=N'+1}^{N',N} & \text{for } a \in (-\infty, a_{N'}], \end{cases} \tag{9}$$

with the weights

$$\begin{cases} q_{i,i-1}^{(0)} = \sum_{j=i}^N Q_j, & i \in [1, N], \\ q_{i,0}^{(1)} = Q_i, & i \in [1, N], \\ q_{i,N}^{(k+1)} = Q_i, & i \in [1, k], \quad q_{i',0}^{(k+1)} = Q_{i'}, \quad i' \in [k+1, N-1], \\ q_{N,0}^{(k+1)} = Q_N + \sum_{j=1}^k Q_j, & k \in [1, N'], \end{cases} \tag{10}$$

where a_k are roots of the Eq. (7).

Detailed proof of the Lemma 1 can be found in [9]. On Fig. 1 it is shown the optimal data transmission graph for the minimum energy problem when $a \in [a_{k+1}, a_k]$ and $k \in [1, N']$.

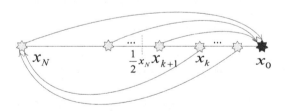

Fig. 1. Optimal data transmission in S_N for $a \in [a_{k+1}, a_k]$

The data transmission cost function

$$E_{i,j}(\bar{\lambda}, \bar{a}) = \sum_n \lambda_n |x_i - x_j|^{\alpha_n}, \quad \forall_n \ \lambda_n \geq 0, \tag{11}$$

which is a sum of monomials (1) with nonnegative coefficients λ_n, for $\forall_n \ \alpha_n \geq 1$ satisfies the inequality (5) and for $\forall_n \ \alpha_n \leq 1$ (6). The following lemma defines conditions under which the solution of (3) for the data transmission cost function (11) is given by (9).

Lemma 2. *For the data transmission cost function* (11), *where* $\forall_n \ \alpha_n \in [a_k, a_{k-1}]$, $k \in [1, N']$ *and* $\lambda_n \geq 0$ *the solution of the minimum energy problem is given by the weighted transmission graph* (9).

Proof. We can split the total energy consumed by the network into summands such, that $E_T(q) = \sum_n \lambda_n E_T^n(q)$, where $E_T^n(q) = \sum_{i=1}^N E_i(q, E^n)$ and $E_{i,j}^n = |x_i - x_j|^{\alpha_n}$. Because for each $\alpha_n \in [a_k, a_{k-1}]$ the minimal energy utilized by the sensors is given by the same transmission graph $T^{(k)}$ from (9), then the optimal data transmission for $E_{i,j}(\bar{\lambda}, \bar{\alpha})$ is also given by $T^{(k)}$.   ◇

3   Solution of the Problem with the Polynomial Cost Function

We note that, the objective function $E_T(q)$ is linear and continuous and for this reason its minima lie on the border of the feasible set defined by the second relation in (3). From this follows, that the data transmitted by each node cannot be split and it must be sent to a single receiver. From the Lemma 2 we know, that when the exponents a, b in the data transmission cost function (2) belong to the same interval $[a_k, a_{k-1}]$, then the optimal transmission graph for $E_{i,j} = |x_i - x_j|^a$ and $E_{i,j} = |x_i - x_j|^b$ is the same and it is also optimal for $E_{i,j} = |x_i - x_j|^a + \lambda |x_i - x_j|^b$. The problem arises when the solutions of (3) for two exponents a and b in $E_{i,j} = |x_i - x_j|^a$ are given by different data transmission graphs T_a and T_b. This is because, the optimal transmission graph for the cost function (2) cannot be sum of the graphs T_a and T_b, unless the transmitted data in the graph $T_a \cup T_b$ are not split.

In this section we consider solutions of (3) in the one dimensional, regular sensor network L_N for which the nodes are located at the $x_i = i$ points of the half line. For the L_N network the data transmission cost matrix (2) has the form

$$E_{i,j} = |i - j|^a + \lambda |i - j|^b. \tag{12}$$

All presented below results are also valid for the non-regular network S_N, but the formulas are cumbersome because of their size and they will not be presented here.

The following lemma describes solution of the the minimum energy problem for the L_N network when the exponents a and b in (12) belong to the neighboring intervals $[a_k, a_{k-1}]$.

Lemma 3. *For the data transmission cost matrix* (12), *when*

$$\begin{cases} a \in [1, \infty), & b \in [a_1, 1], \\ a \in [a_k, a_{k-1}], & b \in [a_{k+1}, a_k], \quad k \in [1, N' - 1], \\ a \in [a_{N'}, a_{N'-1}], & b \in (-\infty, a_{N'}], \end{cases}$$

the solution of the minimum energy problem is given by

$$
\begin{cases}
T^{(0)} & \text{for } \lambda \in [0, \lambda_0], \quad T^{(1)} \quad \text{for } \lambda \in [\lambda_0', \infty), \\
T^{(k)} & \text{for } \lambda \in [0, \lambda_k], \quad T^{(k+1)} \text{ for } \lambda \in [\lambda_k, \infty), \quad k \in [1, N'-1], \\
T^{(N')} & \text{for } \lambda \in [0, \lambda_{N'}], \quad T^{(N'+1)} \text{ for } \lambda \in [\lambda_{N'}, \infty),
\end{cases}
\tag{13}
$$

where

$$
\begin{cases}
\lambda_0 = \frac{2^a - 2}{2 - 2^b}, \\
\lambda_0' = \frac{N^a - |N - N' - 1|^a - (N'+1)^a}{|N - N' - 1|^b + (N'+1)^b - N^b}, \\
\lambda_k = \frac{(N-k)^a + N^a - k^a}{k^b - (N-k)^b - N^b}, \quad k \in [1, N'],
\end{cases}
\tag{14}
$$

$a_0 = 1$, $N' = \frac{N-2}{2}$ *for* N *even and* $N' = \frac{N-1}{2}$ *for* N *odd,* $T^{(k)}$, $k \in [0, N']$ *is a set of data transmission graphs given by* (9).

Proof. If $a \in [a_k, a_{k-1}]$ and $b \in [a_{k+1}, a_k]$ in (12), then we know from Lemma 1 that for a sufficiently small λ the solution of (3) is given by the weighted transmission graph $T^{(k)} = \{T_{i,N}^{(k)}, T_{i',0}^{(k)}\}_{i=1,i'=k}^{k-1,N}$, and for a sufficiently large λ the solution is given by $T^{(k+1)} = \{T_{i,N}^{(k+1)}, T_{i',0}^{(k+1)}\}_{i=1,i'=k+1}^{k,N}$, $k \in [2, N']$. Because of the linearity and continuity of the objective function $E_T(q)$ its minimum lies on the border of the feasible set. This means that the data transmitted by each node cannot be split and the optimal transmission graph for arbitrary value of the λ parameter in (12) cannot be sum of the two graphs $T^{(k)}$ and $T^{(k+1)}$. To find the optimal transmission for any value of λ in (12) we order the transmission graphs in a sequence such that the cost of data transmission along $T^{(k)}$ is less or equal the costs along $T^{(k')}$

$$
E_T(q^{(k)}) \le E_T(q^{(k')}).
$$

The minimal graph $T^{(k')}$ determines the λ_k below which the solution of (3) is given by weight matrix $q^{(k)}$ of the graph $T^{(k)}$. Similarly, the transmission graph $T^{(k'')}$ for which the inequality $E_T(q^{(k+1)}) \le E_T(q^{(k'')})$ is satisfied determines the value of λ_k' above which the solution of (3) is given by $q^{(k+1)}$. We know from the solution (9), (10) and the inequalities (6), (8), that for the k-th node, for any $a \in [a_k, a_{k-1}]$ and $b \in [a_{k+1}, a_k]$, $k \in [2, N']$ between $T^{(k)}$ and $T^{(k+1)}$ there is no other optimal data transmission graphs. For this reason, from the inequality

$$
E_T(q^{(k)}) \le E_T(q^{(k+1)})
$$

we get the values (14) of the parameter λ_k for which the graphs $T^{(k)}$ and $T^{(k+1)}$ are optimal.

When $a \in [1, \infty)$ and $b \in [a_1, 1]$ in (12), then the solution of (3) for a sufficiently small λ is given by the data transmission graph $T^{(0)} = \{T_{i,i-1}^{(0)}\}_{i=1}^{N}$, and for a sufficiently large λ by the transmission graph $T^{(1)} = \{T_{i,0}^{(1)}\}_{i=1}^{N}$. Increasing the parameter λ in (12) we pass by set of data transmission graphs between $T^{(0)}$ and $T^{(1)}$. The next data transmission graph, which requires more energy then the next hop transmission $T^{(0)}$ is the graph $T^{(0+)}$ in which there is an edge $T_{N,N-2}^{(0+)}$ along the N-th node transmits its Q_N of data to the $(N-2)$ node.

The nodes from L_N, $i \in [1, N-1]$ uses the next hop data transmission subgraph with edges $T^{(0)}_{i,i-1}$. Note that, we cannot select an arbitrary edge $T^{(0+)}_{i,i-2}$ for $i \neq N$. This is because we want to transmit the minimal amount of data along the edge $T^{(0+)}_{i,i-2}$ and this is satisfied only for the N-th node. The total energy utilized by the network for $T^{(0+)}$ graph is given by the formula

$$E_T(q^{(0+)}) = E_T(q^{(0)}) - E_{N,N-1} - E_{N-1,N-2} + E_{N,N-2}.$$

Increasing the value of λ in (2) we must pass from the transmission graph $T^{(0)}$ to $T^{(0+)}$. Solving the inequality

$$E_T(q^{(0+)}) \leq E_T(q^{(0)})$$

with respect to the parameter λ, we get the upper bound $\lambda \leq \frac{2^a - 2}{2 - 2^b}$ given in the lemma. When we start decrease the value of λ, above which the data transmission graph $T^{(1)}$ is optimal, then we pass to the graph $T^{(1+)}$ for which the N-th node transmits its Q_N of data along the path which consists of the two edges

$$T^{(1+)}_{N,N-n'-1}, T^{(1+)}_{N-n'-1,0} \in T^{(1+)}.$$

In other words, this is the transmission path which consists of a one hop of the length $n' + 1$ and the second hop of the length $N - (n' + 1)$. For the L_N network and $k \in [1, N']$, where N' is the number of nodes in the first part of the network, i.e., $(0, \frac{1}{2}N)$, $n' = N'$, i.e., $N' = \frac{N-2}{2}$ for N even and $N' = \frac{N-1}{2}$ N odd. The total energy consumed by the network for transmission along the graph $T^{(1+)}$ is given by the formula

$$E_0(q^{(1+)}_N) = E_0(q^{(1)}_N) - E_{N,0} + E_{N,N'+1} + E_{N'+1,0}.$$

Solving the inequality $E_T(q^{(1+)}) \leq E_T(q^{(1)})$, with respect to the parameter λ, we get the lower bound $\lambda'_0 = \frac{N^a - |N-N'-1|^a - (N'+1)^a}{|N-N'-1|^b + (N'+1)^b - N^b}$ for which the optimal data transmission graph is $T^{(1)}$.   ◇

The next two lemmas describe the optimal transmission graphs when the values of a and b in (12) does not belong to the neighboring intervals $[a_k, a_{k-1}]$.

Lemma 4. *Let the exponents of the data transmission cost matrix* (12) *be in the intervals* $a \in [a_k, a_{k-1}]$ *and* $b \in [a_{k'}, a_{k'-1}]$, $k \in [1, N'-1]$, $k' \in [3, N'+1]$, $k' \geq k+2$, *then the optimal transmission graphs to the minimum energy problem are*

$$\begin{cases} T^{(k)} & \text{for } \lambda \in (0, \lambda_k], \\ T^{(k+i)} & \text{for } \lambda \in [\lambda_{k+i-1}, \lambda_{k+i}], \ i \in [1, k'-k-1], \\ T^{(k')} & \text{for } \lambda \in [\lambda_{k'-1}, \infty), \end{cases}$$

where λ_{k+i-1} *is the solutions of the inequality*

$$E_T(q^{(k-1+i)}) \leq E_T(q^{(k+i)}), \quad i \in [1, k'-k].$$

Proof. We know that, for a sufficiently small λ the solution of (3), (12), when $a \in [a_k, a_{k-1}]$ and $b \in [a_{k'}, a_{k'-1}]$, is given by the transmission graph $T^{(k)}$. From the inequality (6) and continuity of the objective function E_T follows that less optimal, the next to $T^{(k)}$ is the transmission graph $T^{(k+1)}$. The value of λ_k above which $T^{(k)}$ is not optimal is determined from the inequality $E_T(q^{(k)}) \leq E_T(q^{(k+1)})$. Similarly, for a sufficiently large λ the solution of (3) is given by the graph $T^{(k')}$. From the inequality (6) it follows that less optimal, the closest to $T^{(k')}$, is the transmission graph $T^{(k'-1)}$. Solving the inequality $E_T(q^{(k'-1)}) \leq E_T(q^{(k')})$ we get the lower bound of $\lambda_{k'-1}$ for which $T^{(k')}$ is an optimal graph. By varying the parameter λ between λ_k and $\lambda_{k'-1}$, when $a \in [a_k, a_{k-1}]$ and $b \in [a_{k'}, a_{k'-1}]$ in (12), we get the various optimal transmission graphs, different from $T^{(k)}$ and $T^{(k')}$. From the inequality (6) and continuity of the objective function E_T follows, that the only solutions of (3) for $\lambda \in [\lambda_k, \lambda_{k'}]$ can be transmission graphs $T^{(k+i)}$, $i \in [1, k'-k-1]$. By solving the set of inequalities $E_T(q^{(k-1+i)}) \leq E_T(q^{(k+i)})$ for $i \in [1, k'-k]$ we get the ordered sequence of λ_{k+i-1}. For any λ in the interval $[\lambda_{k+i-1}, \lambda_{k+i}]$ the optimal transmission graph is $T^{(k+i)}$. $\diamond$

The following lemma defines the optimal transmission graph when $a \in [1, \infty)$ and $b \leq a_1$ in (12).

Lemma 5. *Let the exponents of (12) be in the intervals $a \in [1, \infty)$ and $b \in [a_k, a_{k-1}]$, $k \geq 2$, then the optimal transmission graphs to the minimum energy problem are $T^{(0)}$ for $\lambda \in [0, \lambda_0]$ and $T^{(k)}$ for $k \geq 2$ $\lambda \in [\lambda_k, \infty)$, where λ_0, λ_k are given by (14).*

Proof. This lemma follows from the Lemma 3. To get the upper bound λ_0 of the parameter λ for which the transmission graph $T^{(0)}$ is optimal, we need to solve the inequality $E_T(q^{(0+)}) \leq E_T(q^{(0)})$. To get the lower bound of the parameter λ for which the transmission graphs $T^{(k)}$ are optimal we have to solve the set of inequalities $E_T(q^{(k-1)}) \leq E_T(q^{(k)})$, $k \in [2, N'+1]$ which solution λ_k are given by (14). $\diamond$

The optimal transmission graphs of the minimum energy problem when $a \in [a_1, 1]$ and $b \leq a_3$ in (12) are for the parameter λ.

Lemma 6. *Let the exponents of (12) be in the intervals $a \in [a_1, 1]$ and $b \in [a_k, a_{k-1}]$, $k \geq 3$, then the optimal transmission graphs to the minimum energy problem are $T^{(1)}$ for $\lambda \in [0, \lambda_0']$ and $T^{(k)}$ for $\lambda \in [\lambda_k, \infty)$, $k \in [3, N']$ where λ_0', λ_k are given by (14).*

Proof. This lemma follows from the Lemma 3. To get the upper bound λ_0' of the parameter λ for which the transmission graph $T^{(1)}$ is optimal we need to solve the inequality $E_T(q^{(1+)}) \leq E_T(q^{(1)})$. The lower bound λ_k of the of the parameter λ for which the transmission graphs $T^{(k)}$ are optimal can be determined from the the set of inequalities $E_T(q^{(k-1)}) \leq E_T(q^{(k)})$, $k \in [3, N'+1]$, which solution λ_k are given by (14). $\diamond$

4   Solution of the Problem with SINR Function

In previous sections we defined the minimum energy problem in terms of the data transmission cost matrix $E_{i,j}$ and data flow matrix $q_{i,j}$. In such formalism there is no information in the model about the data transmission rate, the sensors operating time and transmission errors caused by the noise and signal interference. In this section define the minimum energy problem in terms of sensors signal power, data transmission time and capacities of a transmission channels. We show, that for the optimal data transmission of the minimum energy problem in the noisy channel there is no interference of signals. For omnidirectional antennas, when the signal of the transmitting node is heard in the whole network this is equivalent to the sequential data transmission. We prove, that for the point-to-point data transmission utilized by the sensors in the physical layer, when the transmitter adjust the power of its radio signal to the distance to the receiver, the solutions of the minimum energy problem coincide with the solutions discussed in the previous sections.

We assume, that the power of the transmitting signal at the receiver must have some minimal level P_0. This requirement means, that the transmitting node must generate the signal with the strength

$$P_{i,j} = P_0\, \gamma_{i,j}^{-1}, \tag{15}$$

where $\gamma_{i,j} = \gamma(x_i, x_j)$ is the signal gain function between sender and receiver located at the points x_i and x_j of the line. For the transmission model (15) the energy consumed by the i-th sensor is given by the formula

$$E_i(t) = P_0 \sum_{j=1}^{N} \gamma_{i,j}^{-1} t_{i,j}, \tag{16}$$

To get non-trivial solution of the minimum energy problem we must assume that the capacities of the transmission channels are limited, otherwise the minimum energy of each node is reached for zero transmission time $t_{i,j} = 0$. To define the size of the channel capacity we use the Shannon-Hartley formula modified by the Signal to Interference plus Noise Ratio (SINR) function [10,11],

$$C(x_i, x_j, U_{i,j}^n) = \log(1 + s(x_i, x_j, U_{i,j}^n)), \tag{17}$$

where

$$s(x_i, x_j, U_{i,j}^n) = \frac{P_0}{N_o + P_0 \sum_{(k,m)\in U_{i,j}^n} \gamma(x_k, x_m)^{-1} \gamma(x_k, x_j)}$$

is the SINR function and $U_{i,j}^n \subset S_N$ is some set of transmitter-receiver pairs which signal of the transmitters interfere with the signal of the i-th node. For wireless networks in which the nodes use the omnidirectional antennas and the signal is detected by any node of the network, $U_{i,j}^n$ can be defined as a set of node pairs which transmit data simultaneously, i.e.,

$$U_{i,j} = \left\{ (s_{i'}, s_{j'}) \in S_N \times S_N | t_{i,j}^{(s)} = t_{i',j'}^{(s)},\ t_{i,j}^{(e)} = t_{i',j'}^{(e)} \right\}.$$

where $t_{i,j}^{(s)}$ and $t_{i,j}^{(e)}$ is the start and the end of transmission time between i-th and j-th node. By definition $(i,j) \notin U_{i,j}$. The amount of data transmitted by the i-th node to the j-th node during the time $t_{i,j}$ with the transmission rate $c_{i,j}$ is given by the formula

$$q_{i,j} = c_{i,j} \, t_{i,j}. \tag{18}$$

We assume, that the transmission rate $c_{i,j}$ satisfies the inequality $0 \leq c_{i,j} \leq C(x_i, x_j, U_{i,j}^n)$, where $C(x_i, x_j, U_{i,j}^n)$ is given by (17). Because in general a set of sensors can transmit data simultaneously, thus we need to modify the node energy consumption formula (16) to the form

$$E_i(\bar{t}) = P_0 \sum_{j,n} \gamma_{i,j}^{-1} \, t_{i,j}^n,$$

where $\bar{t} = (t^1, \ldots, t^n, \ldots)$ is a tuple of time matrices t^n with elements $t_{i,j}^n$, which define the data transmission time between the i-th and j-th nodes in the presence of transmitters from the set $U_{i,j}^n$. The objective function of the minimal energy problem with SINR function is given by the formula

$$E_T(\bar{t}) = \sum_{i=1}^{N} E_i(\bar{t}) = \sum_{i=1}^{N} P_0 \sum_{j,n} \gamma_{i,j}^{-1} \, t_{i,j}^n.$$

From the data flow constraints, defined by the second formula in (3) and (18), it follows that the minimum energy is consumed by the network when the transmission rate between two nodes is maximal and equals to the channel capacity, i.e. $c_{i,j}^n = C_{i,j}^n$. Taking this into account the minimum energy problem with SINR function can be written in the form

$$\begin{cases} \min_{\bar{t}} E_T(\bar{t}), \\ \sum_{i,n} C_{i,j}^n t_{i,j}^n = Q_i + \sum_{j,n} C_{j,i}^n t_{j,i}^n, \\ t_{i,j}^n \geq 0, \quad Q_i > 0, \end{cases} \tag{19}$$

where $C_{i,j}^n$ is given by (17). The results of the following lemma allows us further reduce the problem (19).

Lemma 7. *The optimal data transmission for the minimum energy problem (19) is the transmission without interference.*

Proof. For the fixed amount of data Q_i generated by each sensor, the transmission times $t_{i,j}^n$ in (19) are minimal when coefficients $C_{i,j}^n$ are maximal. From (17) it follows that maximum value of the transmission rate $C_{i,j}^n$ is achieved when $U_{i,j}^n = \emptyset$, which means that in the network there is no interference of signals. ◇

From the Lemma 7 it follows that to solve the minimum energy problem it is enough to consider only the constant channel capacities $\forall_{i,j} \; C(x_i, x_j) = \log(1 + \frac{P_0}{N_o}) = C_0$. The minimum energy problem for noisy channel with the constant channel capacity can be defined by the following set of formulas

$$\begin{cases} \min_t \sum_{i \in [1,N]} E_i(t), \\ E_i(t) = P_0 \sum_j \gamma_{i,j}^{-1} t_{i,j}, \\ C_0 \sum_i t_{i,j} = Q_i + C_0 \sum_j t_{j,i}, \\ t_{i,j} \geq 0, \quad Q_i > 0. \end{cases} \tag{20}$$

To solve the problem (20) for a given signal gain function $\gamma_{i,j}$ we transform (20) to the minimum energy problem defined in (3). By identifying the variables

$$\begin{cases} q_{i,j} \to P_0 t_{i,j}, \\ E_{i,j} \to \gamma_{i,j}^{-1}, \\ Q_i \to \frac{P_0}{C_0} Q_i, \end{cases}$$

we get the equivalence of the two problems. For the signal gain functions $\gamma_{i,j} = |x_i - x_j|^{-a}$, $\gamma_{i,j} = \frac{1}{|i-j|^a + \lambda |i-j|^b}$, by means of the above transformation we can obtain from (9) and (13) the solutions of (20).

5   Conclusions

In the paper we solved the minimum energy problem in one dimensional wireless sensor networks for the data transmission cost function $E(x_i, x_j) = |x_i - x_j|^a$ with any real value of the exponent a. We showed, how to find the solution of the problem when the data transmission cost function is of the form $E(x_i, x_j) = |x_i - x_j|^a + \lambda |x_i - x_j|^b$ and $a, b \in R$, $\lambda \geq 0$. There are several intervals for the parameter λ for which the optimal transmission graphs are not determined. For example, when $a \in [1, \infty)$, $b \in [a_1, 1]$ in the interval $[\lambda_0, \lambda_0']$ there are transmission graphs which lie between $T^{(0)}$ and $T^{(1)}$ and are solutions of (3). These graphs can be identified by means of the ordering method utilized in the Lemmas 3, 4, 5 and 6. We defined the minimum energy problem in terms of sensors signal power, transmission time and capacities of a transmission channels. We proved, that for the point-to-point data transmission utilized by the sensors in the physical layer, when the transmitter adjust the power of its radio signal to the distance to the receiver, the solutions of the minimum energy problem written in terms of data transmission cost function $E_{i,j}$ and in terms of sensor signal power coincide.

Obtained in the paper analytical solutions of the minimum energy problem for the one dimensional networks can be utilized to solve the problem in two and more dimensions in networks with symmetries. For example, one can easily prove that for the star shaped network, when the sink is located in the center of the network, the solution of the minimum energy problem can be reduced to the solution of the problem discussed in this paper. Similar problems, for the maximum lifetime problem in two-dimensional wireless sensor networks, are discussed in [12]. Presented in the paper classification of solutions of the minimum energy problem, can be applied to study the stability problems of routing algorithms in wireless sensor networks. It is well know, that efficiency of routing algorithms in such networks highly depends on the shape of the network, or more precisely on the cost of data transmission between sensors [13–15]. The reason of instability of the algorithms is that, they do not approximate analytical solution of the routing problem. Thus having classified the analytical solutions of a given problem one can build stability classes of routing algorithms which approximate given solution. Based on the knowledge about stability of the exact solution one can predict stability of routing algorithms in each class and also algorithms which are mixture of algorithms from different classes.

References

1. Chang, J.H., Tassiulas, L.: Energy conserving routing in wireless ad-hoc networks. In: Proceedings of the INFOCOM, pp. 22–31 (2000)
2. Giridhar, A., Kumar, P.R.: Maximizing the functional lifetime of sensor networks. In: Proceedings of the 4th International Symposium on Information Processing in Sensor Networks. Piscataway, NJ, USA. IEEE Press (2005)
3. Acharya, T., Paul, G.: Maximum lifetime broadcast communications in cooperative multihop wireless ad hoc networks: centralized and distributed approaches. Ad Hoc Netw. **11**, 1667–1682 (2013)
4. Li, L., Halpern, J.Y.: A minimum-energy path-preserving topology-control algorithm. IEEE Trans. Wireless Commun. **3**, 910–921 (2004)
5. Rodoplu, V., Meng, T.H.: Minimum energy mobile wireless networks. IEEE J. Sel. Areas Commun. **17**(8), 1333–1344 (1999)
6. Baccelli, F., Blaszczyszyn, B.: Stochastic Geometry and Wireless Networks, vol. 1, 2. Now Publishers Inc., Breda (2009)
7. Garey, M., Johnson, D.: Computers and Intractability: A Guide to Theory of NP-Completeness. Freeman, San Francisco (1979)
8. Steele, M.J.: Probability Theory and Combinatorial Optimization. SIAM, Philadelphia (1997)
9. Lipiński, Z.: On classification of data transmission strategies in one dimensional wireless ad-hoc networks with polynomial cost function. In: Monographs of System Dependability, DepCoS-RELCOMEX, Poland, pp. 85–104 (2012)
10. Gupta, P., Kumar, P.R.: The capacity of wireless networks. IEEE Trans. Inf. Theory **46**(2), 388–404 (2000)
11. Franceschetti, M., Meester, R.: Random Networks for Communication. Cambridge University Press, Cambridge (2007)
12. Lipiński, Z.: On the role of symmetry in solving maximum lifetime problem in two-dimensional sensor networks (2014). arXiv preprint arXiv:1402.2327
13. Akkaya, K., Younis, M.: A survey on routing protocols for wireless sensor networks. Ad Hoc Netw. **3**(3), 325–349 (2005)
14. Cardei, M., Wu, J.: Energy-efficient coverage problems in wireless ad-hoc sensor networks. Comput. Commun. **29**, 413–420 (2006)
15. Lin, J., Zhou, X., Li, Y.: A minimum-energy path-preserving topology control algorithm for wireless sensor networks. Int. J. Autom. Comput. **6**(3), 295–300 (2009)

Detection of Malicious Data in Vehicular Ad Hoc Networks for Traffic Signal Control Applications

Bartłomiej Płaczek$^{(\boxtimes)}$ and Marcin Bernas

Institute of Computer Science, University of Silesia,
Będzińska 39, 41-200 Sosnowiec, Poland
placzek.bartlomiej@gmail.com, marcin.bernas@gmail.com

Abstract. Effective applications of vehicular ad hoc networks in traffic signal control require new methods for detection of malicious data. Injection of malicious data can result in significantly decreased performance of such applications, increased vehicle delays, fuel consumption, congestion, or even safety threats. This paper introduces a method, which combines a model of expected driver behaviour with position verification in order to detect the malicious data injected by vehicle nodes that perform Sybil attacks. Effectiveness of this approach was demonstrated in simulation experiments for a decentralized self-organizing system that controls the traffic signals at multiple intersections in an urban road network. Experimental results show that the proposed method is useful for mitigating the negative impact of malicious data on the performance of traffic signal control.

Keywords: Vehicular networks · Malicious data · Sybil attack · Traffic signal control

1 Introduction

Vehicular ad hoc networks (VANETs) facilitate wireless data transfer between vehicles and infrastructure. The vehicles in VANET can provide detailed and useful data including their positions, velocities, and accelerations. This technology opens new perspectives in traffic signal control and creates an opportunity to overcome main limitations of the existing roadside sensors, i.e., low coverage, local measurements, high installation and maintenance costs. The availability of the detailed data from vehicles results in a higher performance of the traffic signal control [1–3].

The VANET-based traffic signal systems have gained considerable interest in recent years. In this field the various solutions have been proposed that extend existing adaptive signal systems for isolated intersections [4,5]. For such systems the data collected in VANET are used to estimate queue lengths and vehicle delays. On this basis optimal cycle length and split of signal phases are calculated. Similar adaptive approach was also used to control traffic signals at multiple intersections in a road network [6,7]. Particularly advantageous

© Springer International Publishing Switzerland 2016
P. Gaj et al. (Eds.): CN 2016, CCIS 608, pp. 72–82, 2016.
DOI: 10.1007/978-3-319-39207-3_7

for VANET-based systems is the self-organizing signal control scheme, which enables a decentralized optimization, global coordination of the traffic streams in a road network, and improved performance [8].

Effective VANET applications in traffic signal control require new methods for real-time detection of attacks that are based on malicious data. Injection of malicious data can result in significantly decreased performance o the traffic signal control, increased vehicle delays, fuel consumption, congestion, or even safety threats.

This paper introduces a method for the above mentioned applications, which can be used to detect malicious data. The considered malicious data are injected by vehicle nodes that perform Sybil attacks, i.e., create a large number of false vehicle nodes in order to influence the operation of traffic signals. The proposed method detects the malicious data by combining a model of expected driver behaviour with a position verification approach.

The paper is organized as follows. Related works are discussed in Sect. 2. Section 3 introduces the proposed method. Results of simulation experiments are presented in Sect. 4. Finally, conclusions are given in Sect. 5.

2 Related Works

The problem of malicious nodes detection in VANETs has received particular attention and various methods have been proposed so far [9]. The existing solutions can be categorized into three main classes: encryption and authentication methods, methods based on position verification, and methods based on VANET modelling.

In the encryption and authentication methods, malicious nodes detection is implemented by using authentication mechanisms. One of the approaches is to authenticate vehicles via public key cryptography [10]. The methods that use public key infrastructure were discussed in [11]. Main disadvantages of such methods are difficulties in accessing to the network infrastructure and long computational time of encryption and digital signature processing. Public key encryption and message authentication systems consume time and memory. Thus, bandwidth and resource consumption is increased in the public key systems.

In [12] an authentication scheme was proposed, which assumes that vehicles collect certified time stamps from roadside units as they are travelling. The malicious nodes detection is based on verification of the collected series of time stamps. Another similar method [13] assumes that vehicles receive temporary certificates from roadside units and malicious nodes detection is performed by checking spatial and temporal correlation between vehicles and roadside units. These methods require a dense deployment of the roadside units.

Position verification methods are based on the fact that position reported by a vehicle can be verified by other vehicles or by roadside units [14]. The key requirement in this category of the methods is accurate position information. A popular approach is to detect inconsistencies between the strength of received signal and the claimed vehicle position by using a propagation model. According to the method introduced in [15] signal strength measurements are collected when nodes send beacon messages. The collected measurements are used to estimate position of the nodes according to a given propagation model. A node is considered to be malicious if its claimed position is too far from the estimated one. In [16] methods were proposed for determining a transmitting node location by using signal properties and trusted peers collaboration for identification and authentication purposes. That method utilizes signal strength and direction measurements thus it requires application of directional antennas.

Xiao et al. [15] and Yu et al. [17] proposed a distributed method for detection and localization of malicious nodes in VANET by using verifier nodes that confirm claimed position of each vehicle. In this approach, statistical analysis of received signal strength is performed by neighbouring vehicles over a period of time in order to calculate the position of a claimer vehicle. Each vehicle has the role of claimer, witness, or verifier on different occasions and for different purposes. The claimer vehicle periodically broadcasts its location and identity information, and then, verifier vehicle confirms the claimer position by using a set of witness vehicles. Traffic pattern analysis and support of roadside units is used for selection of the witness vehicles. Yan et al. [18] proposed an approach that uses on-board radar to detect neighbouring vehicles and verify their positions.

The modelling-based methods utilize models that describe expected behaviour of vehicle nodes in VANET. These methods detect malicious nodes by comparing the model with information collected from the vehicles. Golle et al. [19] proposed a general model-based approach to evaluating the validity of data collected form vehicle nodes. According to this approach, different explanations for the received data are searched by taking into account the possible presence of malicious nodes. Explanations that are consistent with a model of the VANET get scores. The node accepts data that are consistent with the most scored explanation. On this basis, the nodes can detect malicious data and identify the vehicles that are the sources of such data. Another method in this category relies on comparing the behaviour of a vehicle with a model of average driving behaviour, which is built on the fly by using data collected from other vehicles [20].

In [21] a malicious data detection scheme was proposed for post crash notification applications that broadcast warnings to approaching traffic. A vehicle node observes driver's behaviour for some time after the warning is received and compares it with some expected behaviour. The vehicle movement in the absence of any traffic accident is assumed to follow some free-flow mobility model, and its movement in case of accident is assumed to follow some crash-modulated mobility model. On this basis the node can decide if the received warning is true or false.

A framework based on subjective logic was introduced in [22] for malicious data detection in vehicle-to-infrastructure communication. According to that approach, all data collected by a vehicle node can be mapped to a world-model and can then be annotated with opinions by different misbehaviour detection mechanisms. The opinions are used not only to express belief or disbelief in a stated fact or data source, but also to model uncertainty. Authors have shown that application of the subjective logic operators, such as consensus or transitivity, allows different misbehaviour detection mechanisms to be effectively combined.

According to the authors' knowledge, the problem of malicious data detection for traffic signal control applications has not been studied so far in VANET-related literature. In this paper a malicious data detection method is introduced for VANET-based traffic signal control systems. The proposed method integrates a model of expected driver behaviour with position verification in order to detect the malicious data that can degrade the performance of road traffic control at signalized intersections.

3   Proposed Method

This section introduces an approach, which was intended to detect malicious data in VANET applications for road traffic control at signalized intersections. The considered VANET is composed of vehicle nodes and control nodes that manage traffic signals at intersections. Vehicles are equipped with sensors that collect speed and position data. The collected information is periodically transmitted from vehicles to control nodes. The control nodes use this information for optimizing traffic signals to decrease delay of vehicles and increase capacity of a road network.

In order to detect and filter out malicious data, the control node assigns a trust level to each reported vehicle. The trust levels are updated (decreased or increased) after each data delivery by using the rules discussed later in this section. When making decisions related to changes of traffic signals, the control node takes into account only those data that were collected by vehicles with positive trust level. The data delivered by vehicles with trust level below or equal to zero are recognized as malicious and ignored.

At each time step vehicle reports ID of its current traffic lane, its position along the lane, and velocity. If vehicles i and j are moving in the same lane during some time period and at the beginning of this period vehicle i is in front of vehicle j then the same order of vehicles has to be observed for the entire period. Thus, the following rule is used to detect the unrealistic behaviour of vehicles in a single lane:

$$x_i(t) - x_j(t) > \epsilon_x \wedge x_i(t-\delta) - x_j(t-\delta) < -\epsilon_x \wedge l_i(t') = l_j(t') \; \forall t' : t-\delta \leq t' \leq t, \; (1)$$

where: $l_i(t)$, $x_i(t)$, $v_i(t)$ denote respectively lane ID, position, and velocity of vehicle i at time step t, δ is length of the time period, and ϵ_x is maximum localization error. It is assumed that the frequency of data reports enables recognition of overtaking. In situation when condition (1) is satisfied and both vehicles have positive trust level, the trust level of both vehicles (i and j) is decreased by value α because the collected data do not allow us to recognize which one of the two reported vehicles is malicious. If one of the two vehicles has non-positive trust level then the trust level is decreased only for this vehicle.

According to the second rule (so-called reaction to signal rule), current traffic signals are taken into account in order to recognize the malicious data. If the information received from vehicle i indicates that this vehicle enters an intersection when red signal is displayed or stops at green signal then the trust level of vehicle i is decreased by value α. The following condition is used to detect the vehicles passing at red signal:

$$h_n - x_j(t - \delta) > \epsilon_x \wedge h_n - x_j(t) < -\epsilon_x \wedge s_n(t') = \text{red } \forall t' : t - \delta \leq t' \leq t, \quad (2)$$

where: h_n is position of the stop line for signal n, $s_n(t)$ is the colour of signal n at time t, and the remaining symbols are identical to those defined for rule (1). The vehicles stopped at green signal are recognized according to condition:

$$|h_n - x_j(t')| < \epsilon_x \ \forall t' : t - \delta \leq t' \leq t \wedge s_n(t') = \text{green } \forall t' : t - \delta \leq t' \leq t, \quad (3)$$

where $|\cdot|$ denotes absolute value and the remaining symbols were defined above. In opposite situations, when the vehicle enters the intersection during green signal or stops at red signal then its trust level is increased by α.

Theoretical models of vehicular traffic assume that vehicles move with desired free flow velocity if they are not affected by other vehicles or traffic signals [23]. Based on this assumption, expected velocity of vehicle i can be estimated as follows:

$$\hat{v}_i(t) = \min\left(v_f, \frac{h_i(t) - h_{\min}}{\tau}\right), \quad (4)$$

where: v_f is free flow velocity, $h_i(t)$ is headway distance, i.e., distance between vehicle i and vehicle in front in the same lane or distance between vehicle i and the nearest red signal, $h_{\min}$ is minimum required headway distance for stopped vehicle, τ denotes time which is necessary to stop vehicle safely and leave adequate space from the preceding car or traffic signal. Time τ can be determined according to the two seconds rule, which is suggested by road safety authorities [24].

When the velocity reported by a vehicle differs significantly from the expected velocity then the vehicle is suspected to be malicious. Thus, the trust level of the vehicle is decreased. In opposite situation, when the reported velocity is close to the expected value, the trust level is increased. According to the above assumptions, the trust level is updated by adding value $u_i(t) \cdot \beta$ and $u_i(t)$ is calculated using the following formula:

$$u_i(t) = \begin{cases} 1, & |\hat{v}_i(t) - v_i(t)| < \epsilon_v, \\ -\frac{|\hat{v}_i(t) - v_i(t)|}{v_f}, & \text{else}, \end{cases} \quad (5)$$

where ϵ_v is a threshold of the velocity difference, and the remaining symbols were defined earlier. Threshold ϵ_v was introduced in Eq. (5) to take into account error of velocity measurement and uncertainty of the expected velocity determination.

The last rule for updating trust level assumes that vehicles are equipped with sensors which enable detection and localization of neighbouring vehicles within distance r. In this case each vehicle reports the information about its own position and speed as well as positions of other vehicles in the neighbourhood. The information about neighbouring vehicles, delivered by vehicle j at time t, is represented by set $D_j(t)$:

$$D_j(t) = \{\langle x_k(t), l_k(t) \rangle\}, \tag{6}$$

where $x_k(t)$ and $l_k(t)$ denote position and lane of k-th vehicle in the neighbourhood. The additional data can be utilized by control node for verification of the collected vehicle positions. Position of vehicle i should correspond to one of the positions of neighbours (k) reported by vehicle j if distance between vehicles i and j is not greater than r. Therefore, trust level of vehicle i is increased by α if

$$dist(i, j) \leq r \wedge \exists \langle x_k(t), l_k(t) \rangle \in D_j(t) : dist(i, k) \leq \epsilon_x, \tag{7}$$

where: $dist(i, j)$ is distance between vehicles i and j, r is localization range, and the remaining symbols were defined earlier. The trust level is decreased by α if

$$dist(i, j) \leq r \wedge dist(i, k) > \epsilon_x \forall \langle x_k(t), l_k(t) \rangle \in D_j(t). \tag{8}$$

The symbols in (8) were defined earlier in this section. The above rule is applied only if the trust level of vehicle j is positive.

It should be noted that the parameter of decreasing trust level for the velocity-related rule (β) is different than for the remaining rules (α) because the velocity-related rule can be used after each data transfer, i.e., significantly more frequently than the other rules.

4   Experiments

Simulation experiments were performed to evaluate effectiveness of the proposed method for malicious data detection. The experimental results presented in this section concern percentages of detected malicious data and the impact of these data on vehicle delay at signalized intersections in a road network.

In this study the stochastic cellular automata model of road network, proposed by Brockfeld et al. [25], was used for the traffic simulation. This model represents a road network in a Manhattan-like city. Topology of the simulated network is a square lattice of 8 unidirectional roads with 16 signalized intersections (Fig. 1). The distance between intersections is of 300 m. Each link in the network is represented by a one-dimensional cellular automaton. An occupied cell on the cellular automaton symbolizes a single vehicle. At each discrete time step (1 s) the state of cellular automata is updated according to four steps (acceleration, braking due to other vehicles or traffic light, velocity randomization, and movement). These steps are necessary to reproduce the basic features of real traffic flow. Step 1 represents driver tendency to drive as fast as possible,

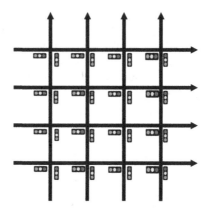

Fig. 1. Simulated road network (Color figure online)

step 2 is necessary to avoid collisions and step 3 introduces random perturbations necessary to take into account changes of vehicle velocity in regions with high density. Finally, in step 4 the vehicles are moved according to the new velocity calculated in steps 1–3. Steps 1–4 are applied in parallel for all vehicles. Detailed definitions of these steps can be found in [25]. Maximum velocity was set to 2 cells per second (54 km/h).

Deceleration probability p for the Brockfeld model is 0.15. The saturation flow at intersections is 1700 vehicles per hour of green time. This model was applied to calculate the stop delay of vehicles. The traffic signals were controlled by using the self-organizing method based on priorities that correspond to "pressures" induced by vehicles waiting at an intersection [26]. The traffic signal control was simulated assuming the intergreen times of 5 s and the maximum period of 120 s. At each intersection there are two alternative control actions: the green signal can be given to vehicles coming from south or to those that are coming from west. The simulator was implemented in Matlab.

Intensity of the traffic flow is determined for the network model by parameter q in vehicles per second. This parameter refers to all traffic streams entering the road network. At each time step vehicles are randomly generated with a probability equal to the intensity q in all traffic lanes of the network model. Similarly, the false (malicious) vehicles are generated with intensity q_F at random locations. The false vehicles move with constant, randomly selected velocity.

During experiments nine algorithms of malicious data detection were taken into account (Table 1). The algorithms use different combinations of the rules proposed in Sect. 3 (4 rules used separately and 5 selected combinations that achieved the most promising results in preliminary tests). Simulations were performed for four various intensities of true vehicles ($q = 0.02, 0.06, 0.10, 0.14$) and false vehicles ($q_F = 0.02, 0.04, 0.06, 0.08$). For each combination of the intensities q and q_F the simulation was executed in 20 runs of 10 min. Based on preliminary results, the parameters used for updating the trust levels were set as follows: $\alpha = 1$ and $\beta = 0.2$.

Table 1. Compared algorithms for malicious data detection

Rule	Algorithm								
	1	2	3	4	5	6	7	8	9
1–Vehicles order	+	−	−	−	+	−	+	−	+
2–Reaction to signals	−	−	+	−	+	+	−	−	+
3–Expected velocity	−	−	−	+	−	−	+	+	+
4–Neighbour detection	−	+	−	−	−	+	−	+	+

Figure 2 shows percentages of correctly detected malicious data and correctly recognized true data for two different intensities of false vehicles. The data were categorized as true or malicious at each one-second interval.

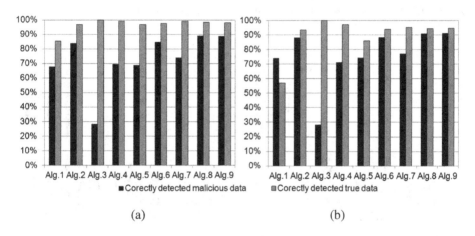

(a)   (b)

Fig. 2. Accuracy of malicious data detection for the compared algorithms: (a) $q_F = 0.02$ veh./s, (b) $q_F = 0.08$ veh./s

Total delay of vehicles for the considered algorithms is compared in Fig. 3. The results in Fig. 3 were averaged for all considered true and false vehicle intensities. Average number of vehicles for one simulation run (10 min) was 384. The best results were obtained for Algorithm 9, which utilizes all proposed rules for detection of the malicious data. This algorithm allows the delay of vehicles to be kept at the low level (close to the value obtained for simulation without malicious data). The delay is increased only by 1 % in comparison to the delay observed when no malicious data are present. Algorithm 9 correctly recognizes 90 % of the malicious data and 96 % of the true data on average. High accuracy was also observed for Algorithm 2, which uses the approach of position verification by neighbouring vehicles without any additional rules. The least satisfactory results were obtained when using the vehicles order rule (Algorithm 1) or the

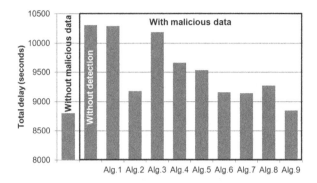

Fig. 3. Average delay of vehicles for the compared algorithms

reaction to signals rule (Algorithm 3). For these algorithms the delay of vehicles is close to that observed when the detection of malicious data is not used.

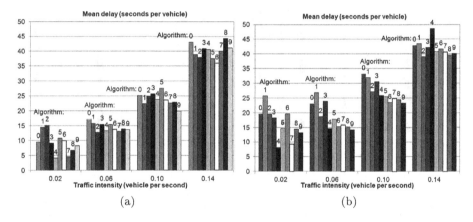

Fig. 4. Mean delay for different traffic intensities: (a) $q_F = 0.02$ veh./s, (b) $q_F = 0.08$ veh./s

Figure 4 shows mean vehicle delays for various intensities of the traffic flow (q) and two different intensities of the false vehicles generation (q_F). Algorithm 0 in Fig. 4 corresponds to the situation when no malicious data detection is implemented. It can be observed in these results that the effectiveness of a particular algorithm strongly depends on the considered intensities. For instance, in case of $q = 0.14$ and $q_F = 0.02$ Algorithm 8 causes a higher delay than those obtained without malicious data detection, while for the remaining intensities Algorithm 8 gives good results. However, for Algorithm 9 the delay is reduced when comparing with those obtained without malicious data detection for all considered intensity settings. This fact confirms that all the proposed rules are useful as they contribute with different degree in various traffic conditions to mitigating the negative impact of malicious data.

5   Conclusion

Sybil attacks can degrade the performance of VANET-based traffic signal control. The proposed approach enables effective detection of malicious data created in VANETs when the Sybil attacks are launched. The introduced detection scheme is based on rules that take into account unrealistic overtaking manoeuvres, expected driver behaviour (reaction to traffic signals and preferred velocity) as well as verification of vehicle position by neighbouring nodes. Effectiveness of this approach was demonstrated in simulation experiments for a decentralized self-organizing system that controls the traffic signals at multiple intersections in an urban road network. The experimental results show that combination of different detection mechanisms allows the malicious data in VANET to be correctly recognized and is essential for mitigating their negative impact on the performance of traffic signal control. Further research is necessary to integrate the method with more sophisticated models of driver behaviours, enable automatic parameters calibration based on collected data, and test the proposed approach in different (more realistic) scenarios with various traffic control algorithms.

References

1. Bajwa, E.J.S., Walia, E.L.: A survey on traffic management systems in VANET. Int. J. Adv. Trends Comput. Appl. 1(4), 28–32 (2015)
2. Płaczek, B.: Efficient data collection for self-organising traffic signal systems based on vehicular sensor networks. Int. J. Ad Hoc Ubiquitous Comput. (in press). http://www.inderscience.com/info/ingeneral/forthcoming.php?jcode=ijahuc
3. Płaczek, B.: A self-organizing system for urban traffic control based on predictive interval microscopic model. Eng. Appl. Artif. Intell. 34, 75–84 (2014)
4. Chang, H.J., Park, G.T.: A study on traffic signal control at signalized intersections in vehicular ad hoc networks. Ad Hoc Netw. 11, 2115–2124 (2013)
5. Kwatirayo, S., Almhana, J., Liu, Z.: Adaptive traffic light control using VANET: a case study. In: Wireless Communications and Mobile Computing Conference (IWCMC), pp. 752–757. IEEE (2013)
6. Maslekar, N., Mouzna, J., Boussedjra, M., Labiod, H.: CATS: an adaptive traffic signal system based on car-to-car communication. J. Netw. Comput. Appl. 36(5), 1308–1315 (2013)
7. Priemer, C., Friedrich, B.: A decentralized adaptive traffic signal control using V2I communication data. In: 12th International IEEE Conference on Intelligent Transportation Systems, ITSC 2009, pp. 1–6. IEEE (2009)
8. Zhang, L., Garoni, T.M., de Gier, J.: A comparative study of macroscopic fundamental diagrams of arterial road networks governed by adaptive traffic signal systems. Transp. Res. Part B: Methodol. 49, 1–23 (2013)
9. Ali Mohammadi, M., Pouyan, A.A.: Defense mechanisms against Sybil attack in vehicular ad hoc network. Secur. Commun. Netw. 8(6), 917–936 (2015)
10. Bouassida, M.S., Guette, G., Shawky, M., Ducourthial, B.: Sybil nodes detection based on received signal strength variations within VANET. Int. J. Netw. Secur. 9(1), 22–32 (2009)
11. Raya, M., Papadimitratos, P., Hubaux, J.P.: Securing vehicular communications. IEEE Wireless Commun. Mag. 13(5), 8–15 (2006). Special issue on inter-vehicular communications

12. Chang, S., Qi, Y., Zhu, H., Zhao, J., Shen, X.: Footprint: detecting Sybil attacks in urban vehicular networks. IEEE Trans. Parallel Distrib. Syst. **23**(6), 1103–1114 (2012)

13. Park, S., Aslam, B., Turgut, D., Zou, C.C.: Defense against Sybil attack in the initial deployment stage of vehicular ad hoc network based on roadside unit support. Secur. Commun. Netw. **6**(4), 523–538 (2013)

14. Guette, G., Ducourthial, B.: On the Sybil attack detection in VANET. In: IEEE International Conference on Mobile Ad Hoc and Sensor Systems, MASS 2007, pp. 1–6 (2007)

15. Xiao, B., Yu, B., Gao, C.: Detection and localization of Sybil nodes in VANETs. In: Proceedings of the 2006 Workshop on Dependability Issues in Wireless Ad Hoc Networks and Sensor Networks, pp. 1–8. ACM (2006)

16. Suen, T., Yasinsac, A.: Ad hoc network security: peer identification and authentication using signal properties. In: Proceedings from the Sixth Annual IEEE SMC Information Assurance Workshop, IAW 2005, pp. 432–433 (2005)

17. Yu, B., Xu, C.Z., Xiao, B.: Detecting Sybil attacks in VANETs. J. Parallel Distrib. Comput. **73**(6), 746–756 (2013)

18. Yan, G., Olariu, S., Weigle, M.C.: Providing VANET security through active position detection. Comput. Commun. **31**(12), 2883–2897 (2008)

19. Golle, P., Greene, D., Staddon, J.: Detecting and correcting malicious data in VANETs. In: Proceedings of the 1st ACM International Workshop on Vehicular Ad Hoc Networks, pp. 29–37. ACM (2004)

20. Raya, M., Papadimitratos, P., Aad, I., Jungels, D., Hubaux, J.P.: Eviction of misbehaving and faulty nodes in vehicular networks. IEEE J. Sel. Areas Commun. **25**(8), 1557–1568 (2007)

21. Ghosh, M., Varghese, A., Gupta, A., Kherani, A.A., Muthaiah, S.N.: Detecting misbehaviors in VANET with integrated root-cause analysis. Ad Hoc Netw. **8**(7), 778–790 (2010)

22. Dietzel, S., van der Heijden, R., Decke, H., Kargl, F.: A flexible, subjective logic-based framework for misbehavior detection in V2V networks. In: 2014 IEEE 15th International Symposium on a World of Wireless, Mobile and Multimedia Networks (WoWMoM), pp. 1–6. IEEE (2014)

23. Płaczek, B.: A traffic model based on fuzzy cellular automata. J. Cell. Automata **8**(3–4), 261–282 (2013)

24. Thammakaroon, P., Tangamchit, P.: Adaptive brake warning system for automobiles. In: 2008 8th International Conference on ITS Telecommunications, ITST 2008, pp. 204–208. IEEE (2008)

25. Brockfeld, E., Barlovic, R., Schadschneider, A., Schreckenberg, M.: Optimizing traffic lights in a cellular automaton model for city traffic. Phys. Rev. E **64**(5), 056132 (2001)

26. Helbing, D., Lämmer, S., Lebacque, J.P.: Self-organized control of irregular or perturbed network traffic. In: Deissenberg, C., Hartl, R.F. (eds.) Optimal Control and Dynamic Games, pp. 239–274. Springer US, New York (2005)

Anti-evasion Technique for the Botnets Detection Based on the Passive DNS Monitoring and Active DNS Probing

Oksana Pomorova$^{(\boxtimes)}$, Oleg Savenko, Sergii Lysenko,
Andrii Kryshchuk, and Kira Bobrovnikova

Department of System Programming,
Khmelnitsky National University, Instytutska, 11, Khmelnitsky, Ukraine
o.pomorova@gmail.com, savenko_oleg_st@ukr.net, sirogyk@ukr.net,
rtandrey@rambler.ru, bobrovnikova.kira@gmail.com
http://www.spr.khnu.km.ua

Abstract. A new DNS-based anti-evasion technique for botnets detection in the corporate area networks is proposed. Combining of the passive DNS monitoring and active DNS probing have made it possible to construct effective BotGRABBER detection system for botnets, which uses such evasion techniques as cycling of IP mapping, "domain flux", "fast flux", DNS-tunneling. BotGRABBER system is based on a cluster analysis of the features obtained from the payload of DNS-messages and uses active probing analysis. Usage of the developed method makes it possible to detect infected hosts by bots of the botnets with high efficiency.

Keywords: Botnet · DNS-traffic · Passive DNS monitoring · Active DNS probing · Botnet's evasion techniques · Cycling of IP mapping · "Domain flux" · "Fast flux" · DNS-tunneling

1 Introduction

Nowadays botnets are one of the most dangerous cyber threats. The vast majority of botnets uses DNS for the purpose of the infected hosts' control [1]. Accessibility, high informative value of the DNS-traffic and its small amount compared with the overall network traffic provide a range of capabilities for botnets detection based on the DNS-traffic analysis. Otherwise, there is a number of evasion techniques based on DNS, which complicate the detection, localization and neutralization of botnets: "fast-flux" service network, "domain flux" and cycling of IP mapping for malicious domain. Also, there is a technique that makes it possible to hide the fact of malicious traffic transmission of command and control, and to mask it as a legitimate DNS-traffic (DNS-tunneling).

2 Related Work

Today there is a number of approaches for botnets detection [2,3], as well as researches aimed for detection of botnets that uses evasion techniques [4–8].

© Springer International Publishing Switzerland 2016
P. Gaj et al. (Eds.): CN 2016, CCIS 608, pp. 83–95, 2016.
DOI: 10.1007/978-3-319-39207-3_8

In [9,10] a detection method of DNS basis botnet communication using obtained NS record history was proposed. Key idea of the approach is that the most of domain name resolutions first obtain the corresponding NS (Name Server) record from authoritative name servers in the Internet, whereas suspicious communication may omit the procedures to hide their malicious activities. The proposed method checks whether the destined name server (IP address) of a DNS query is included in the obtained NS record history to detect the botnet communications.

In [11] an architecture to map network concepts to data stored in relational databases was propose. Based on this architecture, a tool that looks for malicious bot activity, studying, from a unique point of view, DNS traffic from PCAP sources, and TCP connections from IPFIX reports was implemented.

In [4] the detection method for botnets, that use "domain flux", is described. It is based on the analysis of successful and unsuccessful DNS-requests submitted in the time interval that includes successful requests. In order to determine the botnets' IP-addresses, the associated parameters of the domain names' entropy and correlation in time between successful and unsuccessful DNS-requests are analyzed. The disadvantage of this method is the reliance on unsuccessful domain names, as well as the sensitivity to the choice of time interval.

In [5] the detection system of the malicious domains names based on passive DNS monitoring is proposed. It is able to classify the domain names by 4 groups of features that can be extracted from DNS-traffic: (1) time-based features; (2) DNS answer-based features; (3) TTL value-based features; (4) domain name-based features.

In [6] a detection technique, designated as the Genetic-based ReAl-time DEtection (GRADE) system, to identify Fast Flux Service Networks (FFSN) in real time, is proposed. GRADE differentiates between FFSNs and benign services by employing two characteristics: the entropy of domains of preceding nodes for all A-records and the standard deviation of round trip time to all A records. Technique is able to find the best strategy to detect current FFSN trends.

In [7] a technique that uses active probing for identifying the legitimate CDN domains and malicious FFSN domains was proposed. The method is based on identifying characteristics derived from active DNS-probing, namely mapping the domains with the set of IP-addresses that are in different network segments, and taking into account volatility represents the fluctuation of handling time for multiple consecutive time requests for a specific domain name.

The main weakness of the mentioned approaches is that they take in to account a set of features which is insufficient for FFSN detections, as there are a lot of other informative features that are able to indicate the usage of evasion technique fast flux by botnets.

In [8] the known approaches aimed for detection of the botnets, that use of DNS-tunneling, are presented. The two main research directions are presented here: techniques based on analysis of DNS-messages payload and statistical analysis of DNS-traffic.

In [12] authors confirmed that it is effective to check the unknown usage of DNS TXT queries for detecting botnet communication.

The disadvantage of the approaches presented in [8,12] is the concentration on detection of the narrow range of malware, which use specific DNS-tunneling technologies.

The disadvantage of described methods is the concentrating only on the features pointing out the harmfulness of the domain names, and ignoring the behaviour of botnets' bots in DNS-traffic.

Taking into account the wide functionality of the botnets and their usage of evasion techniques, the very important task is to develop a new approach for botnets detection, which would allow combining the benefits of methods for passive and active DNS monitoring in a single technique in order to improve the botnets detection efficiency.

3   Previous Work

In [13] an attempt to solve the botnets detection problem, where bots are using evasion techniques based on DNS, is performed. The method is based on the passive analysis of incoming DNS-traffic and uses cluster analysis of the feature vectors obtained by the analysis of DNS-messages payload. In order to construct the feature vectors based on TTL fields, such incoming DNS-messages are processed: (1) each first captured DNS-message about certain domain name within the TTL-period; (2) each repeated DNS-message received by host within the TTL-period, when the source of the message is non-local DNS-server, and TTL-period referred to this message differs from the rest of TTL-period within which this message have received.

In order to detect the usage of the DNS-based evasion techniques by botnets, such features were used:

- the length of the domain name, L_N;
- the number of unique characters in the domain name, N_U;
- entropy of the domain name, E_N;
- the sign of the usage of uncommon types of the DNS-records (for example, KEY, NULL), or DNS-records that are not commonly used by a typical client (for example, TXT), F_{UR};
- maximum value of entropy of the DNS-records, which are contained in the DNS-messages, E_R;
- maximum size of the DNS-messages about the domain name, L_P;
- the number of A-records corresponding to the domain name in the incoming DNS-message, N_A;
- the number of IP-addresses concerned with the domain name, N_{IP};
- the average distance between the IP-addresses concerned with the domain name, S_{IP};
- number of unique IP-addresses in sets of A-records corresponding to the domain name in the DNS-messages, N_{UA};

- the average distance between the IP-addresses in the set of A-records for the domain name in the incoming DNS-message, S_A;
- the average distance between unique IP-addresses in sets A-record corresponding to the domain name in the DNS-messages, S_{UA};
- number of the domain names that share IP-address corresponding to the domain name, N_D;
- values of TTL-period: mode, T_{mod}; median, T_{med}; average value, T_{aver};
- the sign of success of DNS-query, F_S.

Note. In the approach we are interested in situation, where the average "distance" between each of the IP addresses in the result set is more than 65 535 addresses (equivalent to distinct/16 netblocks) [14].

The method also operates with a dependence function of the DNS-message field entropy f_{Ebn} from its length [15], where n – base of encoding.

Based on mentioned features the feature vectors during the process of the passive monitoring time are formed. Further, the feature vectors are input data of the semi-supervised fuzzy c-means clustering.

The result of clustering is a degree of membership of the feature vectors to one of five clusters, four of which indicates the usage of the evasion techniques (h_1 – cycling of IP mapping, h_2 – "domain flux", h_3 – "fast flux", h_4 – DNS-tunneling, and a cluster h_5 that contains normal queries).

The proposed method demonstrates a high efficiency of detection, but its drawback is an uncertainty of the part of obtained results and a significant level of false positives (at about 4–9 %).

4   Anti-evasion Technique for the Botnets Detection Based on the Passive and Active DNS Monitoring

A possible way to improve the efficiency of detection of botnets in the corporate area networks, that use the evasion techniques, is to develop a new technique, which would involve passive and active DNS monitoring in order to eliminate the disadvantages of the method, described in [13] and increase the botnets detection efficiency.

In this paper, we introduce BotGRABBER, a system that employs a combination of passive and active DNS analysis techniques to detect botnets, which use evasion techniques by its malicious activity in the network.

4.1   Passive DNS Monitoring

In order to increase the technique efficiency, described in Sect. 3, the very first task is to eliminate the uncertainty of the results. For this purpose we propose to involve additional features that are able to indicate the usage of DNS-based evasion techniques by botnets.

One of the important feature, that can be obtained by means of passive monitoring of the DNS-traffic, which indicate malicious activity in the network, is the synchrony of DNS-queries. Therefore, involvement of such feature will separate DNS-queries related to the botnets' functioning, and DNS-queries about the illegal domain names that use evasion techniques based on DNS. We will consider group queries as synchronous, if there is a large number of queries about the same domain name, and they are concentrated in a small time interval (synchronization time of the bots), t_s. If the time interval between the first and the last DNS-response Δt_q for the group query about the same domain name is greater than the length of the time window t_s, then in order to check the synchrony of DNS-queries, we build the vector of the density distribution of DNS-queries in time, which can be described as follows:

$$\overline{W_d} = (\Omega_j)_{j=1}^z, \tag{1}$$

where Ω_j – number of queries within the z-interval; j – number of interval; $z = (t_l - t_f)/((1/3) \cdot t_s)$, where t_l and t_f – the time of the last and the first DNS-responses about the domain name within the TTL-period, or DNS-responses that contained NXDOMAIN error code, or DNS-responses within the interval time when the group flushing of local DNS-caches was fixed [16].

If the maximum number of queries, which belongs to the interval t_s, which consists of three adjacent elements of vector W_d, that describe the DNS-query distribution of continuous interval, exceeds some threshold δ, then the group requests will be considered synchronous [16].

Let us denote the synchrony feature of the group DNS-queries for the domain name, $S_S = N_S/N$, where N_S – number of synchronous group DNS-queries about the domain name; N – the total number of group DNS-queries about the domain name.

Thus, let us present the updated feature vector of the incoming DNS-messages for certain domain name d as follows:

$$\overline{W_d} = (S_S, L_N, N_U, E_N, T_{mod}, T_{med}, T_{aver}, N_A, N_{IP}, S_{IP}, S_A, \\ N_{UA}, S_{UA}, N_D, F_{UR}, E_R, L_P, F_S). \tag{2}$$

The basis of semi-supervised learning of the clusterer is the knowledge about features, which indicate the usage of the evasion techniques based on DNS. Knowledge base can be represented as rules by Algorithm 1.

Based on constructed updated feature vectors, the data matrix is formed. It is the input data for the clusterer. The result of clustering is a matrix of the semi-fuzzy splitting U, where each element of the matrix u_{ij} determines the belonging of i-th element of the set of clustering objects to j-th cluster:

$$U = [u_{ij}] , \quad u_{ij} \in [0,1], \quad i = \overline{1, N_z}, \quad \sum_{j=\overline{1,N_h}} u_{ij} = 1, \tag{3}$$

where N_z – the total number of different domain names requested by network hosts; N_h – the total number of clusters.

Let denote δ and δ' as thresholds that define that the clustering object belongs to the cluster, when the domain name is considered malicious or suspicious respectively. If $u_{ij} \geq \delta$, $j = \overline{1,4}$, then the object belongs to a cluster,

for *all DNS_messages_or_training_data* **do**

    if

    $(S_\text{s} \geq threshold\ and\ T_\text{mod} \in [0, 900]\ and\ T_\text{med} \in [0, 900]\ and\ T_\text{aver} \in [0, 900])$

    then

        if $((N_\text{A} \in (5, \infty)\ and\ S_\text{A} \in (65\,535, \infty))\ or\ (N_\text{UA} \in (8, \infty)\ and\ S_\text{UA} \in$

        $(65\,535, \infty))$ **then**

          | *evasion_technique ← fast_flux*

        end

        if *($F_\text{S} = 0\ and\ N_\text{D} \in [8, \infty])$* **then**

          | *evasion_technique ← domain_flux*

        end

        if $(N_\text{IP} \in (5, \infty)\ and\ S_\text{IP} \in (65\,535, \infty))$ **then**

          | *evasion_technique ← cycling_of_IP_mappings*

        end

        if $((L_\text{N} \in [75, 225]\ and\ N_\text{U} \in (27, 37])\ or\ E_\text{N} \geq f_\text{Eb32}\ or\ (E_\text{R} \geq$

        $f_\text{Eb64}\ or\ E_\text{R} \geq f_\text{Eb256})\ or\ F_\text{UR} = 1\ or\ L_\text{P} > 300))$ **then**

          | *evasion_technique ← DNS_tunneling*

        end

    end

end

Algorithm 1. Rules for semi-supervised learning

which corresponds to one of evasion techniques. The situation where $\delta' \leq u_{ij} < \delta$ means that the object may belong to several clusters and there is an uncertainty of results.

Figure 1(a) and (b) show the plane projection of the set of feature vectors of DNS-messages, which are distributed on clusters. The origin of coordinates is the centroid of the cluster, which contains the set of the feature vectors of incoming DNS-messages about legitimate resources to the hosts of the network. Each marker (dot) in the cluster represents a set of incoming DNS-messages to infected and uninfected hosts of the network about certain domain name.

Figure 1(a) demonstrates that the part of clustering results, described in Sect. 3, is uncertain. Adding the synchrony feature of DNS-query to the feature vector demonstrates the reduction of the results uncertainty that is confirmed experimentally. Studies show that the amount of clustering objects that could not be exactly classified as the malicious or legitimate has been reduced by 30–40 % (Fig. 1(b)).

4.2 Active DNS Probing

Analysis of the outcome of experiments has demonstrated that the part of the results is unable to answer the question whether the domain name is malicious or legitimate.

That's why the next step of the efficiency improvement is the further analysis of the clustering results. We will analyze the requests about the domain names that were placed on the intersection of clusters, and cannot be classified as malicious or legitimate.

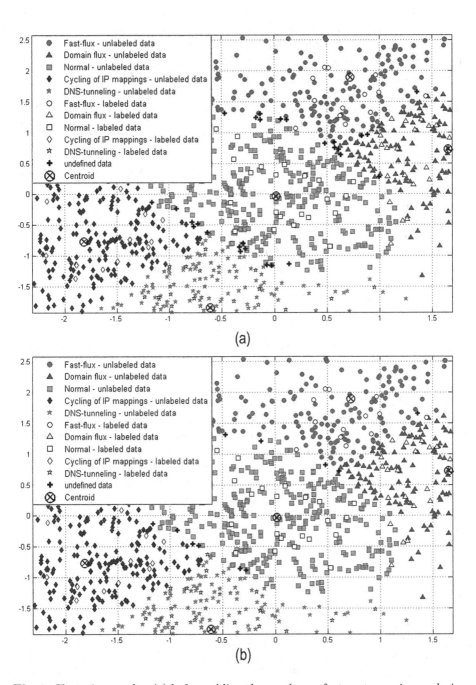

Fig. 1. Clustering results: (a) before adding the synchrony feature to passive analysis; (b) after adding the synchrony feature

Used features are obtained from the passive DNS-traffic monitoring. Therefore, there is no need to initiate specific actions for the purpose of simulation of the malicious activity in order to collect the necessary information regarding to the domain name. However, in a situation when these means are unable to give us result, it is appropriate to involve an active probing analysis.

Such approach will allow us to obtain additional features that may indicate that the domain name is malicious, and it is based on the implementation of NS-request records, A-Records, SOA-records, PTR-records. The approach is applied to refine the detection results of the cycling of IP mapping, "domain flux", "fast flux" evasion techniques.

We also propose to carry out queries for the determination of the autonomous systems numbers (ASN), which include IP-addresses associated with suspicious domain names and their name servers (it will be used for "fast flux" and cycling of IP mapping detection).

Thus, in order to define that the domain name belongs to botnets, which uses evasion techniques, let us use the following features obtained by means of active monitoring: (1) N_{NS} – number of the NS-records in the DNS-response [14]; (2) S_{NS} – the average distance between the IP-addresses for the set of NS-records for the domain name [14]; (3) V_{retry} – value of the fields *retry*, received from the DNS-response by a SOA-request [14]; (4) N_{ASN} – number of different numbers of autonomous systems (ASN), which include IP-addresses associated with server names [14]; (5) N_{ASA} – the number of different numbers of autonomous systems, which include IP-addresses associated with the domain name [14].

Additional Analysis. One of the botnets' feature is coordinated behaviour in the network. That is why the possible way to detect malicious activity of the bots is to find hosts groups by MAC-addresses, which simultaneously have been carrying out DNS-requests about a specific domain name within the interval of time synchronization t_s, and to check them for similarities. This procedure is inapplicable on the stage of the passive monitoring, because it is unuseful on the early botnet's spreading phase. For this purpose, we carry out an analysis of the similarity for hosts' groups, which performed DNS-queries about the domain name.

In situation, if $\delta' \leq u_{ij} < \delta$, $j = 1$ or $j = 3$ (for "fast-flux" or cycling of IP mapping evasion techniques), we perform the determination of the similarity for hosts' groups, which carried out similar requests about the domain names. "Domain flux" or DNS-tunneling evasion techniques provide the frequent change of the domain names related to the botnet's functioning. Therefore, if $\delta' \leq u_{ij} < \delta$, $j = 2$ or $j = 4$ we determine the similarity of groups' hosts, which executed DNS-queries about the same domain names, and the similarity for pairs of group DNS-requests that belong to clusters regarding to evasion techniques.

We will consider hosts similar if $K \geq q$, where K – similarity coefficient, q – threshold value of the similarity. In order to determine the similarity of groups' hosts, that carried out queries about the same domain name, we used the Koch index of dispersity. To determine the similarity for two groups of queries we used the cosine similarity coefficient and define similarity only for those groups that belonged to the same cluster and $\dfrac{a}{\sqrt{ab}} \geq q$, where a and b – amount size of the smaller and larger groups respectively.

4.3 Algorithm of the Anti-evasion Technique of the Botnets Detection Based on the Passive and Active DNS Monitoring

The anti-evasion technique of the botnets detection in the corporate area networks based on the passive and active DNS monitoring (BotGRABBER) is functioning as follows: executing the fuzzy cluster analysis of the feature vectors obtained from passive analysis payload DNS-messages. The result of the clustering is the splitting of the requests about legitimate or malicious domain names into clusters. The next step is to analyze those requests, which were on the intersection of clusters with malicious and normal requests by involvement of active DNS monitoring. Presentation of the method functioning is given as an Algorithm 2.

5 Experiments

In order to test the feasibility of BotGRABBER as a detector botnets which uses evasion techniques in real-life, the campus network of Khmelnitsky National University was used.

In order to hold the experiments, a specialized software was constructed, which was able to make malicious DNS-queries, and had features of the botnets' bots with centralized architecture. The set of constructed software was proportionally divided into groups by its functional properties that related to one of the four evasion techniques – cycling of IP mappings, "domain flux", "fast-flux", DNS-tunneling. Depend on its functionality it executed different types of queries.

For the purpose of the C&C botnets' servers imitation for the duration of experiments the set of the domain names, which we consider malicious, was registered. The C&C botnets' servers made it possible to simulate evasion techniques (such actions as IP-mapping, the domain name changing, cyclically changing of DNS A-records and NS-records for the same domains using round robin algorithm, command and control traffic transfer using DNS-tunneling, etc.).

Developed bots had different sets of features and performed queries about malicious domains that were not previously known to BotGRABBER and had not been used in the training.

For the experiments, a network of 100 hosts was used and they were infected with bots. Also the users activity was simulated and hosts executed legitimate queries. Each botnet carried out the different scenarios of queries and in different time.

The experiment lasted 24 h. It was enough to run all scenarios, embedded in bots. During this time, BotGRABBER detected, analyzed and classified 2369 DNS-responses.

Function passive_analysis
for *all gathered_incoming_DNS_messages* **do**
  for *all first_DNS_messages_within_TTL or*
  all repeated_DNS_messages_within_TTL_from_non_local_DNS_server **do**
   | form *feature_vector_$\overline{W_d}$* // Eq. 2
  end
  form *data_matrix_of_feature_vectors_V*
  data_matrix_of_feature_vectors_V → *set_of_clusters_H* where $V(i) = \overline{W_d}$
  form *fuzzy_splitting_matrix_U*
  if $(u_{ij} \geq \delta)$ **then**
   block
   else
    if $(\delta' \leq u_{ij} < \delta)$ **then**
     | execute *active_analysis*
    end
  end
end
Function active_analysis
for *all $\overline{W_d}$* **do**
  if $(\delta' \leq u_{ij} < \delta)$ **then**
   if $((\overline{W_d} \in H_{\text{cycling\_of\_IP\_mapping}})$ *and* $(\overline{W_d} \in H_{\text{legitimate}}))$ **then**
    if $((T_{\text{mod}} \in [0, 900]$ *and* $T_{\text{med}} \in [0, 900]$ *and* $T_{\text{aver}} \in [0, 900])$
    and $(N_{\text{IP}} \in (5, \infty)$ *and* $S_{\text{IP}} \in (65535, \infty)$ *and* $N_{\text{ASA}} > 2))$ **then**
     | block
    end
   end
   if $((\overline{W_d} \in H_{\text{fast\_flux}})$ *and* $(\overline{W_d} \in H_{\text{legitimate}}))$ **then**
    if $((T_{\text{mod}} \in [0, 900]$ *and* $T_{\text{med}} \in [0, 900]$ *and* $T_{\text{aver}} \in [0, 900])$
    and $((N_{\text{A}} \in (5, \infty)$ *and* $S_{\text{A}} \in (65\,535, \infty))$ *or* $(N_{\text{UA}} \in (8, \infty)$
    and $(S_{\text{UA}} \in (65\,535, \infty))$ *or* $N_{\text{AS}} > 2)$ *and* $(S_{\text{NS}} > 65\,535$
    or $N_{\text{ASN}} > 2$ *and* $N_{\text{NS}} > 3$ *and* $V_{\text{retry}} \in [0, 900]))$ **then**
     | block
    end
   end
   if $((\overline{W_d} \in H_{\text{domain\_flux}})$ *and* $(\overline{W_d} \in H_{\text{legitimate}}))$ **then**
    if $(similarity\_coeficient\_K > threshold\_q$ *or* $N_{\text{D}} \in [8, \infty])$ **then**
     | block
    end
   end
   if $((\overline{W_d} \in H_{\text{DNS\_tunneling}})$ *and* $(\overline{W_d} \in H_{\text{legitimate}}))$ **then**
    if $(similarity\_coeficient\_K > threshold\_q)$ **then**
     | block
    end
   end
  else
   | block
  end
end

Algorithm 2. Algorithm of the anti-evasion technique of the botnets detection based on the passive and active DNS monitoring

Table 1 demonstrates the numbers of detected responses about malicious domains: using the technique which is described in Sect. 3, the technique which used just passive analysis of the DNS-traffic and by BotGRABBER system (using passive and active monitoring). As we can see, after the involvement synchrony feature into passive monitoring process, we obtained better results of detection, but still with significant level of false positives. Finally, usage of the BotGRABBER system demonstrates not only greater detection results, but the decrease of the false positives rage.

Table 1. Experimental results: number of queries carried out by bots, detected responses and false positives

Name of evasion technique	Number of queries carried by bots	Botnets detection technique based on passive monitoring described in [13]	Improved botnets detection technique based on passive monitoring	Botnets detection technique BotGRABBER based on passive and active monitoring
		Detected responses/False positives, %		
Cycling of IP mapping	308	299/2	301/2	301/1
"Domain flux"	1432	1326/3	1406/3	1406/1
"Fast flux"	485	389/3	425/3	425/2
DNS-tunneling	144	142/0	142/0	142/0
Total	2369	2156 (91 %)/8	2274 (96 %)/8	2274 (96 %)/4

Thus, the results of BotGRABBER usage demonstrated the ability of the technology to detect botnets that use evasion techniques up to 96 %, while the level of false positives was about 4 %.

The localization and blocking of the bots' actions on the network hosts are performed by using logs with hosts MAC-addresses that carried DNS-queries and requested their domain names.

As a result, we can make conclusion, that the system BotGRABBER – anti-evasion technique of the botnets detection in the corporate area networks based on combination of the passive and active DNS monitoring can be used as a tool for the detection of infected hosts in the network with high efficiency.

6   Conclusions

Results obtained in experimental studies have demonstrated, that the analysis of the features obtained by passive DNS monitoring in combination with the features extracted by active DNS probing have made it possible to detect botnets, which use evasion techniques. Based on this approach we have constructed an effective botnets detection tool – BotGRABBER, which is able to detect bots, that use such evasion techniques as cycling of IP mapping, "domain flux", "fast flux", DNS-tunneling.

BotGRABBER system is based on a cluster analysis of the features obtained from the payload of DNS-messages and uses active probing analysis.

Usage of the developed system makes it possible to detect infected hosts by bots of the botnets with high efficiency up to 96 % and low false positives at about 4 %.

Acknowledgements. This research was supported by a TEMPUS SEREIN project (Project reference number 543968-TEMPUS-1-2013-1-EE-TEMPUS-JPCP). Additionally, we thank the Khmelnytsky National University for providing access to their DNS-traffic during the early phases of this work.

References

1. DAMBALLA. Botnet Detection for Communications Service Providers. https://www.damballa.com/downloads/r_pubs/WP_Botnet_Detection_for_CSPs.pdf
2. Sochor, T., Zuzcak, M.: Study of internet threats and attack methods using honeypots and honeynets. In: Kwiecień, A., Gaj, P., Stera, P. (eds.) CN 2014. CCIS, vol. 431, pp. 118–127. Springer, Heidelberg (2014)
3. Sochor, T., Zuzcak, M.: Attractiveness study of honeypots and honeynets in internet threat detection. In: Gaj, P., Kwiecień, A., Stera, P. (eds.) CN 2015. CCIS, vol. 522, pp. 69–81. Springer, Heidelberg (2015)
4. Yadav, S., Reddy, A.L.N.: Winning with DNS failures: strategies for faster botnet detection. In: Rajarajan, M., Piper, F., Wang, H., Kesidis, G. (eds.) SecureComm 2011. LNICST, vol. 96, pp. 446–459. Springer, Heidelberg (2012)
5. Bilge, L., Kirda, E., Kruegel, C., Balduzzi, M.: EXPOSURE: finding malicious domains using passive DNS analysis. In: NDSS, pp. 1–17 (2011)
6. Lin, H.T., Lin, Y.Y., Chiang, J.W.: Genetic-based real-time fast-flux service networks detection. Comput. Netw. **57**(2), 501–513 (2013). Elsevier
7. Zhao, Y., Jin, Z.: Quickly identifying FFSN domain and CDN domain with little dataset. In: 4th International Conference on Mechatronics, Materials, Chemistry and Computer Engineering (ICMMCCE 2015), pp. 1999–2004 (2015)
8. Farnham, G., Atlasis, A.: Detecting DNS tunneling. SANS Institute InfoSec Reading Room, pp. 1–32 (2013)
9. Ichise, H., Yong, J., Iida, K.: Detection method of DNS-based botnet communication using obtained NS record history. In: Computer Software and Applications Conference (COMPSAC), 2015 IEEE 39th Annual, vol. 3, pp. 676–677 (2015)
10. Yong, J., Ichise, H., Iida, K.: Design of detecting botnet communication by monitoring direct outbound DNS queries. In: 2015 IEEE 2nd International Conference on Cyber Security and Cloud Computing (CSCloud), pp. 37–41 (2015)
11. Rincon, S.R., Vaton, S., Beugnard, A., Garlatti, S.: Semantics based analysis of botnet activity from heterogeneous data sources. In: Wireless Communications and Mobile Computing Conference (IWCMC), 2015 International, pp. 391–396 (2015)
12. Ichise, H., Yong, J., Iida, K.: Analysis of via-resolver DNS TXT queries and detection possibility of botnet communications. In: 2015 IEEE Pacific Rim Conference on Communications, Computers and Signal Processing (PACRIM), pp. 216–221 (2015)
13. Lysenko, S., Pomorova, O., Savenko, O., Kryshchuk, A., Bobrovnikova, K.: DNS-based anti-evasion technique for botnets detection. In: Proceedings of the 2015 IEEE 8th International Conference on Intelligent Data Acquisition and Advanced Computing Systems: Technology and Applications (IDAACS), IDAAACS-2015, Warsaw, Poland, vol. 1, pp. 453–458, September 2015

14. Nazario, J., Holz, T.: As the net churns: fast-flux botnet observations. In: Conference on Malicious and Unwanted Software (Malware 2008), pp. 24–31 (2008)
15. Dietrich, C.J., Rossow, C., Freiling, F.C., Bos, H., van Steen, M., Pohlmann, N.: On botnets that use DNS for command and control. In: Proceedings of European Conference on Computer Network Defense, pp. 9–16 (2011)
16. Pomorova, O., Savenko, O., Lysenko, S., Kryshchuk, A., Bobrovnikova, K.: A technique for the botnet detection based on DNS-traffic analysis. In: Gaj, P., Kwiecień, A., Stera, P. (eds.) CN 2015. CCIS, vol. 522, pp. 127–138. Springer, Heidelberg (2015)

Measuring Client-Server Anonymity

Rajiv Bagai[✉] and Huabo Lu

Department of Electrical Engineering and Computer Science,
Wichita State University, Wichita, KS 67260-0083, USA
{rajiv.bagai,hxlu}@wichita.edu

Abstract. The primary intent of clients performing anonymous online tasks is to conceal the extent of their already encrypted communication with sensitive servers. We present an accurate method for evaluating the amount of anonymity still available to such clients in the aftermath of an attack. Our method is based upon probabilities arrived at by the attack of possible client-server association levels for being the real one, along with the correctness levels of those associations. We demonstrate how additionally taking correctness levels into account results in more accurate anonymity measurement than the customary approach of just computing the Shannon entropy of the probabilities.

Keywords: Online anonymity · Caching · Infeasibility attacks · Measuring anonymity · Combinatorial matrix theory

1 Introduction

The diversity of activities that can be performed online, such as shopping, job searching, and keeping in touch with friends, has been key to the phenomenal growth enjoyed by the Internet. Of these, activities that people often prefer to perform anonymously were not anticipated when the Internet was designed, thus such activities now require additional support of anonymity systems that provide the desired anonymity. Electronic voting, searching for a partner, whistle-blowing on an authority, are just some examples of online activities that many people prefer to do with the help of an anonymity system that gets the job done while concealing the identity of the performing individuals.

Anonymity systems are subject to a wide range of attacks that undermine the anonymity provided, making it imperative to accurately measure the amount of anonymity the system is actually providing in the aftermath of an attack. An accurate measurement can help understand potency of attacks, compare quality of different anonymity systems, or decide whether or not to employ a particular system for some critical activity, such as for one related to national security.

We develop, in this paper, a technique for measuring anonymity received by individuals, called *clients*, that are performing online tasks via an anonymity system with some target machines, called *servers*. Our technique measures the anonymity remaining after an attack that is capable of narrowing down the possible clients that may have initiated any activity the system is observed to be

© Springer International Publishing Switzerland 2016
P. Gaj et al. (Eds.): CN 2016, CCIS 608, pp. 96–106, 2016.
DOI: 10.1007/978-3-319-39207-3_9

engaged in with the servers. This attack model captures several real-life attacks, including the timing-analysis attack in Adman et al. [1] and the route-length attack in Serjantov and Danezis [2].

Modern anonymity systems often employ data caching for improving the overall system performance and, more importantly, for enhancing the amount of anonymity they provide to clients, as noted in Shubina and Smith [3] and Bagai and Tang [4]. Our technique is tailored especially for such systems. The main characteristic of our technique, however, is that it measures anonymity of *end-to-end* associations, i.e. between clients and servers, not just between the incoming and outgoing messages of the system. Gierlichs et al. [5] and Bagai et al. [6] showed that such end-to-end measurement is essential when clients and servers are involved in sending and/or receiving multiple messages, as is usually the case during anonymous web browsing sessions.

As the system's goal is to hide the actual client-server associations among other possible ones, an adversary strives to expose them as much as possible. The attack model we consider ends up inducing probabilities on each possible association of being factual. Employing Shannon entropy [7], which gives the uncertainty inherent in a probability distribution, is by far the most popular method of measuring anonymity in such situations, as proposed by Serjantov and Danezis [2], and Diaz et al. [8]. We argue that, such an anonymity measure that is based upon just probabilities of events and ignores the correctness levels of those events is, at best, a crude measure. While we also do construct a metric here based on that approach, the main contribution of this paper is another, more accurate anonymity metric, which measures specifically the extent to which the *actual* client-server associations hide. By being sensitive to the correctness levels of other associations, along with their probabilities, our approach results in more accurate measurement and, in that respect, it is an advancement over the current state-of-the-art in anonymity measurement.

The rest of this paper is organized as follows. Section 2 constructs the basic framework of an anonymity system being used by clients to communicate with servers. The system is equipped with data caching abilities, and the clients and servers are capable of sending and/or receiving multiple messages. The attack model we consider is also introduced in this section. Section 3 develops our technique for measuring anonymity of the client-server associations. It shows how an attack essentially results in a probability distribution and an error distribution over the set of all possible associations. It also presents our metric that takes both of these into account, and shows how that results in more accurate measurement than one that ignores the error factor, akin to current practice. Finally, Sect. 4 concludes our work and gives a direction for future work.

2 Anonymity System and Attack Model

This section lays a mathematical framework for an anonymity system used by clients, each of whom may be engaged in carrying out several simultaneous anonymous online activities with some target servers. The clients essentially send requests for some resources to these servers via the anonymity system. The requests may be such as for a particular video clip on youtube.com, or for profiles that meet some characteristics on ashleymadison.com, or even just for some product's page on amazon.com. The anonymity system acts as an intermediary, for providing anonymity to the clients for the tasks they are engaged in. It accomplishes that by first collecting the requests from the clients and then forwarding those requests to the servers in a manner that maintains client anonymity, from both the servers as well as any third-party eavesdroppers.

Let X be the set of requests observed by a passive adversary having been collected by the anonymity system from its clients, and Y be the set of requests observed having been forwarded to the target servers. An important goal of the anonymity system is to prevent the adversary from determining the underlying correspondence it creates between members of X and Y, quite understandably at some acceptable performance cost. It may attempt to hide that correspondence by employing a variety of techniques, such as encryption/decryption of requests to prevent simple content comparison by the adversary, or randomly delaying and forwarding requests in some order other than the one in which they were collected to prevent sequence number associations, etc.

2.1 System Cache

In the interest of mitigating the performance cost introduced by it, the anonymity system is presumed to maintain an internal cache, from which some client requests can be served immediately if their requested content is found within. More importantly, as shown by Bagai and Tang [4], such a cache results in elevating the level of anonymity provided by the system to its clients. As not all incoming client requests need be forwarded to target servers, due to some requests being served from cache, we have that $|X| \geq |Y|$.

The maximum anonymity this system may achieve is when for any particular output request $y \in Y$, each of the input requests in X is a possible candidate to be the one that exited the system as y. We depict this situation by the complete bipartite graph $K_{|X|,|Y|}$ between X and Y, as shown in Fig. 1(a) for example sizes $|X| = 5$ and $|Y| = 3$. Any edge $\langle x_i, y_j \rangle$ in this graph indicates that the incoming request x_i could possibly have been the outgoing request y_j.

An attacker essentially eliminates as many edges as possible from $K_{|X|,|Y|}$ due to their infeasibility revealed by some observable system behavior, with a hope of arriving at a subgraph of $K_{|X|,|Y|}$ that contains all the vertices of $K_{|X|,|Y|}$, but in which the number of edges adjacent to any vertex is as close to one as possible. As an example of an attack, adapted from Adman et al. [1], suppose any request entering the system that is also forwarded by it to its target server is known to the attacker to always exit the system after a delay of between 1 and 4 time units. If five requests enter the system, and three exit, at times shown in Fig. 1(b), then x_1 must be either y_1 or y_2, because y_3 is outside the possible latency window of x_1. Similar reasoning can be performed on all other requests to arrive at the subgraph produced by this attack, shown in Fig. 1(b), whose biadjacency matrix is given in Fig. 1(c).

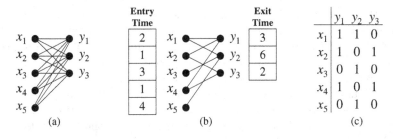

Fig. 1. (a) Complete bipartite graph $K_{5,3}$ depicting full anonymity; (b) Subgraph resulting from a timing analysis attack; (c) Biadjacency matrix of the resulting subgraph

One way to measure the overall amount of anonymity that the system still provides after such an attack, to the *requests that exit*, was proposed by Bagai and Tang [4]. Their method is based upon the number of perfect matchings between Y and any $|Y|$-sized subset of X that are contained in the resulting subgraph. The subgraph of Fig. 1(b) contains 10 such matchings, shown in Fig. 2. As the matching actually employed by the system is still hidden, after the attack, among these 10, the larger this number, the higher the system's anonymity level. By observing the system's outward behavior and eliminating infeasible edges, the attack has succeeded in bringing the number of matchings contained in $K_{|X|,|Y|} = K_{5,3}$, i.e. $|X|!/(|X| - |Y|)! = 5!/(5 - 3)! = 60$, down to 10 contained in the subgraph.

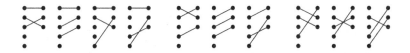

Fig. 2. Perfect matchings contained in the subgraph of Fig. 1(b)

2.2 Request Multiplicities

Gierlichs et al. [5] and Bagai et al. [6] argued that in real-life, clients send multiple requests to the system, destined for the same or different target servers. Moreover, the anonymity system also forwards multiple requests to a server, originating from the same or different users. In such scenarios, the above technique of Bagai and Tang [4], of measuring anonymity as the extent to which the unique perfect matching employed by the system is still hidden among others, is not suitable. As an example, suppose the attacker observes that *David* sent requests x_2 and x_4 of Fig. 1(b), and the outgoing request y_3 went to ashleymadison.com, a web-site that facilitates extra-marital affairs. Consider now the anonymity of the client-server pair $\langle David,$ ashleymadison.com$\rangle$. Five of the perfect matchings in Fig. 2 contain the edge $\langle x_2, y_3 \rangle$, and the other five contain $\langle x_4, y_3 \rangle$. Thus, although the attacker is still unsure of whether y_3 was x_2 or x_4 (each event has 50 % probability), he in fact does not care because he is content to have determined with 100 % certainty that *David* sent one request to ashleymadison.com. Uncovering the exact perfect matching employed by the system between Y and some $|Y|$-sized subset of X is an overkill for the more modest goal of the attacker to just determine *how many* requests any client sent to any server.

Let C be the set of clients sending requests to the anonymity system, and S be the set of servers to which the system forwards some of those requests. While an attack still determines infeasibility of the system's $|X| \cdot |Y|$ *input-output request pairs*, we need a metric to measure the anonymity of $|C| \cdot |S|$ *client-server pairs*, given the externally observable *number* of requests sent by each client and those forwarded to each server. As shown in Fig. 3, for any $i \in \{1, 2, \ldots, |C|\}$, let X_i be the set of requests sent by client i. Similarly, for any $j \in \{1, 2, \ldots, |S|\}$, let Y_j be the set of requests forwarded to server j. We let

$$\overline{C} = \langle |X_1|, |X_2|, \ldots, |X_{|C|}| \rangle \quad \text{and} \quad \overline{S} = \langle |Y_1|, |Y_2|, \ldots, |Y_{|S|}| \rangle$$

be the client and server request multiplicity vectors, respectively. Note that, $\sum_{i=1}^{|C|} |X_i| = |X|$ and $\sum_{j=1}^{|S|} |Y_j| = |Y|$. Clearly,

$$\{X_i \times Y_j : 1 \le i \le |C|, \ 1 \le j \le |S|\}$$

is a partition of $X \times Y$. Any member $X_i \times Y_j$ of this partition is the set of all edges in $K_{|X|,|Y|}$ from client i to server j.

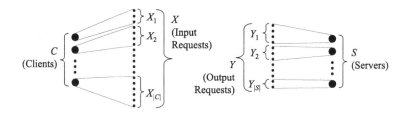

Fig. 3. Clients and servers exchanging multiple requests

3 Measuring Anonymity of Client-Server Associations

We now develop a method to measure the amount of anonymity remaining in the system after an attack that has (a) observed the sending client of each request in X and the receiving server of each request in Y (thus knowing $\overline{C}$ and $\overline{S}$), and (b) determined infeasibility of some input-output request pairs of the system (thus arriving at some $|X| \times |Y|$ biadjacency matrix B, as in Fig. 1(c)).

3.1   Equivalent Perfect Matchings

Let $\mathfrak{P}$ be the set of all $|X|!/(|X|-|Y|)!$ perfect matchings in $K_{|X|,|Y|}$ between Y and some $|Y|$-sized subset of X. The attacker considers two perfect matchings in $\mathfrak{P}$ to be equivalent if they have the same *number* of requests going from each client to each server. This equivalence relation is induced by $\overline{C}$ and $\overline{S}$, as follows.

For any matching $M \in \mathfrak{P}$, we define the *(client-server) association matrix* of M, denoted $\mathbf{A}(M)$, as the $|C| \times |S|$ matrix of nonnegative integers given by:

$$\mathbf{A}(M)_{ij} = |M \cap (X_i \times Y_j)|.$$

Any entry $\mathbf{A}(M)_{ij}$ of this matrix is simply the number of edges (i.e. client-server associations) in M from client i to server j.

Now, any matchings $M_1, M_2 \in \mathfrak{P}$ are *equivalent*, denoted $M_1 \cong M_2$, if they have the same association matrix, i.e. $\mathbf{A}(M_1) = \mathbf{A}(M_2)$.

As an example, in the system of Fig. 1(b), suppose an attacker observed that $X_1 = \{x_1\}$, $X_2 = \{x_2, x_3, x_4\}$, $X_3 = \{x_5\}$, $Y_1 = \{y_1\}$, $Y_2 = \{y_2, y_3\}$, i.e. the system has $|C| = 3$ clients with multiplicities $\overline{C} = \langle 1, 3, 1 \rangle$, and $|S| = 2$ servers with multiplicities $\overline{S} = \langle 1, 2 \rangle$. The two perfect matchings $M_1, M_2 \in \mathfrak{P}$, shown in Fig. 4, are then equivalent to the attacker because they have the same association matrix, also shown in the figure. This example also illustrates that matchings may be equivalent even if they involve different $|Y|$-sized subsets of X.

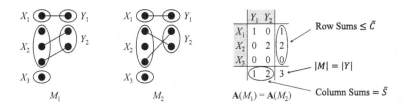

Fig. 4. Example of equivalent perfect matchings and their common association matrix

It is easy to see that for any $M \in \mathfrak{P}$, the sum of all entries in $\mathbf{A}(M)$ is $|M|$, i.e. $|Y|$. While the column-sum vector of $\mathbf{A}(M)$ coincides with $\overline{S}$, the row-sum values, due to caching, can be smaller than the corresponding values of $\overline{C}$, i.e.

$$\sum_{i=1}^{|C|} \mathbf{A}(M)_{ij} = |Y_j|, \ 1 \le j \le |S|, \quad \text{but} \quad \sum_{j=1}^{|S|} \mathbf{A}(M)_{ij} \le |X_i|, \ 1 \le i \le |C|.$$

Thus, for any vector $\overline{R} = \langle r_1, r_2, \ldots, r_{|C|} \rangle$, such that $\sum_{i=1}^{|C|} r_i = |Y|$ and for each $i \in \{1, 2, \ldots, |C|\}$, $r_i \le |X_i|$, any $|C| \times |S|$ matrix of non-negative integers with $\overline{R}$ and $\overline{S}$ as its row- and column-sums vectors, respectively, uniquely represents an equivalence class of $\cong$. Dyer et al. [9] showed that determining the number of such matrices is a #P-complete problem, but several methods for computing their numbers exist, such as in Greselin [10].

We let $\mathcal{CSA}(\mathfrak{P})$ denote the set of all association matrices of matchings in $\mathfrak{P}$.

3.2   Feasible Matchings in an Equivalence Class

The equivalence relation $\cong$ on the set $\mathfrak{P}$ of all possible matchings is induced purely by the client and server request multiplicity vectors, $\overline{C}$ and $\overline{S}$, observed by the attacker. Recall that the attacker additionally arrives at some $|X| \times |Y|$ biadjacency matrix B after determining some edges in $K_{|X|,|Y|}$ to be infeasible. This leg of the attack in fact renders some matchings in $\mathfrak{P}$ infeasible, namely those that contain at least one infeasible edge. In order to measure the resulting anonymity remaining in the system, we need to determine the *number* of feasible matchings left by B in each equivalence class of $\cong$.

Let A be any association matrix in $\mathcal{CSA}(\mathfrak{P})$. Then all matchings in the equivalence class represented by A will have A_{ij} edges between any client i and server j. Let E be the unique sub-matrix of B corresponding to rows X_i and columns Y_j. The rows and/or columns of E may not be contiguous in B. Now, let F be any arbitrary sub-matrix of E, with A_{ij} rows and columns, which are again not necessarily contiguous in E. Clearly, any feasible matching in the subgraph corresponding to F will have exactly A_{ij} edges. It is well known (see, for example Asratian [11]) that there are as many such matchings as the *permanent* of F, given by:

$$\rho(F) = \sum_{\phi \in \Phi} F_{1\phi(1)} F_{2\phi(2)} \cdots F_{n\phi(n)},$$

where $n = A_{ij}$, and Φ is the set of all bijections $\phi : \{1, 2, \ldots, n\} \to \{1, 2, \ldots, n\}$, i.e. permutations of the first n positive integers. For any such F, the number of feasible matchings in the equivalence class that F is involved in is the product of $\rho(F)$ and the feasible matchings between the remaining rows and columns of B. Adding this value over all such F results in the desired count. This technique is employed by the COUNT-FEASIBLE function below to determine the number of feasible matchings in the equivalence class represented by the association matrix A, according to the biadjacency matrix B:

> COUNT-FEASIBLE(A, B)
> > **if** there exist some i and j, such that $A_{ij} \neq 0$
> > > $P = A$
> > > set the entry P_{ij} to 0
> > > E = the sub-matrix of B on rows X_i and columns Y_j
> > > $sum = 0$
> > > **for** each $A_{ij} \times A_{ij}$ sub-matrix F of E
> > > > $Q = B$
> > > > set entries in Q, whose row and/or column is in F, to 0
> > > > $sum = sum + \rho(F) \cdot$ COUNT-FEASIBLE(P, Q)
> > > **return** sum
> > **return** 1

The above essentially computes the number of feasible matchings by multiplying the number of all feasible *partial* matchings with their feasible *extensions*.

3.3   Anonymity Metric

We can now construct a metric for the extent to which the system's client-server communication pattern is still hidden among other such patterns after the attack. Recall that $\mathfrak{P}$ is the set of all $|X|!/(|X| - |Y|)!$ perfect matchings in $K_{|X|,|Y|}$ between Y and some $|Y|$-sized subset of X, as shown in Fig. 5(a). The client and server request multiplicity vectors, $\overline{C}$ and $\overline{S}$, induce a partition of these matchings into equivalence classes, as in Fig. 5(b), and finally, the biadjacency matrix B renders some of the matchings as infeasible, as in Fig. 5(c).

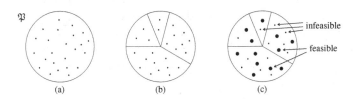

Fig. 5. (a) The set $\mathfrak{P}$ of all $|X|!/(|X| - |Y|)!$ matchings; (b) Equivalence classes on $\mathfrak{P}$ induced by $\overline{C}$ and $\overline{S}$; (c) Feasible/Infeasible categorization of matchings caused by B

The attacker views each equivalence class as one possible scenario, and the fraction of all feasible matchings contained in that class as the probability of the occurrence of that scenario. In other words, an attack induces a probability distribution on the set of all equivalence classes or, alternatively, on $\mathcal{CSA}(\mathfrak{P})$, given the one-to-one correspondence between equivalence classes and association matrices. Let N be the number of all association matrices, for given $\overline{C}$ and $\overline{S}$ vectors, and $A_1, A_2, \ldots, A_N$ be those matrices. The attack thus results in a probability distribution on these matrices:

$$\Pi = \langle \pi(A_1), \pi(A_2), \ldots, \pi(A_N) \rangle,$$

where each $\pi(A_i) = \text{COUNT-FEASIBLE}(A_i, B)/T$, and T is the total number of feasible matchings, i.e. $\sum_{i=1}^{N} \text{COUNT-FEASIBLE}(A_i, B)$. Any value $\pi(A_i)$ is the probability that the *actual* matching in $\mathfrak{P}$ employed by the anonymity system, which the system wants to hide, lies in the equivalence class represented by A_i.

Several methods already exist in the literature for measuring the anonymity of an object that is hiding among some group, when the attacker already has probabilities associated with each member of that group of being the hiding object. Kelly et al. [12] and Bagai and Jiang [13] survey many such methods, which essentially attempt to measure the uncertainty contained in the attacker's probability distribution. Currently, the most popular one is of Serjantov and Danezis [2], improved by Diaz et al. [8], which is based on the Shannon entropy [7] of the distribution. According to their method, the anonymity provided by the system would be:

$$\theta(\Pi) = \frac{-1}{\log N} \sum_{i=1}^{N} \pi(A_i) \cdot \log \pi(A_i).$$

In the above, the base of the logarithm is not important, as its choice only affects the unit of measurement. Also, $0 \cdot \log 0$ is interpreted as 0. Works such as Gierlichs et al. [5] and Bagai et al. [6] employ this approach.

The above metric lies between 0 and 1, and favors uniformity of Π. The more uniform this distribution, i.e. closer to $\langle \frac{1}{N}, \frac{1}{N}, \ldots, \frac{1}{N} \rangle$, the higher the anonymity judged by θ, because higher uniformity of Π essentially results from failure of the attack's biadjacency matrix B to sufficiently expose the equivalence class adopted by the system. While seemingly reasonable on the surface, we argue that this measure is insufficient as it evaluates effectiveness of just one component of the attack, namely B, while overlooking that of the other one, $\overline{C}$ and $\overline{S}$.

In order to arrive at a more accurate measure of anonymity, we propose to consider a *quality* factor of each equivalence class induced by $\overline{C}$ and $\overline{S}$. Let $\widehat{A}$ be the association matrix in $\mathcal{CSA}(\mathfrak{P})$ of the actual matching in $\mathfrak{P}$ employed by the system. Then, the *error* contained in any $A \in \mathcal{CSA}(\mathfrak{P})$ is the amount by which A differs from $\widehat{A}$, given by:

$$\delta(A) = \frac{1}{2\,|Y|} \sum_{i=1}^{|C|} \sum_{j=1}^{|S|} |A_{ij} - \widehat{A}_{ij}|.$$

As the sum of all values in any association matrix is $|Y|$, division by $2\,|Y|$ in the above guarantees the value of $\delta(A)$ to be between 0 and 1. It is also easy to see that $\delta(A) = 0$ iff $A = \widehat{A}$.

If $A_1, A_2, \ldots, A_N$ are, as before, the association matrices induced by $\overline{C}$ and $\overline{S}$, let $\Delta = \langle \delta(A_1), \delta(A_2), \ldots, \delta(A_N) \rangle$ be the attack's error vector. We define the anonymity provided by the system after the attack to be:

$$\Theta(\Pi, \Delta) = \sum_{i=1}^{N} \pi(A_i) \cdot \delta(A_i).$$

The following propositions show that the above is a reasonable anonymity metric. We omit proofs here due to space restrictions.

Proposition 1. $\Theta(\Pi, \Delta) = 0$, *i.e. the system is left with no anonymity, iff the attacker has obtained full information from the attack, i.e.* $\pi(\widehat{A}) = 1$.

Proposition 2. $\Theta(\Pi, \Delta) = 1$, *i.e. the system is still providing full anonymity, iff the attacker has gained no information from the attack, i.e.* $|C| = |X|$, $|S| = |Y|$, *and B has no 0 entries.*

A fundamental departure made by our metric Θ over existing approaches based on θ is in the consideration given by it to the choice of $\widehat{A}$, an aspect that is traditionally ignored. By being sensitive to $\widehat{A}$, our new metric gives a more accurate measure of the extent to which $\widehat{A}$ is still hidden after the attack.

For an illustration of this sensitivity, let us revisit the attack of Figs. 1 and 4. It can be seen that, in this attack, $\overline{C} = \langle 1, 3, 1 \rangle$ and $\overline{S} = \langle 1, 2 \rangle$ partition the 60 matchings in $\mathfrak{P}$ into $N = 8$ equivalence classes whose association matrices, A_1

through A_8, are as in Fig. 6. These are all the matrices whose column-sum vector is $\overline{S}$ and row-sum vector is contained within $\overline{C}$. The attack's biadjacency matrix B, shown in Fig. 1(c), renders 50 matchings in $\mathfrak{P}$ infeasible. The remaining 10 feasible ones, shown in Fig. 2, happen to be evenly distributed, two each, in the classes corresponding to A_1 through A_5. The classes of matrices A_6 through A_8 are not left with any feasible matchings. The probability distribution Π is thus $\langle \frac{1}{5}, \frac{1}{5}, \frac{1}{5}, \frac{1}{5}, \frac{1}{5}, 0, 0, 0 \rangle$, leading to $\theta(\Pi) = \log 5/\log 8 \approx 0.774$.

1	0		0	1		1	0		0	0		0	0		0	1		0	1		0	0
0	2		1	1		0	1		1	2		1	1		1	0		0	1		0	2
0	0		0	0		0	1		0	0		0	1		0	1		1	0		1	0
A_1			A_2			A_3			A_4			A_5			A_6			A_7			A_8	

Fig. 6. Association matrices created by the attack of Figs. 1 and 4

Our new metric Θ takes, in addition, the choice of $\widehat{A}$ into account. For example, if $\widehat{A}$ is A_1, Δ can be seen to be $\langle 0, \frac{2}{3}, \frac{1}{3}, \frac{1}{3}, \frac{2}{3}, 1, \frac{2}{3}, \frac{1}{3} \rangle$, leading to $\Theta(\Pi, \Delta) = 2/5 = 0.4$. The anonymity measure turns out to be the same if $\widehat{A}$ is A_2 or A_3. However, if $\widehat{A}$ is A_4, $\Delta = \langle \frac{1}{3}, \frac{1}{3}, \frac{2}{3}, 0, \frac{1}{3}, \frac{2}{3}, \frac{2}{3}, \frac{1}{3} \rangle$, producing $\Theta(\Pi, \Delta) = 1/3 \approx 0.33$, a different anonymity measure. The choice of A_5 for $\widehat{A}$ results in the same anonymity level, while the choices A_6 through A_8 are not possible because their probabilities in Π are 0 and the attacker is assumed to employ only observable system behavior to conduct the attack, thus making it impossible to incorrectly render any edge between X and Y as infeasible, a condition that is necessary to reduce the probability of $\widehat{A}$ to 0.

4 Conclusions

Accurate measurement of the amount of anonymity received by any clients communicating anonymously with some servers in the wake of an attack is essential for evaluating the robustness of the underlying anonymity system. It is already identified, as in Gierlichs et al. [5] and Bagai et al. [6], that an attack essentially generates a probability distribution over the set of all possible client-server associations for being the real one. The customary approach for measuring remaining anonymity after the attack is by determining the uncertainty inherent in this distribution with the Shannon entropy technique [7]. This has been the state-of-the-art ever since the proposals of Serjantov and Danezis [2], and Diaz et al. [8].

We showed that if, in addition to this probability distribution, the correctness levels of all possible associations are also taken into account, a more accurate measure of anonymity can be arrived at. We developed an anonymity metric that incorporates the correctness levels, thereby resulting in an improvement over existing metrics.

Arriving at the probability distribution requires computing permanents of matrices, a problem already shown by Valiant [14] to be #P-complete. Thus, the attack itself, as well as all metrics mentioned above, namely the existing ones and ours, are hard to compute. Development of an efficient and reasonably accurate heuristic for our metric is therefore a useful future undertaking.

Acknowledgments. This work was partially supported by the United States Navy Engineering Logistics Office contract number N41756-08-C-3077.

References

1. Edman, M., Sivrikaya, F., Yener, B.: A combinatorial approach to measuring anonymity. In: IEEE International Conference on Intelligence and Security, pp. 356–363. Informatics, New Brunswick, USA (2007)
2. Serjantov, A., Danezis, G.: Towards an information theoretic metric for anonymity. In: Dingledine, R., Syverson, P.F. (eds.) PET 2002. LNCS, vol. 2482, pp. 41–53. Springer, Heidelberg (2003)
3. Shubina, A.M., Smith, S.W.: Using caching for browsing anonymity. ACM SIGEcom Exch. **4**(2), 11–20 (2003)
4. Bagai, R., Tang, B.: Data caching for enhancing anonymity. In: 25th IEEE International Conference on Advanced Information Networking and Applications (AINA), pp. 135–142. Singapore (2011)
5. Gierlichs, B., Troncoso, C., Diaz, C., Preneel, B., Verbauwhede, I.: Revisiting a combinatorial approach toward measuring anonymity. In: 7th ACM Workshop on Privacy in the Electronic Society, pp. 111–116. Alexandria, VA, USA (2008)
6. Bagai, R., Tang, B., Khan, A., Ahmed, A.: A system-wide anonymity metric with message multiplicities. Int. J. Secur. Network. **10**(1), 20–31 (2014)
7. Shannon, C.: A mathematical theory of communication. Bell Syst. Tech. J. **27**, 379, 623–423, 656 (1948)
8. Díaz, C., Seys, S., Claessens, J., Preneel, B.: Towards measuring anonymity. In: Dingledine, R., Syverson, P.F. (eds.) PET 2002. LNCS, vol. 2482, pp. 54–68. Springer, Heidelberg (2003)
9. Dyer, M., Kannan, R., Mount, J.: Sampling contingency tables. Random Struct. Algorithms **10**(4), 487–506 (1997)
10. Greselin, F.: Counting and enumerating frequency tables with given margins. Statistica Applicazioni **1**(2), 87–104 (2003)
11. Asratian, A., Denley, T., Häggkvist, R.: Bipartite Graphs and Their Applications. Cambridge University Press, Cambridge (1998)
12. Kelly, D., Raines, R., Baldwin, R., Grimaila, M., Mullins, B.: Exploring extant and emerging issues in anonymous networks: a taxonomy and survey of protocols and metrics. IEEE Commun. Surv. Tutorials **14**(2), 579–606 (2012)
13. Bagai, R., Jiang, N.: Measuring anonymity by profiling probability distributions. In: 11th IEEE International Conference on Trust, Security and Privacy in Computing and Communications (TRUSTCOM), pp. 366–374. Liverpool, UK (2012)
14. Valiant, L.: The complexity of computing the permanent. Theoret. Comput. Sci. **8**(2), 189–201 (1979)

Probabilistic Model Checking of Security Protocols without Perfect Cryptography Assumption

Olga Siedlecka-Lamch[1]([⊠]), Miroslaw Kurkowski[2], and Jacek Piatkowski[1]

[1] Institute of Computer and Information Sciences,
Czestochowa University of Technology, Częstochowa, Poland
{olga.siedlecka,jacekp}@icis.pcz.pl
[2] Institute of Computer Sciences,
Cardinal Stefan Wyszynski University in Warsaw, Warsaw, Poland
mkurkowski@uksw.edu.pl

Abstract. This paper presents the description of a new, probabilistic approach to model checking of security protocols. The protocol, beyond traditional verification, goes through a phase in which we resign from a perfect cryptography assumption. We assume a certain minimal, but measurable probability of breaking/gaining the cryptographic key, and explore how it affects the execution of the protocol. As part of this work we have implemented a tool, that helps to analyze the probability of interception of sensitive information by the Intruder, depending on the preset parameters (number of communication participants, keys, nonces, the probability of breaking a cipher, etc.). Due to the huge size of the constructed computational spaces, we use parallel computing to search for states that contain the considered properties.

Keywords: Computer networks security · Security protocols verification · Perfect cryptography assumption · Probabilistic analysis

1 Introduction

Security protocols are a key point of security systems in computer networks. They are concurrent algorithms which are described as sequences of actions performed by a few (two or more) entities (sides, servers), allowing for realization of the intended purpose. Some actions of those protocols are performed using cryptographic algorithms. As correct functioning of the security protocol we mean achievement of the required goals: unilateral or mutual authentication, confidentiality, message integrity or new session key distribution.

Nowadays, increasingly more data need to be encrypted, for their safety and to comply with legislation. The problem is how to manage such a large number of keys (sometimes hundreds of thousands of keys). Their ongoing generation and decryption is a burden for a server. The threat is not only the breaking of them

© Springer International Publishing Switzerland 2016
P. Gaj et al. (Eds.): CN 2016, CCIS 608, pp. 107–117, 2016.
DOI: 10.1007/978-3-319-39207-3_10

but also an ordinary takeover. Therefore, is a perfect cryptography assumption today fully justified?

The problem of protocol correctness can be analyzed on many levels. There is the aspect of its implementation (both hardware and software), or technical components. However, the essence of protocols are algorithms, hence the most important form of verification are formal methods that sometimes can prove whether an investigated protocol has the necessary security property or not.

There are at least three main groups of formal methods of protocol verification: inductive [1], axiomatic (deductive) [2] and model checking [3]. The last of them, on which our research is focused, involves building a proper mathematical model of protocol executions. The model checking method shows in a formal way (preferably automated) that all the considered executions possess the desired properties. There are several high-profile projects linked with model checking of security protocols such as Avispa [4], SCYTHER [5] or native VerICS [6].

Our approach is based on a much simpler and more efficient technique of constructing chains of states for a given protocol, and it also follows works on the VerICS system [7,8]. These structures encode all the significant actions of the protocol and execution interlaces occurring in real network communications. The tree of real executions of the protocol, built on this basis, allows one to automatically search for the path of an attack. In [7,12] we showed that the parallel approach not only significantly decreases the time of analyzing and searching constructed spaces, but also allows one to check the properties of protocols with a greater number of considered parameters (number of communication participants, keys, nonces, the probability of breaking a cipher, etc.) or in the case of more complicated protocols.

So far, all these approaches use the main assumption called *the perfect cryptography assumption*. That means that it is impossible to get any information about the content of an encrypted message without knowing the key, which is necessary to decrypt that message. In many situations it seems reasonable to somewhat reduce the level of key security. Suppose that the information should be kept as a secret only for a very short period of time. It seems that a good example of this situation is communication between a landing airplane and the control tower when the exchanged information should be protected only until the plane has landed. Another situation is when the information contained in the messages does not really matter (gossip on a communicator in a local network). In these situations we can consider that the power of the used cryptography need not be very strong. Therefore we suppose that in these cases there is a very small but measurable probability of breaking the key. Thus the system administrator has the ability to match the level of security to his needs.

To our previous studies, we have added the probability of breaking a key. In this paper we present the analysis of possible executions of the protocol in this case. In order to do this, we generate probabilistic trees that model these executions. Our work is based on our previous structures – chains of protocol states that were presented in [7,8]. We illustrate our consideration using as an example the well-known NSPKL protocol (from the names of the authors Needham,

Schroeder, Lowe [9,10]), for various configurations of participant communication. The tool constructed for this purpose generates and explores different executions of the protocol in terms of the probability of a leak of confidential information, due to the preset parameters: the keys, nonces, number of parties, the probability of breaking a key, etc. In view of the experimental results, the method is ideally suited for parallelization (already at the level of generating performances, or probabilistic trees, etc.) and thanks to it we obtained significant acceleration.

Our paper is divided into five main sections: the introduction, the section describing the idea of chains of states, the section about probabilistic analysis of protocol executions with the assumed probability of breaking a key, the section that contains the experimental results, together with a description of parallel algorithms, and finally the summary and plans for future work.

2   Chains of States

In the following section the idea of chains of states will be presented. As will be shown, these chains describe all the important behaviors of the system from the modeling and checking point of view, and can be used for formal specification of the considered protocol.

In our approach we define four types of states:

- States that represent specific executions of the protocol steps. States of this type will be determined further by: S_j^i, where parameter i specifies the number of steps in the corresponding execution of the protocol, and the j parameter specifies the execution number.
- States representing the generation of confidential information (nonces, encryption keys) by users. These states will be denoted by G_U^X, which means the generation of secret X by user U (for example, $G_a^{n_a}$ means that user A generates number n_a).
- States representing learning about the different elements of the message (which may be ciphertext) by the receiver. These states are denoted by K_U^X – acquiring knowledge by user U about X (for example $K_a^{n_b}$ – obtaining by a number n_b).
- States representing the need of a user to have knowledge about the information necessary to compose and send a message in the next step. Those states will be marked by P_U^X – user U must possess knowledge about X. For example $P_a^{k_b}$ means that user a must know the public key of b.

Consider the following example of the NSPKL protocol [9] written in Common Language:

$$
\begin{aligned}
A \to B \quad & \langle i(A), N_A \rangle_{K_B} \\
B \to A \quad & \langle N_A, N_B, i(B) \rangle_{K_A} \\
A \to B \quad & \langle N_B \rangle_{K_B}.
\end{aligned}
\tag{1}
$$

In the first step of protocol execution, participant A creates random number (nonce) N_A, adds its identifier and sends this message which is encrypted by public key B to participant B. The next step of NSPK protocol execution starts from the generation of its own random number N_B by participant B. Next, it creates the message consisting of nonces of both communication participants and its identifier, encrypts this message by means of public key K_A, and sends it to participant A.

After that A runs the decrypting operation of the received message, and the operation of comparing the N_A number, which was received from B with number N_A prepared by itself. If both numbers are correct, A considers B as authenticated. In the last step of the protocol execution, A sends nonce N_B encrypted by the K_B key to B. After decrypting and comparing the numbers, participant B can consider participant A as authenticated. Executing this protocol should guarantee both participants the identity of the communicating person.

The method of automatic generation of chains of states will be as follows: if we consider the k-th step in execution number i – inside the sequence we put state S_k^i. Before this state we place type $P_{Send_k^i}^X$ (needed) and Gen_k^i (generated) states. After state S_k^i in a constructed chain – we place type K_U^X states, for all the information that the recipient can extract from messages using his own keys. The Intruder generates as many strings as sets of generators there are for the received message. Details of the algorithm can be found in [7].

For the NSPKL protocol we distinguish states representing the execution steps: S_1^1, S_2^1, S_3^1. The method generates the following chains of states:

$$\alpha_1^1 = (G_A^{N_A}, S_1^1, K_B^{N_A}),$$
$$\alpha_2^1 = (P_B^{N_A}, G_B^{N_B}, S_2^1, K_A^{N_B}), \quad (2)$$
$$\alpha_3^1 = (P_A^{N_B}, S_3^1).$$

Denote as $PreCond(S)$ those states before executions of the steps of the Protocol and $PostCond(S)$ states that follow them. Using such a notation NSPKL execution can be modeled as:

$$\alpha_1^1, \alpha_2^1, \alpha_3^1 = (G_A^{N_A}, S_1^1, K_B^{N_A}, P_B^{N_A}, G_B^{N_B}, S_2^1, K_A^{N_B}, P_A^{N_B}, S_3^1). \quad (3)$$

Let us note that following conditions hold:

$$PreCond(S_2^1) = \{P_B^{N_A}, G_B^{N_B}\}, \quad PostCond(S_2^1) = \{K_A^{N_B}\}. \quad (4)$$

It would be useful to define a sequence of states representing a real protocol execution on networks.

Let Π be a space consisting of a defined number of users and their attributes (nonces, identifiers, cryptographic keys, etc.). Moreover, in this space we have to consider all the executions of a protocol and all the chains of states representing all these executions. In the set of all the chains of states we can define the states which correspond to the runs in the computational structure from [6].

Definition 1. *The chain of states* $\mathfrak{s} = s_1, s_2, \ldots, s_p$ *will be called a* **correct chain of states** *if and only if the following conditions hold:*

1. if $s_i = S_j^k$ for some $j, k \leq p$, then $j = 1 \vee \exists_{t<i}(s_t = S_{j-1}^k)$ and $PreCond(S_j^k) \subseteq \{s_1, \ldots, s_{i-1}\} \wedge PostCond(S_j^k) \subseteq \{s_{i+1}, \ldots, s_p\}$,
2. if $s_i = G_U^X$, then $\forall_{t \neq i}(s_t \neq G_U^X)$,
3. if $s_i = P_U^X$, then $\exists_{t<i}(s_t = G_U^X \vee s_t = K_U^X)$.

The first point guarantees the correct order of the steps of the given execution. The further two points provide appropriate knowledge dependencies (the user can send only what he has generated or what he has learned). In this simple way we can represent all the possible protocol executions, and during the building of a tree of runs – find a possible attack.

3 Probabilistic Approach

The idea of chains of states has been successfully used to verify many protocols, and the results of these studies can be found in [7,8]. The research was performed with the Dolev-Yao (DY) Intruder model [3] and mentioned before the *perfect cryptography assumption*. However, we can investigate protocols using four different types of Intruder model. Apart from DY we can consider its restricted version where the Intruder knows the information sent only to him. We can also consider the model called the Lazy Intruder model where the Intruder does not create new data, and it is a restricted version.

As we mentioned before, we investigate a situation in which the Intruder is trying to break the key. We assume that there is a certain minimal, but measurable probability that he will succeed. In continued research, in this way we can assume the same or a different probability of breaking each key. One can also consider a different Intruder computing power. The time dependence between executions of the following steps of a protocol can be another parameter. In the considerations presented below we have assumed that the probability of breaking keys is the same, the Intruder has a certain computing power and the protocol steps are performed at the same time intervals. As in other approaches we do not consider the delays on networks when transmitting further information. Another parameter is the size of the space specified by the number of communication participants, keys and nonces.

As an example, we present the consideration of one execution of the NSPKL protocol between two honest users A and B presented in Eq. 1. Having assumed the parameters: communication sites, used keys and their number, the number of generated nonces, the probability of breaking a key; we can analyze the possible situations and the knowledge that the Intruder will gain in the following steps, depending on whether he breaks one of the keys, combinations of them, or none. In Fig. 1 we used notations of chains $P_U^X; G_U^X; K_U^X; K_I^X$: needed; generated; learned by an honest user; learned by the Intruder.

Firstly, the Intruder tries to break key K_B used in the first sent message $(\langle i(A), N_A \rangle_{K_B})$. He will succeed with probability p_{1K_B}, and in this case he will know key K_B and nonce N_A. Otherwise, in the next step, he will still be trying to break key K_B, or another key K_A which appears within message

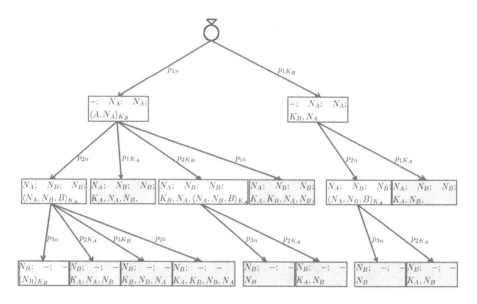

Fig. 1. Probabilistic tree for NSPK protocol

$\langle N_A, N_B, i(B) \rangle_{K_A}$, or both keys. The operating principle is similar in subsequent steps of the protocol. In Fig. 1, the grey nodes indicate the states in which the Intruder has discovered all of the sensitive contents of the information sent during protocol execution (in our example there are nonces N_A and N_B).

For all the states, we can easily calculate the overall probability of achieving a given place. This information will give the user knowledge: the importance of set encryption strength for correct execution of the specified protocol.

As one can see from the example above, in this Protocol the Intruder can focus on breaking only one key to get all the secret information. He does not have to divide his computing power to break both keys used in the protocol. The present tree indicates that the Intruder should concentrate on breaking key K_B, since he has the most time for it and that is why the likelihood of breaking this key is the greatest. In the case of the K_A key, the Intruder also obtains all the secret information, but he has less time for breaking it.

The presented tree shows only one of the scenarios of protocol executions possible to generate. As mentioned above (and described in works [7,8]) we can generate many more scenarios and thereby take into account different user behaviors, especially the Intruder, in real computer networks.

On the basis of the results returned by the tool one can match the power of cryptographic protection of keys whose breaking/gaining will give the highest probability of gaining the secret knowledge.

4 Experiments

In this section, we will present a description of the tool which automatically generates the discussed probabilistic trees and designated states (situations) in which the Intruder can gain secret knowledge with a certain probability. This tool can be used by network administrators to investigate what cryptographic power for security protocols they must ensure to get the proper probability of confidentiality of transmitted information. We will also present experimental results for several versions of the NSPKL protocol.

The experiments are based on our previous experience. Firstly, the protocol is specified in ProToc specification language [11]. Then, its executions are generated in a specified, space given by the user, defined by the preset parameters (number of communication participants, keys, nonces). Executions are encoded as chains of states [7,8]. As was shown in [7], the parallel approach is ideal for the first two phases of verification. Subsequently, probabilistic trees can also be parallelly constructed and tested.

As already mentioned, to build probabilistic trees we must have a set of chains of states and a set of executions of a given protocol. These collections represent the input for the tree-building algorithm. Each node of the tree must provide access to information being the appropriate subset of each of the afore-mentioned sets. We do not want to duplicate existing data, so it is sufficient to store addresses to the relevant objects. The node must keep a collection of additional individual information – calculated on the construction stage of the tree. In the practical implementation of this issue we defined the following classes:

```
class prnode{
    prnode*          prev;   // address of the previous node
    vector<prnode*>  next;   // vector of addreses to following nodes
    chain_position   know;   // address to the part of knowledge
    prob_supp_node   supp;   // supplement
    //...
};
class prob_supp_node{
    wsk_vector not_used_keys;
    vector<double> probability; //probabilities of breaking the key
    wsk_vector intru_know;       //Intruder's knowledge
    //...
};
```

where chain_position is a constant iterator for the chain vector and wsk_vector is teh pointer vector for an abstract base class, after which inherit classes representing various types of processed data, such as keys, nonces, id and so on.

To facilitate control and analysis of generated probabilistic trees we defined the class that maintains the entire tree structure (addresses of its nodes) and addresses of input data. Dispatching the addresses of each tree node offers easy extraction of information collected at the stage of generating its branches – creating a path of information increment, identifying sets of keys untapped by the Intruder or probability values of breaking a key.

```
class pr_bush{
  typedef vector  < p_prnode > one_generation;
  typedef vector  < one_generation > tree;
  chain_position origin;
  substitution_position sub_pos;
  tree bush;
  //...
};
```

where substitution_position is a constant iterator for the execution vector. In this class we also defined several methods that allow, among others, one to determine tree shapes, trace chosen knowledge paths or record results to the output stream. We defined the class responsible for grouping tasks within a single thread, and its functor that calls the method responsible for initializing construction of the tree.

```
class ThreadFunctor{
  vector<pr_bush*>::iterator b_pos;
  size_t trees_per_thread;
public:
  ThreadFunctor():trees_per_thread(0){}
  ThreadFunctor(const bush_position& ref, const size_t& a2):
            b_pos(ref), trees_per_thread(a2){}
  void operator()()const{
    for (size_t i=0; i< trees_per_thread; ++i)
      (*(b_pos+i))->grow_up();
  }
};
```

Functors of the ThreadFunctor class are passed as arguments for the constructor of the std::thread() class.

In the discussed trees, the maximum number of levels is at most equal to the number of steps of the analyzed protocol, whereby a given branch of finish nodes, in which the Intruder gained complete information sent during protocol execution (gray nodes in Fig. 1). The process of determining the nodes for each j-th level of the tree is carried out according to the following algorithm:

Algorithm 1. Generation of one level of the probabilistic tree
1 : Establish necessary keys to break ciphertext in given step of protocol
2 : **for** $i = 1$ *to sizeof*$(N[j])$ **do**
3 :   find Intruders knowledge in current node (A)
4 :   **if** (! Intruder has all needed information (A))
5 :   **if** (! information encrypted by Intruder key)
6 :   Add to current node W_i^j new k-th node W_1^{j+1} in following level $j + 1$
7 :   Add gained knowledge (cryptogram) to W_1^{j+1}
8 :   Add undecrypted keys to W_1^{j+1}
9 :   Each element of a set of keys assigns a probability equal to 1 (for W_1^{j+1})
10 :   **end if**
11 :   Find unused keys and their combinations without repetition (B_i^j)
12 :   **for** $k = 1$ *to sizeof*(B_i^j) **do**

13 :   Find Intruder's knowledge after breaking keys (C_k)
14 :   Add to current node W_i^j new k-th node W_1^{j+1} in following level $j + 1$
15 :   Add knowledge from set C_k to W_1^{j+1}
16 :   Add knowledge derived from W_1^{j+1}
17 :   Add any unused single keys to W_1^{j+1}
18 :   Each element of a set of keys assign a probability equal to p
19 :   **end for**
20 : **end if**
21 : **end for**
where $N[j]$ – thevector of tree nodes at level j.

The depth of the constructed trees will not usually be large, but the number of nodes at all the subsequent levels grows exponentially (this depends on the number of all possible combinations of keys that the Intruder must break). The pessimistic time complexity of teh algorithm is exponential because of all the parameters of the tested protocol: the number of steps, executions, users and objects used by the users. The sequential building of trees is time consuming and resource intensive, hence the use of parallel techniques is fully justified. Trees for individual executions are calculated in parallel using standard library `std::threads` in the C++11 standard.

In the case of the NSPKL protocol, by using parallel computing we obtained triple acceleration of calculation. The parallel approach allowed us to obtain the results shown in Table 1 for several variants of the NSPKL protocol (for two users – NSPKL; for two users, server and need of sending two public keys NSPKL_server2; for two users, server and need of sending one public key NSPKL_server1; for further communication – NSPKL_non2). The table shows the generation time and the size of the constructed space.

Table 1. Parallel expermimental results for NSPKL

Protocol	Number of nodes	Computing time [s]
NSPKL	1 112	0.011
NSPKL_server2	17 004	0.55
NSPKL_server1	58 442	2.53
NSPKL_non2	2 691 920	183

The calculations were made on a computer with Windows 8.1 and Quad-Core Intel I7-4710HQ and 8 GB of RAM.

To conclude, the proposed algorithm is easy to parallelize, which enables accelerating calculations and studying protocols with a greater number of considered parameters (number of communication participants, keys, nonces, etc.) or simply more complex protocols.

5 Summary

In this paper we have presented the analysis of possible executions of the protocol without perfect cryptography assumption. In order to do this we have generated probabilistic trees that model protocol executions. We illustrate our consideration using as an example the NSPKL protocol, with various configurations of communication participants. The tool constructed for this purpose generates and explores different executions of the protocol in terms of the probability of a leak of confidential information, due to the preset parameters: the keys, nonces, number of parties, the probability of breaking a key, etc. According to the experimental results: the method is ideally suited for parallelization (already at the level of generating performances, or probabilistic trees, etc.) and thanks to it we obtained significant acceleration.

The implemented tool allows us to determine for the input protocol which keys are most important (breaking/gaining them guarantees the fastest interception of confidential information), and hence the correct choice of encryption strength or security. It also highlights the keys that are not so important and their cryptographic power can be reduced – thus relieving the server.

References

1. Paulson, L.: Inductive Analysis of the Internet Protocol TLS, TR440. University of Cambridge, Computer Laboratory (1998)
2. Burrows, M., Abadi, M., Needham, R.: A logic of authentication. Proc. R. Soc. Lond. A **426**, 233–271 (1989)
3. Dolev, D., Yao, A.: On the security of public key protocols. IEEE Trans. Inf. Theory **29**(2), 198–207 (1983)
4. Armando, A., et al.: The AVISPA tool for the automated validation of internet security protocols and applications. In: Etessami, K., Rajamani, S.K. (eds.) CAV 2005. LNCS, vol. 3576, pp. 281–285. Springer, Heidelberg (2005)
5. Cremers, C.J.F.: The Scyther tool: verification, falsification, and analysis of security protocols. In: Gupta, A., Malik, S. (eds.) CAV 2008. LNCS, vol. 5123, pp. 414–418. Springer, Heidelberg (2008)
6. Kurkowski, M., Penczek, W.: Verifying security protocols modeled by networks of automata. Fundamenta Informaticae **79**(3–4), 453–471 (2007). IOS Press
7. Kurkowski, M., Siedlecka-Lamch, O., Szymoniak, S., Piech, H.: Parallel bounded model checking of security protocols. In: Wyrzykowski, R., Dongarra, J., Karczewski, K., Waśniewski, J. (eds.) PPAM 2013, Part I. LNCS, vol. 8384, pp. 224–234. Springer, Heidelberg (2014)
8. Kurkowski, M., Siedlecka-Lamch, O., Dudek, P.: Using backward induction techniques in (timed) security protocols verification. In: Saeed, K., Chaki, R., Cortesi, A., Wierzchoń, S. (eds.) CISIM 2013. LNCS, vol. 8104, pp. 265–276. Springer, Heidelberg (2013)
9. Lowe, G.: Breaking and fixing the Needham-Schroeder public-key protocol using FDR. In: Margaria, T., Steffen, B. (eds.) TACAS 1996. LNCS, vol. 1055, pp. 147–166. Springer, Heidelberg (1996)
10. Needham, R.M., Schroeder, M.D.: Using encryption for authentication in large networks of computers. Commun. ACM **21**(12), 993–999 (1978)

11. Kurkowski, M., Grosser, A., Piatkowski, J., Szymoniak, S.: ProToc - an universal language for security protocols specification. In: Wiliński, A., El Fray, I., Pejaś, J. (eds.) Soft Computing in Computer and Information Science. AISC, vol. 342, pp. 237–248. Springer, Heidelberg (2015)
12. El Fray, I., Hyla, T., Kurkowski, M., Maćków, W., Pejaś, J.: Practical authentication protocols for protecting and sharing sensitive information on mobile devices. In: Kotulski, Z., Księżopolski, B., Mazur, K. (eds.) CSS 2014. CCIS, vol. 448, pp. 153–165. Springer, Heidelberg (2014)

High-Interaction Linux Honeypot Architecture in Recent Perspective

Tomas Sochor$^{(\boxtimes)}$ and Matej Zuzcak

University of Ostrava, Ostrava, Czech Republic
tomas.sochor@osu.cz
http://www1.osu.cz/home/sochor/en/

Abstract. High-interaction honeypots providing virtually an unlimited set of OS services to attackers are necessary to capture the most sophisticated human-made attacks for further analysis. Unfortunately, this field is not covered by recent publications. The paper analyses existing approaches and available open source solutions that can be used to form high-interaction honeypots first. Then the most prospective approach is chosen and best applicable tools are composed. The setup is tested eventually and its usefulness is proven.

Keywords: Honeypot high-interaction · Internet threat · SSH service · Linux · Honeypot low-interaction · HonSSH · Bifrozt

1 Introduction to High-Interaction Honeypots

At present, cyberattacks are becoming more and more sophisticated and elaborated. To perform their detailed analysis, it is necessary to have a tool for recording such attacks. Honeypots represent a useful tool to do so. A honeypot is a closed system that can (and should) be attacked and every attacker's step is monitored while confining any undesired attacker's activity. Most honeypots are so-called low-interaction honeypots, where some most specific services of the corresponding real system are emulated. No more activity is usually allowed here. Low-interaction honeypots usually serve primarily for automated attack detection.

While there are some recent publications concentrating on low-interaction honeypots and their application [1–4] as well as associated security techniques [5,6], similar publications focusing to high-interaction honeypots are missing completely. Therefore, this paper offers a unique overview of existing approaches to high-interaction honeypot implementation, proposes a feasible composition of open-source software tools and verifies it in a practical network.

In honeypots, it is often necessary to make a further step upwards from service emulation: the ultimate goal for honeypots should be to create an environment for attackers that is as realistic (i.e. similar to real systems) as possible. The attacker operating in such a system should not feel any limitation, but required security measures must be met (i.e. primarily the attacker's activities

© Springer International Publishing Switzerland 2016
P. Gaj et al. (Eds.): CN 2016, CCIS 608, pp. 118–131, 2016.
DOI: 10.1007/978-3-319-39207-3_11

must be confined to avoid threatening of the laboratory or third party infrastructure). The primary goal of the presented research is the monitoring of human attacker's activities when attempting to compromise the monitored system. i.e. the honeypot.

1.1 Availability for Attackers Versus Monitoring and Restrictions

Surveillance on attacker's activities inside the closed system of a high-interaction honeypot should follow the two main antagonistic principles:

- To let attackers do virtually anything they want to in the honeypot.
- To monitor the attacker's activity in the honeypot in details, and to avoid any harmful influence on other computers from the honeypot.

Both principles are antagonistic in nature because any perfect monitoring and confinement of harmful results inevitably eventuates in steps or measures that are easier to be recognized by an attacker. On the other hand, when some monitoring activities are confined, a loss of certain important information about the attack could happen, or even damages could ensue. Researchers should therefore find a trade-off between the freedom for attackers and the honeypot operation safety and sufficient monitoring, i.e. they find the maximum possible level of monitoring and honeypot protection that can provide sufficient information about the attack.

1.2 Principles of High-Interaction Honeypot Implementation

In addition to the two main high-interaction honeypot principles listed above there are some implementation aspects that should be taken into account in any high-interaction honeypot implementation as follows.

- To create a certain way of communication in order to control the honeypot and send the gathered data about attacks into a (remote) processing node.
- Presentation part of the system located outside the honeypot system.

All the above aspects imply just from the philosophy of high-interaction honeypots and their operation and should be present in any of the following categories of high-interaction honeypot. Nevertheless, practical implementation of the above aspects differs across the following categories.

2 Existing Approaches in High-Interaction Honeypot Implementation

There are several feasible approaches to high-interaction honeypot design, layout and implementation differing significantly in technology. In this section the major existing approaches are classified.

2.1 In-the-Box Solutions

The philosophy of this approach is based on the implementation of a monitoring tool directly into the honeypot system (usually in the form of a rootkit). Here it is described using the best-known representative of this category, Sebek capture tool [7]. Sebek honeypot is split into a server and a client part. The client part is installed directly on the system to be monitored (i.e. the honeypot) where it tries to operate invisibly like a rootkit. Therefore it modifies the system kernel. Its primary role consists in monitoring of certain types of system calls. So its task is to gather data and send it to the Sebek server subsequently. The server can collect data from two types of sources – from a live transmission (Sebek packets about activities in the monitored system), and/or from tcpdump files. Sebek itself cannot filter or suppress any dangerous activity forwarded from the system so it should be combined with a honeywall, e.g. Roo.

Sebek was software with huge opportunities thanks to its perfect conception. Unfortunately, it is completely useless nowadays. This is not only because its versions are limited to outdated operating systems infrequently used (supported are Linux up to 2.6.26, Windows up to XP SP2). Neither ENISA [8] nor designers recommend its use. Moreover, Sebek has also become easily detectable by common anti-rootkit techniques and security tool at present, even without privileged permissions, using techniques like [9,10], or using a passive monitoring of network communication. Therefore, it is well known among attackers [11]. Even worse, it can be easily eliminated after its detection.

Similar problems are common for virtually any tool implemented directly into the host system serving as a honeypot. In fact, this is a variant of a cat-and-mouse game in a way similar to the case of malware detection.

2.2 Tools Limited Just to the First Stadium of Attacks

There are various tools that focus only on the analysis of exploitation stadium of attacks, i.e. to study methods how an attacker enters the system [8]. They do not analyze further activities and therefore they do not provide any opportunity to operate the system to the attacker any further. The best-known tool in this category is Argos Secure System Emulator[1]. In practice, it operates so that it analyses the network data coming to the system. In order to store the network data into RAM, the Dynamic Taint Analysis [12] technology is applied. When an attack is detected, the analysis is performed with the output to the log file. The support of this tool is limited, however, with no significant community. Another significant aspect is that the tool does not provide data about other activities in the attack in addition to the system penetration stadium. For many research activities it seems to be insufficient because further information is required.

2.3 Out-of-the-Box VM-Based Solutions

This category covers tools directly related to virtualization and external software. This high-interaction honeypot category uses a virtualization tool, e.g. QEMU,

---

[1] http://www.few.vu.nl/argos/.

Oracle VirtualBox, Xen, VMWare etc., and the specific honeypot that cooperates on the API level with the specific virtualization software and analyses and reports actions that the virtualization tool can provide.

In addition to better hiding of monitoring beyond the honeypot system, The fact that the system calls are monitored from outside the system lowers the risk of monitoring disclosure by an attacker. In this case the driver is not implemented inside the monitored system but it is a part of the host system. Such a tool allows monitoring of all types of instructions that allow system calls directly assisted by the virtualization software. Obviously, for the subsequent analysis it is necessary to select only suitable system calls. Another option is that in some system only selected internal operations could be monitored, not all of them.

In overwhelming majority of projects falling into this category the VMI (Virtual machine introspection) technology is applied that is used by IDS systems as well [13]. Their basic idea is to move the monitoring component beyond the virtualized system used as a honeypot. The monitoring component is often integrated into so-called virtual machine monitor (VMM) that is provided by virtualization software. VMM runs on a host system and processes all requests of the guest virtualized system. In the case of an attack, VMM generates an overview of the attacker's activity that can be provided subsequently for the purpose of further processing – e.g. into an analysis of internal events of the virtualized honeypot.

One of representatives of this category is Qebek honeypot [14] based on Qemu virtualization tool. It is a successor of Sebek honeypot. Qebek itself monitors virtualized honeypot using the VMM module and it is installed in the host system. Qebek project had improved high-interaction honeypot monitoring significantly. But however, it has not been developed for long period and no community exist and no support is available. Moreover, it supports only one single operating system. i.e. relatively outdated Windows XP SP2, as a guest system. For this reason it is not suitable for our research.

VMI approach is applied in a range of other projects that are not publicly available and were described just as prototypes like VMScope [15] and VMWatcher [16]. Some interesting aspects are present in the project VMI-Honeymon [17] that uses the LibVMI[2] interface. LibVMI presents a library providing resources for monitoring the virtualized system and it can be used for building other VMI-based systems. It is used for Windows system by seemingly presently the only publicly available analytical tool that is solely VMI-based called DRAKVUF [18].

The following two categories are often considered to form a single group. Because of the fact that the paper presents their differences, too, and finally the out-of-the-box approach is to become a basis for further conception and its modification will be presented, the groups are distinguished here.

---

[2] http://libvmi.com/.

2.4 Out-of-the-Box Solutions

Solutions in this category are characterized by monitoring the honeypot activity only from outside with no direct cooperation with host system virtualization software. The system itself used as a honeypot does not contain any modifications in this direction. The monitoring is not only based on the network activity as in the case of network-based approaches but certain monitoring of the honeypot activity happens also on the honeywall (e.g. Linux shell monitoring). Such a solution is extremely transparent for an attacker but a "semantic gap" problem emerges here [19,20]. This issue is present even in association with virtualization but here it is even more significant. The monitoring is fully transparent for the attacker here but the researcher does not collect enough data necessary for a detailed analysis, e.g. memory contents, file system or registry modifications, used system calls etc. Such problems happen when an attacker leaves the system shell and uses a binary file, or encrypts the network communication. This problem cannot be resolved (i.e. the valuable semantic reconstruction of the attack cannot be obtained) without an erosion of the transparency for attackers.

The most significant representative of this category is Honeywall Roo. Honeywall Roo[3] is not just a honeypot; it is rather a purposely-modified firewall intended to protect honeypots from abuse. Its most frequent application is as a component of a high-interaction honeynet. It consists of a honeypot securing, network traffic limitation and its logging into a MySQL database, as well as its basic analysis. The captured network traffic can be obtained in the pcap format as well. For detection of attacks to be further confined, the Snort-inline module is used. Unfortunately, the development of Roo honeywall does not continue and the honeywall is not supported any more. Other solutions belonging into this category are HonSSH[4] and Bifrozt[5], which are used in the research further presented in this paper.

2.5 Network-Traffic-Monitoring-Only Approach (Network-Based Honeypots)

Solutions in this category monitor only the network traffic in a passive way. One of examples frequently referred to in this category is tcpdump tool capturing the network traffic. The captured data can be processed and visualized further. This approach is applied in network based IDS (NIDS), too. A significant issue of such solutions is the absence of any information about internal system events. For example, an attacker can use an encryption resulting in unavailability of any data for further analysis. Therefore this approach is insufficient despite the maximum transparency for attackers. They have no way how to recognize that their activities are monitored.

---

[3] http://www.honeynet.pk/honeywall/roo/index.htm.
[4] https://github.com/tnich/honssh/.
[5] https://github.com/Bifrozt/ALPHA.

Table 1. Summary of main aspects of honeypot approaches. Question mark means that the specific feature is dependent on the specific software and no generalization is possible.

Approach	Transparency/ detectability	Internal activities	Network activities	Potential for attack analysis	Linux monitoring support (guest)	Recency/ community/ support	Usefulness for present research
In-the-box	🙁	🙂	😐	😐	🙂	🙁	🙁
1st attack stadium	🙂	🙁	😐	😐	🙁	😐	🙁
Out-of-the-box-VM-based	🙂	🙂	😐	😐	🙁	🙁	🙁
Out-of-the-box	🙂	😐	🙂	🙂	🙂	🙂	🙂
Network-based	🙂	🙁	🙂	?	🙂	?	🙁

A simple summary of important aspects of the approaches described above is shown in Table 1. The symbols used there were inspired by similar indications in the study [8] by ENISA.

3   Selection of the Appropriate Approach

Not all approaches described above, however, have been feasible. The "in-the-box" approach has been impeached by numerous research papers mentioned in Sect. 2.1 and it could be considered as out-dated for now. The second approach is limited to the exploitation stadium of the attack and tools using this approach usually focus just to the RAM memory content analysis. Therefore, they can provide data for exploitation analysis, i.e. vulnerability abuse can be monitored this way but no subsequent attacker's activity could be monitored.

The out-of-the box VM-based approach represents an efficient way of the honeypot monitoring only if very close cooperation with the VM manager can be reached. This is, however, rather unusual in existing virtualization tools. Therefore, solutions in this category presenting themselves as sandboxes (e.g. Limon project [21]) or Cuckoo for Windows[6] install their agents on a virtualized system; such agents try to load covertly into the system kernel. There is a high risk that a human attacker accessing a system modified in this way can easily detect those modifications using various commands or tools. Therefore, they can be categorized (at a higher abstraction level) rather to the in-the-box category. Moreover,

---

[6] https://cuckoosandbox.org.

these solutions are not real honeypots, rather sandboxes. There is another disadvantage consisting in the fact that they can monitor system calls and internal system operations perfectly but their operation is based on the analysis of a specific binary file or script (some solutions are able to analyze even a given web page) so that they cannot process any activity that could be considered as human-attacker's interaction with the system in real time. Therefore, they are combined with out-of-the box approach in this case.

Solutions in the last category (network-based honeypots) offer just a network traffic dump for analysis; this is completely insufficient for the analysis of any further attacker's activity in the honeypot.

As the discussion of the approaches above has shown, the most feasible approach seems to be in the out-of-the box category. As already mentioned, this category often merges into a single group with the last category of network-based honeypots. Nevertheless, the conception presented in this paper is built on the solution that performs certain analysis even on the honeywall. The present research is devoted solely to the Linux system. Therefore SSH protocol is considered to be the ingress attack channel. Recording of any activity performed by SSH is made by HonSSH system with several modifications.

4 Conception

The aim of our analytical tool is to capture more sophisticated attacks (ideally those performed by human attackers). In order to analyze botnet and/or script-kiddies activity, low-interaction honeypots are sufficient [3]. For the purpose of elimination of automated or very simple attacks, an attacker is provoked to perform an analytical activity preceding the attack against the system itself. Therefore, in addition to the SSH shell input monitoring, also the web application seeming to be an ordinary production webpage is monitored. This webpage is intentionally vulnerable and monitored in detail using Shadow Daemon[7] and by the CMS itself, too. This web application completes the idea of a complex system presenting an added value for potential attacking against it. The layout of the system to be attacked is shown in Fig. 1.

What is extremely important is that the communication should seem to be fully transparent for an attacker. A local network is emulated inside the honeypot using Honeyd tool in order to improve the plausibility of the system. The potential attacker can then continue in attacking other targets and their behavior could be analyzed in more details. This part of the system is composed of low-interaction honeypots that serve as a LAN emulation behind the high-interaction honeypot. Because of the fact that these emulated LAN nodes are not intended as a primary targets of attack, the honeyd was chosen. The attacker has little opportunities to find they are honeypots because logging in them is not allowed.

---

[7] https://shadowd.zecure.org/.

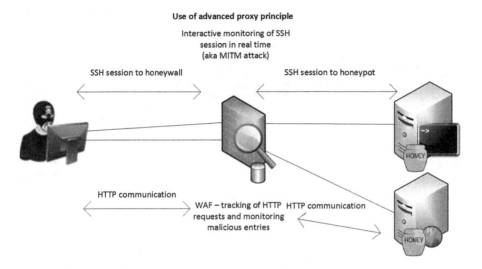

Fig. 1. Layout of the system to be attacked as it is presented to potential attackers

5   Technical Implementation

The whole conception of the solution presented here is shown in Fig. 2. The first issue to resolve was which IP address to choose for such a system. To assign an address from the CESNET academic network did not seem to be reasonable because a more sophisticated attacker could become suspicious about the system purpose [3] after an easy whois request. Nevertheless, the research system has been prepared in the data centre of the university so it was necessary to obtain a commercial virtual hosting server for camouflage purpose. This hosting server operates as a proxy server from where a transparent VPN is configured to the research system.

5.1   Implementation Details

All necessary systems were virtualized using Oracle VirtualBox. The VPN tunnel was configured using OpenVPN using L2 TAP interface. The TAP adaptor forms a single network connection for the first virtualized system – honeywall on the honeynet perimeter. Honeywall operates as a proxy server as well; it is perfectly transparent for the attacker. The transparency has been obtained using a suitable iptables configuration as well as thanks to the temporary NAT[8]. Behind the honeywall, there are two more virtualized systems that seem as a single one from the outside.

On the first system all SSH sessions from HonSSH are forwarded. This is a virtualized Linux Debian 8 system not containing any other modifications.

---

[8] Using Advanced Networking tool available at https://github.com/tnich/honssh/wiki/Advanced-Networking.

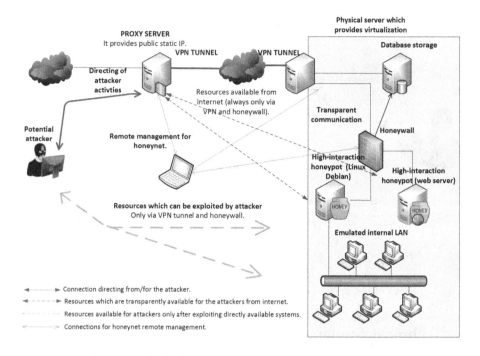

Fig. 2. Layout of the high-interaction honeynet designed in the research

However, it contains a lot of dummy data, MySQL database and other content. If an attacker starts scanning the local network after penetration into the system, they observe that there are several other workstations there. There are some Windows and Linux systems and a single Cisco router. All these devices are emulated using Honeyd and contain single scripts responding to incoming requests, mostly taken from the Honeyd project[9].

Requests to port 80 are forwarded to the other virtualized system from the honeywall. This system provides web server services, seemingly on the same IP address. The web server is monitored by a high-interaction honeypot/Web Application Firewall (WAF) and uses intentionally vulnerable older version of the CMS system Drupal 7[10]. Drupal 7 includes a robust method for assigning permissions to individual users. The strategy of the system is that any user can register into the system obtaining a limited access into CMS administration automatically as well. Vulnerabilities intentionally contained in the web application do not allow permission escalation so the attacker cannot, e.g., browse log files or change the system settings. However, the attacker gains access to the stored contents (its potential use is mentioned below). All attacker's activities like logging in, executing commands, running scripts etc. together with attacker's IP address are stored into the MySQL database from where the data can be visualized. The architecture layout is shown in Fig. 3.

---

[9] https://github.com/DataSoft/Honeyd.
[10] http://www.drupal.org.

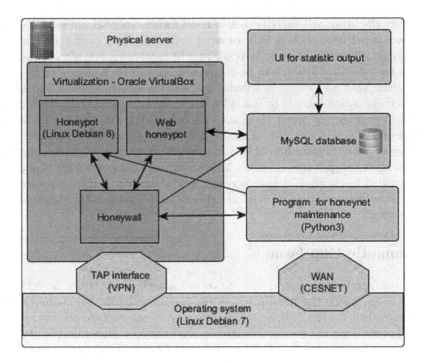

Fig. 3. Architecture layout of the honeypot system

5.2   Proposed Solution Sustainability, Protection and Analysis

After the connection to the honeypot (valid SSH session) was created, it is inevitable to allocate certain period of "free" activity to the attacker. During this period the attacker may freely investigate the system, continue its exploitation, execute scripts or malware, introduce a backdoor etc. Determination of this period has been a subject of some discussions but finally it has been decided to be set to 30 min for the pilot implementation period. It is expected, however, that this period length setting is going to be subject to change according to first results (first attack captured and analyzed) because it significantly influences the scope of attacker's activities that can be captured and analyzed.

After a successful login and opening an SSH session, the honeypot administrator is notified via e-mail. If required, the administrator can connect to the honeywall to observe the attacker's behavior in real time. For the purpose of the attacker's free period countdown, a Python script was written counting down the 30-minute period, and after the period elapses, the attacker will be disconnected, a snapshot of the system will be made and the system will be restored from scratch to be prepared to capture other potential attacks. If the attacker disconnects earlier, the Python script makes a snapshot and restores the system immediately.

During the attacker's activity, not only all inputs via SSH are recorded but any downloaded files, e.g. via SCP or wget commands, are captured. The files can be analyzed later using sandboxes. The VM snapshot made after the attack allows a later detailed analysis to be performed, or possibly to renew the system from the snapshot and allow the attacker to log in the system again. The decision between these options should be made by the honeypot administrator that is also entitled to prolong the preset 30-minute period to longer one when he/she follows the attack and comes to a conclusion that this specific case is so interesting that it deserves it. Suppression of the adverse effects of the attack (e.g. incorporation of the honeypot into a botnet and performing DDoS attacks) is done by the honeywall using iptables that is configured according to Bifrozt project. In addition, the automated capture of all network traffic has been implemented on the honeywall.

6   Semantic Gap Issue

The above-described system is able to record attacker's activities in detail as they are done inside the monitored system (honeypot) through connection to the remote shell via the SSH protocol. Similarly, it can capture all downloaded files and control any network traffic leaving the honeynet. However, the attacker can, for instance, run its own shell on other port (except the monitored port 22). For this purpose, e.g., the netcat tool can be used. When this method is used, the attacker passes by the HonSSH monitoring element (proxy server) but its communication is reliably captured by the network traffic monitoring on the honeywall. When the attacker runs a binary file or script that performs its activities apart from SSH protocol and the communication is encrypted, the situation is even worse from the point of view of monitoring. This is due to the fact that monitoring of system calls on the honeypots cannot be performed; in such a situation the only option is to perform an offline dynamic analysis. This can be done, however, only after the "free" period for the attacker has elapsed, using the two following methods. The first is running the downloaded binary file or script in Limon offline sandbox operating similarly to Cuckoo (see Footnote 6) sandbox (designed for Windows) by loading its agent into the monitored system. The other method is the direct application of the snapshot made after the attacker has logged out. Inside the system restored from the snapshot, the Sysdig[11] system is loaded that performs modifications on the kernel level in a similar way.

The latter method represents a concept that establishes the required trade-off between the two principles mentioned in Sect. 1.1. The tradeoff consists in the avoidance of immediate (and therefore permanently present and visible) system kernel modification (like in Limon or Cuckoo tools already mentioned) on one hand, and in later obtaining detailed information comparing common network-based approach as described is Sect. 2.5. As a result, the attacker will not have any opportunity to find any modifications of the system used as honeypot but at

---

[11] http://www.sysdig.org/.

the same time its behavior can be completely analyzed later. The analysis has to be split into two parts – the first is online monitoring of the attacker's activity though SSH access, and the other is an offline analysis of software resources used. The latter requires active work done by an analyst. However, this possibility of the honeypot administrator is still vital in order to avoid undesirable effects of sophisticated attacks.

7   Practical Implementation and Use Scenario

The software honeypot describer in Sect. 5 has been implemented in the authors' academic network and has been operation for a short period so far. However, the operation demonstrated at least the feasibility of the proposed solution so far. Because of the short period of operation and lacking promotion, no sophisticated human-made attacks have been recorded. The following paragraphs summarize the measures planned to be implemented in the near future in order to obtain some interesting results.

In the first step, automated attacks from botnets, robots and frequently script-kiddies were avoided. This has been reached simply by configuring a password that is strong enough to resist simple dictionary or brute-force attacks. In order to attract human attackers, several optional vulnerabilities are going to be prepared for the pilot phase of the research to promote the honeynet described above as well as to control the access to it. The weaknesses are listed as follows:

- Creation of a production company webpage with intentionally retained vulnerabilities allowing the attacker to access the contents by abusing the web application errors; the contents includes a webpage with login data for the honeypot.
- Creation of several logins for administration the web system with weak passwords (e.g. admin/toor, root/toor) where the attacker can log in and as a result they get the webpage with SSH logins.
- After self-registration of a new user permission allowing to access the webpage with SSH logins are automatically assigned.
- The webpage with SSH logins is publicly accessible (without the necessity to register as a user), e.g. for the purpose of indexing by search engines.
- Easy-to-guess login data (e.g. name and password root/Toor12345) were configured for SSH access to the shell so as to avoid automated attacks but it is easy to break for a human attacker.
- The access to the shell via SSH is intentionally controlled so that certain small percentage (e.g. 3 %) of connection attempts is allowed to login with any login name and password (percentage computed on 24-hour basis).
- And eventually, a small delusive promotion campaign will be run using certain commercial web presentations to attract human attackers.

The above strategies are subject of success evaluation in terms of attracting successful attacks at present due to quite a short period elapsed from the implementation of the above steps (some of them are just planned).

8 Conclusions

The paper proposes a conception of high-interaction honeypot intended to analyze human-made highly sophisticated attacks via SSH to a Linux shell. The research of existing approaches and their detailed comparison has been done, and a new approach to solve their issues is proposed. The new approach is a balanced trade-off between the detection mechanism observability by human attackers and the level of recorded details about attackers' activities. The new solution evaluation as well as suppression of possible actions threatening the rest of the infrastructure outside the honeypot is included.

Acknowledgment. The paper was supported by the project *Application of fuzzy methods for system analysis, description, prediction and control* No. SGS02/ AVAFM/16 of the Student Grant Competition of the University of Ostrava.

References

1. Kheirkhah, E., et al.: An experimental study of SSH attacks by using honeypot decoys. Indian J. Sci. Technol. **6**(12), 5567–5578 (2013)
2. Sokol, P., Andrejko, M.: Deploying honeypots and honeynets: issues of liability. In: Gaj, P., Kwiecień, A., Stera, P. (eds.) CN 2015. CCIS, vol. 522, pp. 92–101. Springer, Heidelberg (2015)
3. Sochor, T., Zuzcak, M.: Study of internet threats and attack methods using honeypots and honeynets. In: Kwiecień, A., Gaj, P., Stera, P. (eds.) CN 2014. CCIS, vol. 431, pp. 118–127. Springer, Heidelberg (2014)
4. Sochor, T., Zuzcak, M.: Attractiveness study of honeypots and honeynets in internet threat detection. In: Gaj, P., Kwiecień, A., Stera, P. (eds.) CN 2015. CCIS, vol. 522, pp. 69–81. Springer, Heidelberg (2015)
5. Pomorova, O., Savenko, O., Lysenko, S., Kryshchuk, A., Nicheporuk, A.: A technique for detection of bots which are using polymorphic code. In: Kwiecień, A., Gaj, P., Stera, P. (eds.) CN 2014. CCIS, vol. 431, pp. 265–276. Springer, Heidelberg (2014)
6. Pomorova, O., Savenko, O., Lysenko, S., Kryshchuk, A., Bobrovnikova, K.: A technique for the botnet detection based on DNS-traffic analysis. In: Gaj, P., Kwiecień, A., Stera, P. (eds.) CN 2015. CCIS, vol. 522, pp. 127–138. Springer, Heidelberg (2015)
7. The Honeynet Project: Know Your Enemy: Sebek - A kernel based data capture tool. Honeynet.org. (2003). http://old.honeynet.org/papers/sebek.pdf
8. Grudziecki, T. et al.: Proactive Detection of Security Incidents Honeypots. In: Polska, C., ENISA (eds.) ENISA (2012). https://www.enisa.europa.eu/publications/proactive-detection-of-security-incidents-II-honeypots
9. Dornseif, M., Holz, T., Klein, C.N.: NoSEBrEaK - Attacking Honeynets (2004). http://arxiv.org/abs/cs/0406052
10. Corey, J.: Local Honeypot Identification (2003). http://www.phrack.org/unofficial/p62/p62-0x07.txt
11. Quynh, N.A., Takefuji, Y.: A novel stealthy data capture tool for honeynet system. In: Proceedings of the 4th WSEAS International Conference on Information Security, Communications and Computers, Tenerife, pp. 207–212 (2005)

12. Portokalidis, G., Slowinska, A., Bos, H.: Argos: an emulator for fingerprinting zero-day attacks for advertised honeypots with automatic signature generation. In: ACM SIGOPS Operating Systems Review, vol. 40(4), pp. 15–27. ACM (2006). http://www.few.vu.nl/argos/papers/p15-portokalidis.pdf

13. Floeren, S.: Honeypot-architectures using VMI techniques. In: Proceeding zum Seminar Future Internet (FI), Innovative Internet Technologien und Mobilkommunikation und Autonomous Communication Networks, vol. 17, pp. 17–23 (2013)

14. Song, C., Ha, B., Zhuge, J.: Know your tools: Qebek-conceal the monitoring. In: Proceedings of 6th IEEE Information Assurance Workshop. The Honeynet Project (2015)

15. Jiang, X., Wang, X.: "Out-of-the-Box" monitoring of VM-based high-interaction honeypots. In: Kruegel, C., Lippmann, R., Clark, A. (eds.) RAID 2007. LNCS, vol. 4637, pp. 198–218. Springer, Heidelberg (2007)

16. Jiang, X., Wang, X., Xu, D.: Stealthy malware detection through vmm-based out-of-the-box semantic view reconstruction. In: Proceedings of the 14th ACM Conference on Computer and Communications Security, pp. 128–138. ACM (2007)

17. Lengyel, T.K., Neumann, J., Maresca, S., Payne, B.D., Kiayias, A.: Virtual machine introspection in a hybrid honeypot architecture. In: CSET (2012)

18. Lengyel, T.K., Maresca, S., Payne, B.D., Webster, G.D., Vogl, S., Kiayias, A.: Scalability, fidelity and stealth in the DRAKVUF dynamic malware analysis system. In: Proceedings of the 30th Annual Computer Security Applications Conference. ACM (2014)

19. Chen, P.M., Noble, B.D.: When virtual is better than real. In: Proceedings of the Eighth Workshop on Hot Topics in Operating Systems HOTOS 2001. IEEE Computer Society (2001)

20. Dolan-Gavitt, B., Leek, T., Zhivich, M., Giffin, J., Lee, W.: Virtuoso: n arrowing the semantic gap in virtual machine introspection. In: Security and Privacy, pp. 297–312. IEEE (2011)

21. Monnappa, K.A.: Automating Linux Malware Analysis Using Limon Sandbox. https://www.blackhat.com/docs/eu-15/materials/eu-15-KA-Automating-Linux-Malware-Analysis-Using-Limon-Sandbox-wp.pdf

About the Efficiency of Malware Monitoring via Server-Side Honeypots

Mirosław Skrzewski[✉]

Institute of Computer Science, Silesian University of Technology,
Akademicka 16, 44-100 Gliwice, Poland
miroslaw.skrzewski@polsl.pl

Abstract. Gathering information on malware activity is based on two sources of information: trap systems (Honeypots) and program agents in the AntiVirus tools. Both of them deliver only fragmentary picture of malware population, visible from trap systems or from users systems on corporate or home networks. Due to this fragmentation, there is no uniform overall picture of malware state, and various sources present different, often quite different approximations thereof, depending on the their ability of gathering samples of various types of threats and operating locally malware. Another question is how complete is this picture and whether the tools used do not lose some important informations. The paper compares current available informations about malware with data gathered by a set of honeypot systems and discusses usability of some types of malware traps at current state of malware expansion.

Keywords: Malware monitoring · Network activity · Honeypot performance

1 Introduction

Creating an image of malware activity is based on two main information channels: capturing malware through honeypot systems and capturing malware via software agents of anti-virus. Both channels deliver only fragmentary picture, visible from trap systems or from user computers on corporate networks. Resulting from these sources image of malware state is complemented with informations from analysis of hacking into servers, stealing confidential information from companies and government agencies.

Malware infection methods are changing very quickly, and there is no guarantee that these changes are just as evident in the various monitoring systems. Most systems monitoring the presence of malware is run by the corporations security departments or antivirus companies. Aggregating data from its own sources, they form a total picture of the activity of recorded versions of the malware. As a result, images of individual firms differ, like malware they are based on.

The question arises whether these differences correspond only to differences in the prevalence of various types of malware, or they are in some part a reflection

© Springer International Publishing Switzerland 2016
P. Gaj et al. (Eds.): CN 2016, CCIS 608, pp. 132–140, 2016.
DOI: 10.1007/978-3-319-39207-3_12

of ability to raise new copies of the malware by various types of honeypot systems and software agents.

The paper presents an analysis of the malware captured by server-side honeypots systems during nearly four years of their activity on a research network and compares results with the presented on-line quantitative and qualitative picture of current threats.

Remaining parts of the paper are organized as follows: Sect. 2 describes the environment of a research network and installed trap systems. Analysis of types and quantity of recorded threats is presented in Sect. 3, Sect. 4 discusses the global malware state and Sect. 5 discusses the differences between local and global threats picture and required changes in methods of obtaining malware samples.

2   Malware Research Network

Research was carried out on a separate class C subnet with direct VPN connection to Internet access router, bypassing any packet filtration. Access router works also as DHCP server, assigning IP addresses to systems on research network and as a DNS server. Most of the address space was used to monitor incoming traffic to the vacant IP addresses.

On one of the final IP addresses on system with quad core processor and 8 MB RAM was installed Xen 3.1 hypervisor and CentOS 5.9 operating system as dom0. The system was configured as hardware platform for machine virtualization, on which operated honeypot systems [1] and network traffic monitoring system with packet argus [2].

As the primary source for collecting malware was installed virtual Fedora system (domU) with dionaea honeypot [3], initially there were also tested other versions of honeypots [4] like honeyd [5] or multipot. Dionaea emulated on incoming requests operation of various network services and enabled the upload of malware code. Captured copies of malware dionaea stored as files with names corresponding to their MD5 hash values.

The captured malware samples were then copied to another VM system with installed antivirus software (NOD32 in this case), which attempt to identify them, determine their type and name or in the absence of a signature ignored them, treating it as a harmless. These unidentified malware samples MD5 hashes were also checked on Internet via google search engine and quite big part of the samples – about 20–25 % were unknown on Internet at all. All its operation and communication with the environment dionaea recorded in the local SQLite database, what facilitates further detailed analysis.

Dionaea honeypot started it's operation from March 2011 and operated till the end of 2013, then again from third quarter of 2014 till now (the end of 2015). Results of the first year of its operation was presented in [6]. At the middle of 2014 there were started another copies of dionaea honeypots, operating at selected places of network address space.

3 Results of Honeypot Operation

Results of honeypot operation vary with time, from the very start it capture a relatively large number of copies of a variety of malware, registering also a lot of network traffic related to honeypot communication. Individual monthly or quarterly periods differed, but the overall intensity of the honeypot related traffic and appearance of malware was quite high, and ranged in initial two years from 20000 to 90000 connections per quarter. Figure 1 presents quarterly aggregate amount of incoming connections to the honeypot system throughout the entire period of system activity.

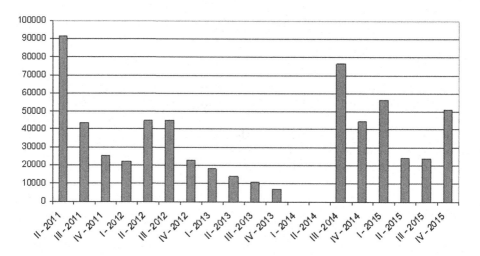

Fig. 1. Quarterly numbers of connections to the dionaea honeypot

The volume of incoming connections remain on nearly constant level, but its results vary significantly with time. Figure 2 shows the quarterly number of failed (refused) external connections to the honeypot system. Clearly the number of failed connections begins to increase from mid-2013.

Most of successful connections to the honeypot system led to attempts to infect the system and register malware samples. Initially (years 2011–2012) captured malware represented many different types of malware families, and appear in many versions [6]. With time, the number of recorded malware has visibly reduced. Table 1 shows the number of samples from different malware families registered in selected quarters of system honeypot activity.

Attempts to identify the copies of malware by antivirus program often ends assigning them names like "variant of the program", indicating that malware belong to a group having a similar behavior. Figure 3 presents graphs of the number of registered copies of malware divided on malware families and variants of various families.

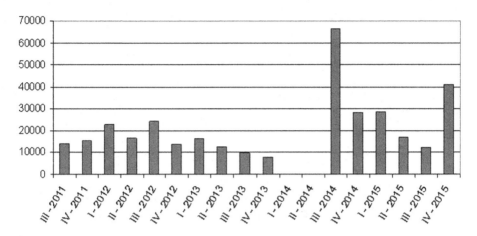

Fig. 2. Quarterly amounts of rejected connections to honeypot

Table 1. Quarterly number of registered samples of malware from different families

Malware family name	2-2011	3-2011	1-2012	3-2012	1-2013	3-2013	4-2014	2-2015
IRC/SdBot	49							
MSIL/Injector	26							
Win32/Agent	2	1	4	3				1
Win32/Allaple	51	38	15	10	7	16	28	21
Win32/AutoRun	265	88	6	17	5		2	1
Win32/AutoRun.IRCBot	323	660	59	52	19	18	40	11
Win32/Dorkbot	5							
Win32/Hatob	5		5					
Win32/Injector	161	148	3					
Win32/Kryptik	77	9	4				4	5
Win32/Pepex	26	6	4	4	3	4	9	11
Win32/Rbot	5	87	173	171	107	6		
Win32/Spy.Agent	15		3	5	1	4		
Win32/TrojanDownloader	14							
Win32/Conficker		1		3387				
Win32/Rozena		1						
Win32/Lethic		10						

To get a more detailed view of the operation of honeypot were compared malware activity from three quarters with a similar number of external connections to the honeypot i.e. the third quarter of 2011, third quarter of 2012 and the fourth quarter of 2014 years. The Table 2 presents a quantitative comparison of those quarters.

For selected quarters were analyzed the most frequently encountered malware programs and remote hosts attempting to infect the honeypot. The Table 3 shows

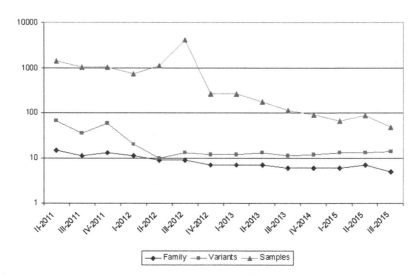

Fig. 3. Quarterly number of malware families, variants and samples

Table 2. Comparison of malware activity

Compared quarters:	III_2011	III_2012	IV_2014
Number of all external connections	43 614	44 980	44 343
Number of accepted connections	30 029	40 962	8 020
Number of named malware samples	919	3 649	83
Number of all samples	1 056	4 140	92
Number of unique malware samples	108	136	45
No. of remote_hosts seen two or more times	67	51	11

from when showed up particular versions of the malware, how long was seen and how many samples of the malware honeypot registered, and also which remote host and what method delivers the malware. Compiled statistics allow to evaluate what has changed in the network communication of honeypot with the environment and how is changing the nature of existing threats.

The most noticeable difference between the 2012 and 2014 quarters is a marked reduction in the number of connections from emerging malware. Top sender (most frequently occurring) malware in 2014 recorded up to 12 connections in three months, many others do not exceed two. The previous quarters were dominated by bot programs, generating dozens of connections to honeypot. In the last quarter there appear on the list rare version of such programs, but provide very little malware samples.

Most bots' infections came from the network 83.x.x.x that dominated address space of remote hosts. In the last quarter on the list of systems providing most malware appeared also other numbers of IP network. Analyzing the addresses

Table 3. Comparison of malware activity in selected quarters 2011–2014 years

First_day	Interval	Malware_name	Count	Remote_host	Download_url
2011-08-27	116	Win32/AutoRun.Delf.AG	84	83.230.192.106	http://83.230.192.106:30517/x
2011-08-04	21	Win32/AutoRun.IRCBot.FC	22	83.229.119.19	ftp://119.188.6.227:5809/tyf.jpg
2011-08-08	168	unknown	19	83.230.3.117	http://146.185.246.139/g.exe
2011-08-12	18	Win32/AutoRun.IRCBot.FC	17	83.230.244.28	http://46.45.164.228/s.exe
2011-09-09	349	Win32/Rbot	16	83.218.175.189	tftp://83.218.175.189/upds.exe
2011-08-20	10	Win32/Injector.JUV	11	83.157.16.146	emulate://
2011-09-29	25	Win32/Injector.KBU	11	83.177.156.36	emulate://
2011-08-09	167	Win32/AutoRun.IRCBot.FC	9	83.228.103.78	http://83.228.103.78:16254/x
2011-08-20	75	Win32/Injector.KQH	9	83.229.48.222	http://146.185.246.53/XZ44.exe
2011-08-08	12	Win32/Injector.ITJ	6	83.228.104.228	emulate://
2012-07-05	1	Win32/Conficker.AA	1461	83.230.117.101	http://83.230.117.101:5543/saekqry
2012-07-03	20	Win32/Rbot	37	83.218.170.222	tftp://83.218.170.222/udps.exe
2012-07-04	54	Win32/AutoRun.IRCBot.FC	32	83.230.180.226	http://188.132.163.17/nn.exe
2012-07-01	17	Win32/AutoRun.IRCBot.FC	29	83.222.58.166	http://83.222.58.166:21284/x
2012-08-02	36	Win32/AutoRun.IRCBot.FC	27	83.222.58.162	http://83.222.58.162:7521/x
2012-09-12	48	Win32/AutoRun.IRCBot.FC	24	83.222.58.104	http://83.222.58.104:30575/x
2012-08-27	104	Win32/AutoRun.KS	14	83.229.243.102	http://146.185.246.67/aa.exe
2012-08-31	75	Win32/Allaple	9	83.230.229.171	emulate://
2012-07-19	10	Win32/AutoRun.IRCBot.FC	6	83.222.59.118	http://83.222.59.118:33134/x
2012-08-06	6	Win32/AutoRun.IRCBot.GQ	5	83.222.173.109	emulate://
2014-11-12	118	unknown	12	84.111.156.198	smb://84.111.156.198
2014-10-05	344	Win32/AutoRun.IRCBot.DI	8	83.14.193.236	http://83.14.193.236:4590/x
2014-10-07	8	Win32/Pepex.F	5	186.74.238.213	smb://186.74.238.213
2014-09-04	0	unknown	2	197.242.253.131	smb://197.242.253.131
2014-09-13	21	Win32/Allaple	2	212.9.29.105	emulate://
2014-10-26	10	Win32/AutoRun.IRCBot.CX	2	83.228.103.198	http://83.228.103.198:28477/x
2014-10-28	0	unknown	2	88.2.241.167	smb://88.2.241.167
2014-11-07	134	Win32/Allaple.D	2	200.47.225.147	emulate://
2014-11-22	0	INF/Autorun.C.Gen	2	200.253.159.90	smb://200.253.159.90
2014-11-25	0	INF/Autorun.C.Gen	2	83.39.136.196	smb://83.39.136.196

Table 4. Remote hosts delivering Win32/Pepex variant malware

First_day	Count	MD5_hash	Remote_host	Locality
2015-06-22	2	fd46e75a29737c709b6914ca2a3a6727	200.35.110.250	Venezuela, Caracas
2014-10-07	1	dfbe916ddefffa3c2e62735ca43a983b	82.166.241.18	Israel, Hazafon
2014-12-09	1	9288a0069f0c43b5308662ecf7515462	24.167.5.14	Texas, Brownsville
2015-01-09	1	f6e4e93828c920d55f71152de02573b7	81.198.249.146	Latvia, Riga
2015-02-24	1	c90a4585af4f89cd52c7af544908b879	173.70.35.7	New Jersey, Tinton Falls
2015-03-12	1	760a8166b88bffe81ef2bec906f90d31	213.60.184.42	Spain, A Coruna
2015-04-14	1	4606b26f51650ff64d4e41e24047978f	69.214.198.185	Illinois, Chicago
2015-06-01	1	79834645d783bd5f23a227cabc73d183	146.60.77.187	Germany, Kleve

of hosts which send different versions of the malware, for example Win32/Pepex one can find addresses from all over the world (Table 4).

Starting from 2013 can be notice a marked decrease of the amount of captured samples of the malware. Total number of connection attempts to honeypot remained at a similar level or even increased, but the number of copies decreased significantly, also did not appear any new families of malware. Actually appear only additional copies classified as variants of the same well-known several programs, which over the past few years trying to infect systems. It appeared also many copies of malware that were not detected by the antivirus program. This decrease in the number of recorded malware is shown in Fig. 4.

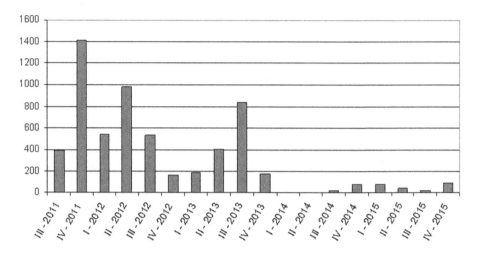

Fig. 4. Quarterly number of captured malware samples

The observed results stands in contradiction to the presented in various corporate studies informations of almost constant virtually exponential increase in the number of circulating malware copies.

4   Malware Threats in Online Reports

Many corporate security departments [7,8] and antivirus companies [9–11] periodically publish papers with information about the current state of network threats, presenting the currently active malware programs, amounts of malware samples registered in some period of time and similar information, citing a corporate systems dedicated to monitoring threats on the Internet.

Basic information repeated in these publications are data on quantities of registered new malware samples in a given period and their total accumulated quantities of hundreds of millions of copies [12,13]. Rarely, however there is the information, how many different types of malware programs correspond to this numbers, and what size was the address space the data are collected from.

The figures are often overwhelming, companies [14,15] are reporting growth in amount of malware samples of 10 million during the year, with the aggregated resources of the order of 200 million unique malware samples. Looking at these figures one should remember that the global image of malware activity is created as an aggregation of data from many sources, and its quantitative dimension can be directly due to the quantities of sources, but can also be the result of ambiguities in the classification and counting of malware samples.

In [16] the author analyzes quantities of signatures provided by antivirus company and comes to the conclusion that 17.5 million signature really corresponds to about 270 different types of malware, constantly slightly modified by programs packing malware code for transmission and by the processes of malware self-modification after infection. In effect, he explains the quantitative relationship between malware family, malware variants of the basic version and the recorded malware samples identified by different values of MD5 hash. This explains the number of samples reported in the reports, but the fact remains continued trend of increase in the quantity of emerging malware families circulating on the Internet.

5   Conclusions

Comparing the changes over time in the numbers of malware collected by the server-side honeypots (e.g. dionaea) and included in the reports informations about increase in number of malicious software, it becomes obvious that the classic versions of server-side honeypot ceased to serve its purpose of delivering new malware.

Analyzing the published reports it can be seen a clear change in the nature of distribution channels of the malware – the place of independently migrating malware programs (internet worms) occupied the versions of malware delivered to systems by various Internet channels – infection through web sites, using the errors in popular application (pdf, flash), using social engineering methods directing users to the prepared sites distributing malware. These new channels make use of the user activity, and are not visible to the system traps of server-side type (providing hosted services).

One can clearly see that this change in the trend occurred near 2013 and now the traps register only next auto-modifications of previous generations of well-known malware running on still infected systems on the network. It is evident that the server-side honeypot systems, such as dionaea reached the end of their usefulness as one of the main sources of information about new threats. This state is indirectly confirmed by the fact that within the last few years has not appeared in the literature any new ideas or projects of systems of this type, and the existing ones are not further developed.

To capture new versions of contemporary malware are necessary new type of trap systems well imitating behavior of careless user's, browsing eagerly various websites and receiving numerous email correspondence from social networking sites, which is something that can be called active client-side honeypot system.

References

1. http://www.honeynet.org/
2. http://qosient.com/argus/

3. http://dionaea.carnivore.it/
4. Skrzewski, M.: Monitoring malware activity on the LAN network. In: Kwiecień, A., Gaj, P., Stera, P. (eds.) CN 2010. CCIS, vol. 79, pp. 253–262. Springer, Heidelberg (2010)
5. http://www.honeyd.org/
6. Skrzewski, M.: Network malware activity – a view from honeypot systems. In: Kwiecień, A., Gaj, P., Stera, P. (eds.) CN 2012. CCIS, vol. 291, pp. 198–206. Springer, Heidelberg (2012)
7. IBM X-Force Threat Intelligence Quarterly, 4Q 2015. http://www-01.ibm.com/common/ssi/cgi-bin/ssialias?subtype=WH&infotype=SA&htmlfid=WGL03099USEN&attachment=WGL03099USEN.PDF
8. Verizon 2014 data breach investigation report. https://dti.delaware.gov/pdfs/rp_Verizon-DBIR-2014_en_xg.pdf
9. Symantec, Internet security threat report 2014. http://www.symantec.com/content/en/us/enterprise/other_resources/b-istr_main_report_v19_21291018.en-us.pdf
10. McAfee Labs, 2016 Threats Predictions. http://www.mcafee.com/us/resources/reports/rp-threats-predictions-2016.pdf
11. Emm, D.: The threat landscape. http://media.kaspersky.com/en/business-security/kaspersky-threat-landscape-it-online-security-guide.pdf
12. https://www.av-test.org/en/statistics/malware/
13. Symantec, Internet security threat report appendices, ISTR20. https://www4.symantec.com/mktginfo/whitepaper/ISTR/21347932_GA-internet-security-threat-report-volume-20-2015-social_v2.pdf
14. McAfee Labs, Threats report Fourth Quarter 2013. http://www.mcafee.com/mx/resources/reports/rp-quarterly-threat-q4-2013.pdf
15. http://landing.damballa.com/state-infections-report-q4-2014.html
16. Bott, F.: The malware numbers game: how many viruses are out there? http://www.zdnet.com/article/the-malware-numbers-game-how-many-viruses-are-out-there/

Algorithms for Transmission Failure Detection in a Communication System with Two Buses

Andrzej Kwiecień, Błażej Kwiecień, and Michał Maćkowski[✉]

Institute of Computer Science, Silesian University of Technology, Gliwice, Poland
{akwiecien,michal.mackowski}@polsl.pl, blazej.kwiecien@gmail.com

Abstract. Designing systems with parallel transmission for industrial purposes, requires first defining the types of a failure, and next detecting it. Another step is to execute an appropriate procedure that can ensure the continuity of transmission. The issue of a failure detection can be a problem in itself, but is can also be a component of data transmission via dual bus. To make the transmission system work correctly, apart from creating the scenario of exchanges, it is necessary to solve the problem of a failure occurrence so that to maintain the transmission continuity. The paper presents the methods of failure detection and algorithms used when such failure occurs.

Keywords: Distributed control system · Redundancy · Dual bus communication · Master-slave · Industrial networks · Distributed real-time systems · Medium access · Detection of transmission failure

1 Introduction – The Redundant Network Model Fundamental Problems

In present automation system the essential elements are the IT systems operating in the real time (RT) mode. Their task is to organize and process the data that can ensure the correct work of a device or a group of devices. The fundamental feature of RT systems is that they are able to respond correctly to events, to analyze them and to react (generating a response) in a defined amount of time. Currently, it is common to use RT systems together with other RT system, creating distributed real time system (DRT). In that case, it is necessary to apply such type of communication model, which would guarantee the right communication exchange and time determinism [1,2]. DRT systems are built based on commonly known models of communication exchange i.e., Master-Slave, token passing, producer-distributor-consumer [3–7] and their modifications or hybrids [8–11].

A very numerous group are also protocols whose aim is to obtain time determinism in Ethernet networks. These protocols include: Modbus/TCP, EtherCAT, Profinet, EthernetIP, Ethernet PowerLink, HSE Fieldbus Fundation, CC-Link-IE [6,7,12–15].

© Springer International Publishing Switzerland 2016
P. Gaj et al. (Eds.): CN 2016, CCIS 608, pp. 141–153, 2016.
DOI: 10.1007/978-3-319-39207-3_13

Taking into consideration the reliability of RT systems, in particular the ones based on Master-Slave model of exchange, a very specific solution may be applying, in the broad sense, the bus and interface redundancy [16–21]. Redundancy can be particularly use for improving the transmission time parameters [22]. It can be achieved by dividing the tasks between two buses, conversely to the classical redundancy [1,23] where the same information is transmitted via the main (primary) bus and backup (secondary) bus. Here, it is suggested to divide the scenario of transmission so that on each bus would be executed various transactions, and the condition influencing the division would be the duration of a communication exchange. It is worth to mention here, that the correct form of the scenario of exchanges in Master-Slave networks is essential in terms of minimizing the number of transactions i.e., shortening the duration of data exchange cycle. Such form of the system enables to shorten the exchange cycle providing that all devices perform correctly. In case of a failure, it is necessary to do some steps to switch the transactions to the well-functioning part of the system [1,24,25]. While designing the real-time industrial systems, it is important to take into consideration also the fact that all transactions are supposed to be executed in a defined time T_{RES} (to ensure the time determinism). The worst case e.g., the bus failure, need to be considered. Thus, the failure and its influence on the entire transmission time must be taken into consideration. Hence, in case of a failure the parameters will worsen, but the rule of determinism will not be broken. To put it in other words, the lack of a failure means better parameter of the system, and the occurrence of a failure means worse parameters, though the rule of time determinism is not broken.

There is no doubt that applying two buses for communication exchange is beneficial for the shortening of the duration exchange. It is beyond the question that two parallel data streams are capable to transmit more information than one. However, in a system with such configuration there is a problem with detection of a failure in transmission line. In most communication systems a failure is recognized as the lack of response. A particular parameter for the situation above is *Timeout*, which in case of industrial networks is a configurable parameter. Therefore, in most communication systems the failure is correspond to the situation where in a defined time (*Timeout*) a polled station will not send the response message (network's gape). As the research shows [26–28] *Timeout* introduces a serious delay for switching the communication between the buses of a redundant system. The remaining values of delays are negligble small in comparison to the defined *Timeout*.

There are many techniques for shortening the system reaction. One of them can be constant monitoring the duration of communication exchange. Knowing the analytic dependencies, times of transmission exchange and the maximum response time for each transaction, it is possible to start sending the control frame in advance (within *Timeout*) to find the reasons of a failure. Such approach can shorten the time for a system reaction when a failure of a bus or a node occurs. The failure detection is the foundation of efficiency in the presented method in parallel transmission via redundant buses [1].

When using the redundant system for parallel data transmission via two independent buses, it is important to know that it can be possible when two lines work correctly. When one of the communication lines fails, then the parallel transmission need to be stopped and instead of two independent data streams, only one stream should be transmitted. To achieve this goal, appropriate algorithms ought to be developed. They should be used for constant diagnosis of communication, failure detection and its classification, and what is more they can start the process of switching the data transmission to one bus. Thus, the two problems have to be faced:

1. How to organize the data transmission via two buses having only the scenario of all cyclic exchanges in the system?
2. How to detect the bus failure, and then instead of two channels link organize one channel link?

It should be added that these two issues must respect the rules of the real time system. Therefore, applying the two-channels transmission in current systems can results in saving time, provided that there are methods for failure detection and algorithms controlling the data flow.

The issues referring to the first problem presented above were described in details in [1], whereas the current paper concerns the second problem. The authors presented here three situations that can result in incorrect system work:

- subscriber failure,
- failure of subscriber interface,
- damage of communication medium.

The next sections present the methods of a failure detection and algorithms used when a failure occurs in communication system.

2 Failure Types and Detection Methods

The authors in the paper applied Master-Slave network as an exemplary model. In case of a correct transmission, the data is transmitted and exchanged among subscribers via two communication buses (Fig. 1).

As it is known, using the redundancy allows to increase the bandwidth and usable efficiency of the network. However, it is important to check the system behavior in case of a failure, since the system is then interrupted and requires a service (switching to the one-channel communication, detection of previously transmitted data). The amount of time needed for the error service can be estimated based on the rule of "the worst case". Calculating or even estimating the value of delay is important, because a big delay can disorder the system work. Nonetheless, the result depends chiefly on the type of a failure. What is more, it can be also noticed the fall of efficiency and bandwidth rate, but this is temporary and without any significant meaning to the system work. In the present paper the authors included three operation states:

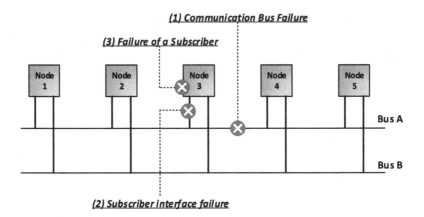

Fig. 1. The redundancy of communication medium: (1) Bus failure, (2) Subscriber interface failure, (3) Subscriber failure

- correct system work – both buses are efficient,
- communication medium failure (Fig. 1 Point (1)),
- failure of subscriber interface (Fig. 1 Point (2)),
- subscriber failure (Fig. 1 Point (3)).

When a failure is detected, the following procedures are advised to be executed, no matter what type of a failure it is, which:

- allow detecting a failure and determining its type in the shortest possible time,
- allow stabilizing the communication system,
- can start the appropriate transmission procedures recognizing the failure type,
- enable the further work, excluding the damaged part of the system.

The most important elements that determine choose of an algorithm are the duration of failure detection and define the failure type. In consequence, the right choice of an algorithm allows the further correct system work. As stated before, a failure can be a defect of I/O module, disruption of a communication bus, damage of a subscriber unit (e.g., PLC – Programmable Logic Controller) or even wrong execution of an algorithm implemented in the subscriber (an error in program).

2.1   Failure of Communication Medium (Bus) or Subscriber Interface

When a communication system works correctly, the data is exchanged among subscribers via two communication buses. However, to ensure the security and the data cohesion it is necessary to prepare some procedures that can protect the communication system in case of a failure of one of the buses. For example, if the subscriber 3 (Node 3 in Fig. 2) is not responding to the request, some steps need to be taken. The simplest way to detect the reason of such failure is to send a control frame via the same communication bus to another subscriber

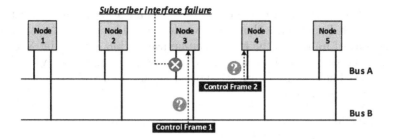

Fig. 2. Transmission of the control frames in case of subscriber interface failure

(Node 4 in Fig. 2). If the subscriber responses, it means that a failure occurs in Node 3 or its interface. The problem appears when there is no response from other subscribers. This can mean that:

- communication medium is disrupted,
- a failure occurs in other subscribers.

There is of course a possibility to poll the subscribers one after another to find out if the communication medium is really damaged. But the question is, how much time does it take and whether the time requirements for communication system will be met.

According to the authors the best solution is such approach, which may enable the immediate detection of a failure type. If the subscriber (Node 3), for some reasons, does not exchange the information via one of the buses (Bus A), then one control frame should be sent to Node 3 using the other bus (Bus B), and the second control frame should be sent (via Bus A) to another subscriber (Node 4 in Fig. 2). Concluding that the simultaneous failure in both buses is unlikely to happen, it is possible to determine quickly the source of a failure (Subscriber 3, interface of Node 3 or a communication bus).

When both subscribers (Node 3 and Node 4) response to the control frame, it means the Node 3 interface failure. In contrast, if the response comes from the subscriber (Node 3) polled by the second bus (Bus B), it means the failure of the entire communication link (Bus A).

2.2 Subscriber Failure

A very similar situation is when a failure occurs at one of the subscribers (Fig. 1 Point (3)). When a system works correctly, and suddenly one of the subscribers (Node 3) stops exchange the information in the communication system, then two hypothesis need to be made:

- communication medium has been damaged,
- the subscriber interface has been damaged or the failure occurs at the subscriber.

Detection of such failure, as in case of communication medium disruption, has to be carried out in the shortest possible time. The simplest method is to send the control frame via the same communication bus but to the Node 4. If the subscriber responses, it means the failure of Node 3. Similarly as in case of communication medium failure, if another subscriber (Node 4) also does not response, this can mean that:

– communication medium is disrupted,
– a failure occurs in other subscribers.

An alternative method used for detection of such type of a failure, as in case of communication medium failure, is to send a control frame to the same subscriber (Node 3) but via the other bus (Bus B) (Fig. 3). If the subscriber will not confirm it received the control frame, then it means that a failure occurs at a subscriber in a node, and the communication medium is efficient (providing that the failure in both buses is unlikely to happen).

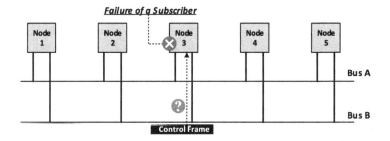

Fig. 3. Sending the control frame

Recognizing the type of a failure should not take a lot of time, however a very essential is how the communication system should react in case of a failure. While the case with damaged medium does not give a lot of opportunities because of the physical reason, it is not possible to parallelize the process of information transmission, the case concerns the failure at a subscriber seems to be very attractive. Here, the communication system will not be functioning (no subscriber). Whereas, if the subscriber failure for some reason is not critical for the entire system, the work can be continued using redundant communication medium. This can ensure the parallel exchange of information in the system.

3   Algorithms for Failure Detectionm

To ensure the basic functionalities, such as security and data cohesion of redundant systems, it is important to prepare appropriate procedures, which can protect the system in case of a failure of one of the buses, subscriber failure or interface failure.

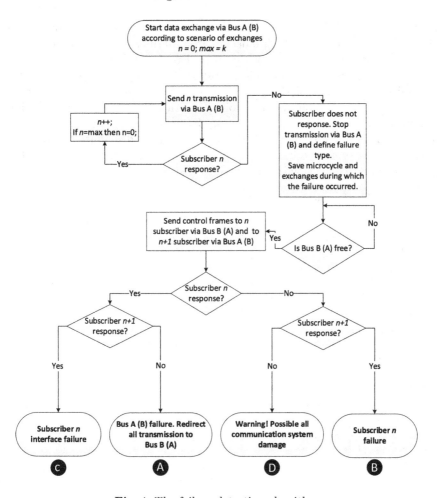

Fig. 4. The failure detection algorithm

The developed algorithm is used for detecting the failure in a communication medium, the entire subscriber failure, the failure of subscriber interface or a failure in the whole communication system. The algorithm presented in Fig. 4 is used to control permanently the correct work of the network. The system collects the information each time any transaction included in the scenario of exchange, is executed. Such transactions are the only source of information for algorithm and enable to detect a failure in a communication medium, at a subscriber or his interface.

The first symptom of incorrect system work is the appearing of so-called *Timeout*. It is worth to mention that it is one of the most important parameters determining the reaction time of each communication system, when a failure occurs. The problem of *Timeout* is not the main issue of the present paper, however it is the basic indicator that informs about the failure of a bus, a subscriber or interface. If is estimated incorrectly, there are delays in a system [29]. In the presented algorithm, *Timeout* is crucial and it occurs once in case of interface failure, and twice when the failure occurs at subscriber or in a bus [26–28].

The task of the presented algorithm is to detect and define the type of a failure: a failure of Bus A or B (Point A in Fig. 4), a failure of the subscriber (Point B), a failure of the one of the subscriber interfaces (Point C), the entire failure of the communication system (Point D). An effective detection of failures in operating system allows continuous updating the status, which permits to take another steps connected with the transactions in distributed system. The fragment of the source code, written in C language that shows the procedures made after failure detection, is presented in Listing 1.1.

Listing 1.1. The fragment of source code presenting procedures made after failure detection type.

```
if(ERROR_BUS_1) { //or ERROR_BUS_2
    count_error_Bus_1++;
    send_Bus_2(i);
    Send_Exchange_Bus_1 = false;

    if(count_error_Bus_1 == limit) { //the limit of the counter
        count_error_Bus_1 = 0;
        STEP_Bus_1 = 0;
        STEP_Bus_2 = 0;
        ERROR_BUS_1 = 0;
        ERROR_BUS_2 = 0;
    }}
    else {
    if(error_Slave) {
        count_error_Slave++;
        send_Bus_1(i+1);

        if(count_error_Slave == limit) { //the limit of the counter
            count_error_Slave = 0;
            error_Slave = false;
            STEP_Bus_1 = 0;
            STEP_Bus_2 = 0;
        }}
        else {
        send_Bus_1(i);
    }}
```

Figures 5, 6, 7 and 8 present flow charts of algorithms executed after the failure detection in one of the buses, at subscriber or his interface and the total breakdown of the system. The algorithms use the matrices of exchanges for Bus A and Bus B. Let's have an exemplary matrix L in which the number of rows is L_M (number of microcycles), and the first column of matrix defines the number of exchanges in each microcycles. Each row of matrix describes one microcycle. Only one row of the matrix will be completed totally with the numbers of transactions. This is because the whole transactions of exchanges are not executed in every microcycle. In microcycles in which only part of the transactions is executed the sequence T_{TN_i} is completed with the value 0 till the end

$$
L = \begin{bmatrix}
L_{T_1} & T_{TN_i} & \cdots & T_{TN_n} \\
L_{T_2} & T_{TN_j} & \cdots & T_{TN_m} \\
\vdots & \vdots & \vdots & \vdots \\
L_{T_{LM}} & T_{TN_{LM}} & \cdots & T_{TN_z}
\end{bmatrix}
$$

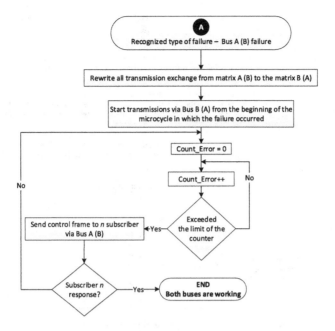

Fig. 5. Algorithm in case of a bus failure

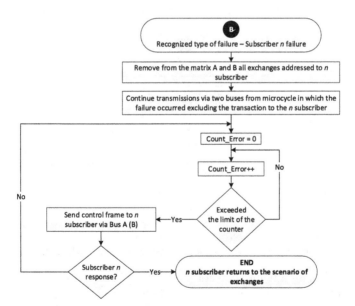

Fig. 6. Algorithm in case of subscriber failure

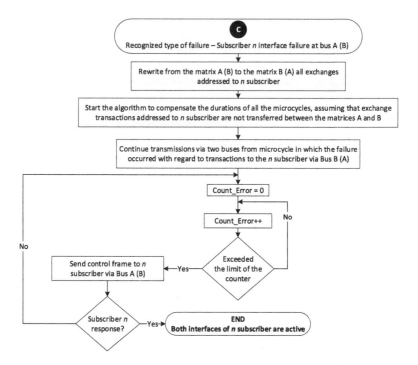

Fig. 7. Algorithm in case of subscriber interface failure

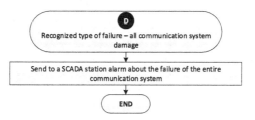

Fig. 8. Algorithm in case of the entire communication system failure

where:

L_{T_i} – the number of transactions in the i microcycle,
seque.$\{T_{TN_i}\}$ – the sequence of numbers of transactions within the i microcycle.

The algorithm for a failure detection precedes whether after a certain number of cycles (Failure Counter) the causes of a failure are removed. To achieve this goal, a control frame is sent. If it turns out that a failure disappeared, then the transmission scenario is changed into the original one and it executes the transmission in a totally efficient communication system.

4 Conclusion

The presented idea of applying redundant bus for increasing the frequency of exchanges in distributed real time systems must be strongly related to the methods of failure detection. The results clearly indicate that the presented solution is effective. The developed algorithm allows to detect a failure of interface, link or the whole subscriber in a redundant network that is based on Master-Slave model of exchanges. Failure detection of interface or bus makes the system to change the operation mode by reconfiguring the scenario of exchanges and switching the communication to other bus.

The research results of applying the above algorithms has been presented in another authors' paper [30]. The paper discussed in more details the failure influence on parameters of real-time system with two buses. Conducted tests enable to present the thesis that discussed algorithm allows not only to detect properly a failure, but also to realize the exchanges scenario correctly (which cannot be realized via the corrupted bus). Despite the system is not a standard redundant system, it enables the proper operation in case of failure of communication bus, even if in some failures the periodicity of micro cycle realization may be exceeded.

Acknowledgments. This work was sponsored from BK-263/2015.

References

1. Kwiecień, A., Kwiecień, B., Maćkowski, M.: Automatic scenario selection of cyclic exchanges in transmission via two buses. In: Gaj, P., Kwiecień, A., Stera, P. (eds.) CN 2015. CCIS, vol. 522, pp. 150–161. Springer, Heidelberg (2015)
2. Gaj, P., Jasperneite, J., Felser, M.: Computer communication within industrial distributed environment - a survey. IEEE Trans. Industr. Inf. **9**(1), 182–189 (2013)
3. Miorandi, D., Vitturi, S.: Analysis of master-slave protocols for real-time-industrial communications over IEEE 802.11 WLAN. In: 2nd IEEE International Conference on Industrial Informatics 2004, pp. 143–148 (2004)
4. Conti, M., Donatiello, L., Furini, M.: Design and analysis of RT-ring: a protocol for supporting real-time communications. IEEE Trans. Industr. Electron. **49**(6), 1214–1226 (2002)
5. Raja, P., Ruiz, L., Decotignie, J.D.: On the necessary real-time conditions for the producer-distributor-consumer model. In: IEEE International Workshop on Factory Communication Systems 1995, WFCS 1995, pp. 125–133 (1995)
6. Modbus-IDA. Modbus Application Protocol Specification V1.1b3, December 2006. http://modbus.org/docs/Modbus_Application_Protocol_V1_1b3.pdf
7. Modbus-IDA. Modbus Messaging on TCP/IP Implementation Guide V1.0b, October 2006. http://modbus.org/docs/Modbus_Messaging_Implementation_Guide_V1_0b.pdf
8. PACSystems Hot Standby CPU Redundancy. GE Fanuc Intelligent Platforms, doc. no: GFK-2308C (2009)
9. PACSystems - Hot Standby CPU Redundancy. GE Fanuc Automation, Programmable Control Products, doc. no: GFK-2308a (2006)

10. PROFIBUS Nutzerorganisation e.V. (PNO), PROFIBUS System Description - Technology and Application, Order number 4.332 (2010)
11. Genius I/O System and Communications. GE Fanuc Automation, doc. no: GEK-90486f1 (1994)
12. Decotignie, J.: Ethernet-based real-time and industrial communications. Proc. IEEE **93**(6), 1102–1117 (2005)
13. SIEMENS: SIMATIC PROFINET description of the system. Siemens, doc. no: A5E00298288-04 (2009)
14. Automation, P.: EtherNet/IP Specification: ACR Series Products (2005)
15. Ethernet POWERLINK Standardisation Group: EPSG Draft Standard 301, Ethernet POWERLINK, Communication Profile Specification, Version 1.20, EPSG (2013)
16. Kirrmann, H., Hansson, M., Muri, P.: IEC 62439 PRP: bumpless recovery for highly available, hard real-time industrial networks. In: IEEE Conference on Emerging Technologies and Factory Automation, ETFA 2007, pp. 1396–1399, September 2007
17. Kirrmann, H., Weber, K., Kleineberg, O., Weibel, H.: Seamless and low-cost redundancy for substation automation systems (high availability seamless redundancy, HSR). In: Power and Energy Society General Meeting, 2011 IEEE, pp. 1–7 (2011)
18. Neves, F.G.R., Saotome, O.: Comparison between redundancy techniques for real time applications. In: Fifth International Conference on Information Technology: New Generations, ITNG 2008, pp. 1299–1300 (2008)
19. Wisniewski, L., Hameed, M., Schriegel, S., Jasperneite, J.: A survey of ethernet redundancy methods forreal-time ethernet networks and its possible improvements. In: Proceedings of 8th International Conference Fieldbuses Networks Industrial; Embedded Systems (FET 2009), vol. 8, pp. 163–170 (2009)
20. IEC 62439, Committee Draft for Vote (CDV): Parallel Redundancy Protocol. In: Industrial Communication Networks: High Availability Automation Networks, chap. 6 (2007)
21. IEC 62439, Committee Draft for Vote (CDV): Media Redundancy Protocol based on a ring topology. In: Industrial communication networks: high availability automation networks, chap. 5 (2007)
22. Kwiecień, A., Sidzina, M.: Dual bus as a method for data interchange transaction acceleration in distributed real time systems. In: Kwiecień, A., Gaj, P., Stera, P. (eds.) CN 2009. CCIS, vol. 39, pp. 252–263. Springer, Heidelberg (2009)
23. Kwiecień, A., Stój, J.: The cost of redundancy in distributed real-time systems in steady state. In: Kwiecień, A., Gaj, P., Stera, P. (eds.) CN 2010. CCIS, vol. 79, pp. 106–120. Springer, Heidelberg (2010)
24. Kwiecień, A., Maćkowski, M., Stój, J., Sidzina, M.: Influence of electromagnetic disturbances on multi-network interface node. In: Kwiecień, A., Gaj, P., Stera, P. (eds.) CN 2014. CCIS, vol. 431, pp. 298–307. Springer, Heidelberg (2014)
25. Kwiecień, A., Sidzina, M., Maćkowski, M.: The concept of using multi-protocol nodes in real-time distributed systems for increasing communication reliability. In: Kwiecień, A., Gaj, P., Stera, P. (eds.) CN 2013. CCIS, vol. 370, pp. 177–188. Springer, Heidelberg (2013)
26. Kwiecień, B., Sidzina, M.: Research of failure detection algorithms of transmission line and equipment in a communication system with a dual bus. In: Kwiecień, A., Gaj, P., Stera, P., et al. (eds.) CN 2013. Communications in Computer and Information Science, pp. 166–176. Springer, Heidelberg (2013)

27. Sidzina, M., Kwiecień, B.: The algorithms of transmission failure detection in master-slave networks. In: Kwiecień, A., Gaj, P., Stera, P. (eds.) CN 2012. CCIS, vol. 291, pp. 289–298. Springer, Heidelberg (2012)

28. Kwiecień, B., Sidzina, M., Hrynkiewicz, E.: Industrial implementation of failure detection algorithm in communication system. In: Kwiecień, A., Gaj, P., Stera, P. (eds.) CN 2014. CCIS, vol. 431, pp. 287–297. Springer, Heidelberg (2014)

29. System, A., S7–400 CPU Specifications. Siemens Reference Manual, doc. no: 6ES7498-8AA04-8BA0 (2006)

30. Kwiecień, A., Kwiecień, B., Maćkowski, M.: A failure Influence on parameters of real-time system with two buses. In: Gaj, P., Kwiecień, A., Stera, P. (eds.) 23th Conference on Computer Networks, CN 2016, Brunów, Poland. CCIS, vol. 608. Springer International Publishing Switzerland (2016)

A Failure Influence on Parameters of Real-Time System with Two Buses

Andrzej Kwiecień, Błażej Kwiecień, and Michał Maćkowski[✉]

Institute of Computer Science, Silesian University of Technology, Gliwice, Poland
{akwiecien,michal.mackowski}@polsl.pl, blazej.kwiecien@gmail.com

Abstract. The paper refers to the time parameters of transmission in industrial systems that use two buses. Applying the systems with two buses makes sense only if it is possible to control the transmission all the time, and make modifications and reconstruction of transmission scenario in case of a failure. The paper presents the results of empiric research into testing software algorithm for failure detection of transmission line and network node in industrial communication system. After implemented this algorithm in PLC the results referring to measurements of duration of basic transaction in a system and duration of failure detection on communication buses were presented. The authors tried to clarify whether a failure detection in two buses transmission can have an influence on the delays of transmission. The paper consists of description of a test bench for time parameters measurements, the test results and conclusion.

Keywords: Distributed control system · Redundancy · Dual bus communication · Master-slave · Industrial networks · Distributed real-time systems · Medium access · Detection of transmission failure

1 Introduction

The main goal of redundancy is to ensure a correctly operating system. In case of a system failure, appropriate prevention must be implemented to protect the system against the loss of important information (e.g. losing the object state in a real-time system) [1–3]. The principle aim of using a redundant system is to increase the reliability of a communication system. Therefore, the idea of transmitting data via the unused bus (different than via the primary bus) obtained by dividing tasks between the redundant buses, seems to be an interesting issue [4–7].

Taking into consideration the reliability of the system (which can be increased thanks to the redundancy) or network overload (permanent or temporary) there is a problem how to increase the flexibility of network connections in order to improve the time of data exchange. The mentioned redundancy can also be used as a way of improving the transmission time parameters, and it also may increase the bandwidth [8–10].

The current paper is related to the previous authors' article: "Algorithms for transmission failure detection in a communication system with two buses" [11].

© Springer International Publishing Switzerland 2016
P. Gaj et al. (Eds.): CN 2016, CCIS 608, pp. 154–167, 2016.
DOI: 10.1007/978-3-319-39207-3_14

It should be reminded that a fundamental feature of the developed method described in [11,12], is not only the ability to transmit data via two buses, but also the possibility to continue the transmission even if a failure occurs within a defined amount of time [12].

The theoretical analysis and the research presented in the further part of the paper were based on Master-Slave model, and for the measurements GeFanuc controllers and Modbus RTU network were used [13–15]. A type of network is not significantly important, because the authors intended mainly to confirm the practical correctness of algorithm and to compare it to the measurements of theoretical quantities, namely the delays and their influence on the requirements of Real Time system.

The further part of the paper comprises equations for timing calculations (delays), which are used for measuring within a real system. In order to determine the time for realization of a failure detection algorithm in a real time environment and its influence on the time parameters of a real time system, a test bench was designed and prepared and appropriate tests were conducted.

2   A Failure Influence on the System Time Parameters

Since the next part of the paper based on Master/Slave Model so it is important to define some basic parameters. Figure 1 presents the scenario of exchanges in transmission via a single bus. Particular symbols in Fig. 1 stands for:

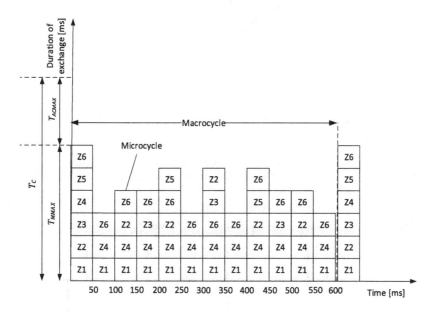

Fig. 1. Model of cyclic transmission (scenario of exchanges)

$Z1 - Z6$ – symbolic sign of a single exchange (e.g. read/write from/to any
subscriber). The heights of particular squares that represent the exchanges
correspond to their duration; here: all exchanges have the same durationis.

T_C – cycle duration of an exchange in the network.

T_{AC} – the remaining time for executing acyclic exchanges.

T_{MMAX} – maximum time of microcycle execution within a macrocycle.

T_{ACMAX} – maximum acceptable time for acyclic exchange.

In the abscissa in Fig. 1, the periodicity of exchanges is marked. The Fig. 1
shows that the minimal periodicity is 50 ms. Hence, the following microcycles
are executed in every 50 ms.

$$T_{MICR} = \sum_{i=1}^{l} T_{Ei} \tag{1}$$

where:

T_{Ei} – the time of the i request-response transmission,

l – the number of exchanges in the i microcycle.

A failure of a communication system influences in different ways the parameters of the real time system. These can be noticed as continuous delays in
realization of the scenario of exchanges (a total breakdown of one of the buses).
The authors in the paper applied Master-Slave network as an exemplary model.
In case of a correct transmission, the data is transmitted and exchanged among
subscribers via two communication buses (Fig. 2).

In such case, the complete time determined within the scenario of exchanges,
equals:

$$T_{CA1} = T_{CW1} + T_{CW2} \tag{2}$$

where:

T_{CW1}, T_{CW2} – time of exchange in particular buses in the efficient system,

T_{CA1} – complete time of network cycle when one bus is completely disrupted.

The realization of the scenario of exchanges can be shortened as a result of
removing some exchanges in it. This happens in case of a subscriber failure. The

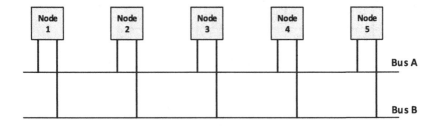

Fig. 2. The redundancy of communication medium

consequence of it is of course the lack of communication with subscriber. Then, in a determined time:

$$T_{CA2} = T_{CW} - \sum_{i=1}^{ns} T_{AWi} \qquad (3)$$

where:

T_{CA2} – the total time of the network cycle in case of the subscriber failure,
T_{CW} – the the total time of exchange in the efficient system,
T_{AWi} – transmission time of i transaction removed from the scenario of exchange,
ns – the number of transactions removed from the scenario because of subscriber failure.

Moreover, the damage of one of the subscriber interfaces influences the load of one of the buses, since the transmission to this subscriber is switched to the second bus.

Finally, the entire communication system can collapse, and then the only thing that can be done is to generate an appropriate alarm e.g., activating a special output.

When a failure is detected, no matter what failure type it is, some temporary disturbances and delays take place; after the transmission is stabilized they disappear. This can be referred to as transient state. On the other hand, by stabilized transmission it should be understood the end of a failure detection process and reconfiguration of the scenario of exchanges.

2.1 Delays Resulting from the Duration of Subscriber Failure Detection

In case of a failure of one of the subscribers, a delay on the bus connected to the disrupted subscriber resulting from a failure detection and change of a scenario of exchanges is:

$$T_{OP1} = T_{PR2} + T_{TR2} + T_{DE2} + 2 * T_{OD} + T_{ALG1} \qquad (4)$$

where:

T_{OP1} – delay resulting from the detection of a disrupted subscriber,
T_{OD} – maximum response time for request frame (timeout),
T_{PR2} – time for preparing the control frames that are sent to subscribers,
T_{TR2} – transmission time of control frames sent to subscribers,
T_{DE2} – time for control frames detection,
T_{ALG1} – time for executing the algorithm used for detecting a failure at subscriber and the scenario change.

From the point of view of the real time system requirements, this case is the safest, because after the failure is detected and the network is stabilized, the total time of executing the macrocycle is shortened (the number of exchanges will be reduced). Certainly, it is possible to determine the duration of a transient state for the worst case [11]. Taking into account the time for realization of the longest microcycle, its maximum realization time is:

$$T_{\text{MIKRmax}} = \max\left(T'_{\text{MIKR}i}, T''_{\text{MIKR}i}\right) + T_{\text{OP1}} \tag{5}$$

where:

T_{MIKRmax} – the maximum time of microcycle realization, in case of a failure, $\max\left(T'_{\text{MIKR}i}, T''_{\text{MIKR}i}\right)$ – the maximum time of microcycle realization in particular buses when a failure occurs.

To weight up the influence of the occurred failure the following formula can be used:

$$T_{\text{C}} > T_{\text{MMAX}} + T_{\text{WWMAX}} \tag{6}$$

where:

T_{C} – time of network cycle (time for realizing the whole exchanges within the one macrocycle),
T_{MMAX} – maximum time for realizing a microcycle within the macrocycle,
T_{WWMAX} – maximum acceptable time of triggered exchange.

Bearing in mind that if a failure occurs, then T_{MMAX} takes value T_{MIKRmax}, and by keeping the value T_{C}, the time for realization of triggered exchanges will be shortened. However, if it is unacceptable from the technological point of view, then the only possible solution is increasing the value T_{C}. It can be done at the designing phase of a communication system. The realization of some triggered exchanges can be abandoned by making some modifications in the main program (in this approach the value T_{C} is not changed).

The precise measurement of the delay time will be possible after having the authors' algorithm implemented on the exact processing unit.

2.2   Delay Resulting from a Failure Detection Time in a Bus

When a failure occurs in one bus, then the delay on the second, correctly performing bus, resulting from the failure detection and changes in the scenario of exchanges, is:

$$T_{\text{OP2}} = T_{\text{PR2}} + T_{\text{TR2}} + T_{\text{DE2}} + 2 * T_{\text{OD}} + T_{\text{ALG2}} \tag{7}$$

where:

T_{OP2} – delay caused by a failure detection in a bus,
T_{ALG2} – time for executing the algorithm of a bus failure detection and modifications in the scenario of exchanges.

This case is the most difficult from the point of time analysis, since it requires to go back to the original form of the matrix L [11,12]. Thus, the values such as efficiency or useful bandwidth takes the original form of their quantities. The transient state will introduce a delay leading to:

$$T_{\text{MIKRmax}} = \max\left(T'_{\text{MIKR}i}, T''_{\text{MIKR}i}\right) + T_{\text{OP2}} \tag{8}$$

which must be considered at the system designing phase. The time for executing the algorithm for detection a medium failure and the switching the transmission scenario into one bus, can be precisely measured after having the algorithm implemented in an exact processing unit.

2.3 Delay Resulting from a Detection Time of a Subscriber Interface Failure

In case when a subscriber interface is disrupted, a delay in a bus, which the damaged interface was connected to, resulting from a failure detection and the change of the scenario of exchanges equals:

$$T_{\text{OP3}} = T_{\text{PR2}} + T_{\text{TR2}} + T_{\text{DE2}} + 2 * T_{\text{OD}} + T_{\text{ALG3}} \tag{9}$$

where:

T_{OP3} – delay caused by a failure detection in a subscriber interface,
T_{ALG3} – time for executing the algorithm used for detection a subscriber interface failure and the changes in the scenario of exchanges.

Considering the requirements of the time analysis, this case is very difficult, because the scenario of exchanges needs to be edited entirely and thoroughly [12]. And again, as previously the threads of maintaining the rules of the real time system come from the following dependency:

$$T_{\text{MIKRmax}} = \max\left(T'_{\text{MIKR}i}, T''_{\text{MIKR}i}\right) + T_{\text{OP3}}. \tag{10}$$

As before, a delay can influence the time parameters of the real time system, but only in the transient state. After the transmission is stabilized, the minimal modifications in the structure of macrocycles can be noticed. This happens, because some exchanges normally addressed according to the order determined by the algorithms of creating matrices A and B [12], will be transmitted in an imposed manner via the bus not connected to the disrupted interface. Therefore, it will not have any special influence on the dependency (4), and the designing conclusions will be almost the same as in the two previous cases. The time for executing the algorithm of interface failure detection as well as the changes of transmission scenario can be precisely measured after having the algorithm implemented in an exact processing unit.

3   Test Bench and Research Procedure Based on PLC

In order to determine the time for realization of a failure detection algorithm in a real time environment and its influence on the time parameters of a real time system, a test bench was designed and prepared (Fig. 3) and the following tests were conducted:

1. Measurement of the duration of exchange in Bus A.
2. Measurement of the duration of exchange in Bus B.
3. Measurement of the duration of exchange in case of Bus A failure, Bus B failure, subscriber failure and subscriber interface failure for various values of timeout parameter.
4. Measurement of the duration for failure detection algorithm.

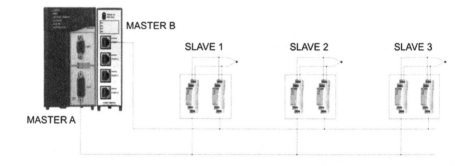

Fig. 3. Test bench based on PLC (GE Intelligent Platforms)

The authors decided also to prepare an additional test bench based on microcontroller, thanks to a completely new concept and opportunity to build a multiprotocol industrial node [15]. This model has also been tested, but the received results are far beyond the current paper issue.

The paper includes the measurements of the duration of algorithm designed for failure detection of a bus (subscriber interface), subscriber or the entire communication system.

The research was carried out for industrial control-measuring circuit based on Modbus/RTU protocol. The discussed conception required applying solutions used in industrial systems. The measurement of failure detection was conducted on a test bench, which was prepared straight for this purpose, and was provided with special software. The measurements of software were conducted on test bench consisting of several devices. The most essential is to mention that basic assumption of the system was its work in a redundant system. As Master station, the authors used PAC RX3i Intelligent Platforms controller. The central unit of controller was equipped with RS-485 interface. In order to double interface, it could be possible to complement the device with RS-232/RS-485 converter for port 1 or additional network coprocessor. However, in the discussed case the PAC controller was equipped with additional network coprocessor CMM004 with RS-485 interface. This two coprocessors have different functionalities. Coprocessor with RS-485 interface is in-built in PLC and has implemented communication driver to Modbus RTU. Driver is distributed by PLC producer. To start communication it is necessary to implement driver function in PLC application, each data exchange has to be declared by programmer. Second coprocessor CMM004 is a module with microcontroller which realizes data exchanges. This coprocessor is independent device and programmer has to declare data exchanges scenario during configuration. This device operates in the main cartridge of PAC controller, and the information can be realized into three ways. The test procedures include:

- Measurement of the exchange duration in Bus A and B.
- Measurement of the exchange duration in case of a failure in Bus A or B (including also the failure detection time).
- Measurement of the exchange duration, including also a failure detection time, in case of a subscriber failure.
- Measurement of the exchange duration, including also the failure detection time, in case of a subscriber interface failure.

The measurements of the duration were conducted by using a special application that operates in Master Station. The program started the internal clock while the transmission was being initialized. The clock was stopped when a word of transmission status, confirming its realization, was read. The other important values used for measurement tests, were:

V_{T} – transmission speed – 9600 bits/s,
L_{BT} – the number of bits per one character equals 8,
L_{BS} – the number of control bits (start and stop bits) equals 3,
T_{PRZ} – time for preparing the request frame in Master station 5.5–6 ms (manufacturer standard),
T_{DT} – time for frame detection, equal to the time of transmission of 3.5 characters,
T_{PRO} – time for preparing the response frame from Slave module, 10 ms (manufacturer standard).

Moreover, according to the specification of Modbus protocol:

- The request transmission frame contains 8 bytes,
- The response transmission frame contains 6 bytes,
- The used function was function with code 02.

For such determined values, and that transmission frame is described by the below dependency:

$$T_{\mathrm{TR}i} = \frac{(L_{\mathrm{BT}} + L_{\mathrm{BS}})}{V_{\mathrm{T}}} = \frac{(8+3)*8}{9.6} = 9.16\,[\mathrm{ms}] \tag{11}$$

the time for request transmission (T_{TRZ}) frame (from Master to Slave) is:

$$T_{\text{TRZ}} = T_{\text{DT}} + T_{\text{TR}i} + T_{\text{PRZ}} = 4 + 9.16 + 6 = 19.16\,[\text{ms}] \qquad (12)$$

and the time for response transmission (T_{TRO}) frame (from Slave to Master) equals:

$$T_{\text{TR}j} = \frac{(8+3)*6}{9.6} = 6.88\,[\text{ms}] \qquad (13)$$

$$T_{\text{TRO}} = T_{\text{DT}} + T_{\text{TR}j} + T_{\text{PRO}} = 4 + 6.88 + 10 = 20.88\,[\text{ms}]. \qquad (14)$$

Thus, the total time of i exchange:

$$T_{\text{W}i} = T_{\text{TRZ}} + T_{\text{TRO}} = 40.04\,[\text{ms}]. \qquad (15)$$

4   Research Results

The received transmission times in both buses (Figs. 4 and 5) are the same as the calculated ones in the previous point. As the charts show, the maximum transmission times for data exchange $T_{\text{W}i}$ ("request-response") respectively for Bus A and B were: 46.9 ms (A) and 47.2 ms (B), and their minimum values are 41.2 ms (A) and 40.8 ms (B). The differences of transmission times in Bus A and B result from applying two different coprocessors the buses were connected to; and in consequence, they have different transmission times.

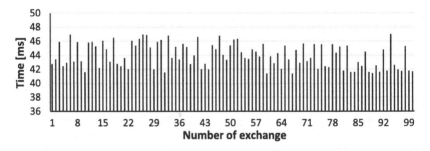

Fig. 4. Measurement of duration of exchange transaction in Bus A – PLC

To ensure the basic functionalities, such as security and data cohesion of redundant systems, it is important to prepare appropriate procedures, which can protect the system in case of a failure of one of the buses, subscriber failure or interface failure. The average duration of transaction exchange in Bus A, B and standard deviation are presented in Table 1.

Standard deviation of about few milliseconds results from the cyclic work of a controller. The cycle phase of a controller determines the moment of network coprocessor service, and at the same time, it defines the time after which the transmitted data can be accessible for executing application [15]. Having looked

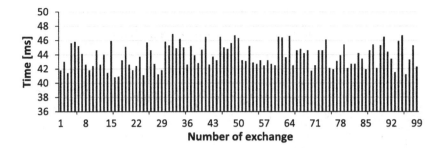

Fig. 5. Measurement of duration of exchange transaction in Bus B – PLC

Table 1. The summary of durations of transaction exchanges in Bus A and B

	Bus A	Bus B
Average time of exchange [ms]	43.95	44.21
Standard deviation [ms]	1.71	1.74

at Figs. 4 and 5, it can be said that the time of a central unit cycle while the measurement were conducted was:

$$T_{WiMAX} - T_{WiMIN} \cong 5.7\,[\text{ms}] \tag{16}$$

which was confirmed by observations.

Figures 6 and 7 present the durations of transaction exchange while a failure occurs in one of the buses, at subscriber or a subscriber interface. When a failure is detected in one of the buses, the implemented algorithm changes the scenario of exchanges so that the remaining transactions could be switched to the efficient bus. In case of a subscriber interface failure, the scenario of exchanges is modified so that the exchanges to this subscriber could be transmitted via bus connected to the efficient interface. Next, a failure of a subscriber results that the exchanges to this subscriber are removed from the scenario of transmission. Additionally, it is checked every 1000 cycles of a controller if a failure has not been removed by sending the control frame. Figures 6 and 7 clearly show how many times, according to the algorithm, Timeout was counted – respectively, twice for a bus failure and subscriber and one for an interface failure (the value of Timeout parameter was set at 100 ms). In consequence, the transaction duration in case of an interface failure is shorter than in case of a bus or subscriber failure. Figure 8 presents the delays resulting from the time of a failure detection in a bus, at subscriber or his interface. It is clear that these delays meet the below dependency:

$$T_{OP2} < T_{OP1} < T_{OP3}. \tag{17}$$

It results from the analysis of algorithm for a failure detection. When a bus failure occurs it is enough to switch all transactions from one bus to another, and this takes a very little time (T_{OP2}). In case of a subscriber failure, additionally

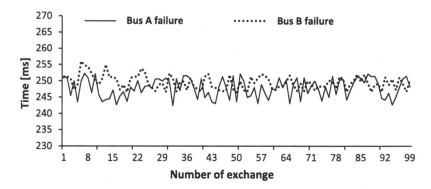

Fig. 6. Measurement of transaction duration in case of Bus A and Bus B failure including the time of a failure detection – PLC

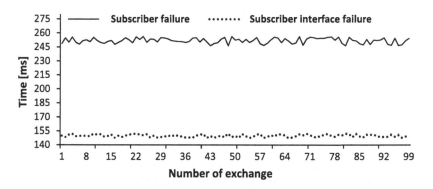

Fig. 7. Measurement of transaction duration in case of subscriber and subscriber interface failure including the time of a failure detection – PLC

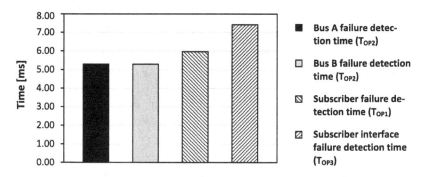

Fig. 8. Delays resulting from the time of a failure detection

Table 2. Summary of times in case of Bus A and B failure

	Bus A failure	Bus B failure
Average time of exchange [ms]	247.91	249.45
Average time of failure detection and changing the scenario of exchange [ms]	5.29	5.29

Table 3. Summary of times in case of a subscriber and a subscriber interface failure

	Subscriber failure	Subscriber interface failure
Average time of exchange [ms]	249.83	151.12
Average time of failure detection and changing [ms] the scenario of exchange	5.96	7.42

it is advised to search the whole scenario of exchanges in order to remove the transactions addressed to a faulty subscriber [12] – this delay is T_{OP1} long. The detection of an interface failure takes the most amount of time, because it is necessary to modify the whole scenario of exchanges – T_{OP3}.

The Tables 2 and 3 present the average times of exchange for failures in Bus A and B, subscriber failure and subscriber interface failure. Moreover, they show an average time of a failure detection with the duration of changes in the scenario of exchanges. Time of a failure detection and the time for changes in the scenario of exchanges were measured by using appropriate, commonly known time functions.

The research results show the time dependencies occurring in the presented algorithm that was developed for a failure detection in a bus, at a subscriber, in its interface or even in the entire communication redundant system that is based on Master-Slave model. The differences in an average time of a failure detection, depending on a failure type, are the result of various steps the algorithm executes. In case of a bus failure, all transactions are switched to an efficient bus; whereas in case of a subscriber failure or its interface failure, the entire scenario of exchanges must be modified so that not to send another exchanges. It is worth to notice that the time for algorithm realization is very little (a few milliseconds) and it does not introduces a significant delay into real time system.

It should be noticed that the biggest delay is the result of the value of Timeout parameter. Its value influences significantly the time of switching the communication between buses of the redundant system. If the time of maximum response is wrongly selected, then all procedures related to improving the system reliability can be lost.

The presented results clearly indicate the efficiency of the discussed solution. The developed algorithm [11] allows to detect a failure of an interface, bus or subscriber in a redundant network based on Master-Slave model of exchange. During the tests, the algorithms each time defined correctly the type of occurred failure.

5 Conclusion

The aim of the conducted research presented in this paper was to detect various types of failures in the industrial communication system. The presented idea of applying redundant bus for increasing the frequency of exchanges in distributed real time systems must be strongly related to the methods of failure detection. The presented approach is based on redundant communication bus, but the data exchange model is different than in e.g. PRP (Parallel Redundancy Protocol) where the same information are transmitted via parallel/redundant buses.

If tested algorithm are omitted, then one should expect the system failure. It seems that mechanism of interface failure detection is particularly useful. Failure detection of interface or bus makes the system to change the operation mode by switching the communication to other bus.

The conducted tests allow to make an argument that the described algorithm [11] enables to detect a failure and what is more, allows to execute the scenario of exchanges that cannot be executed via disrupted bus. Although the system is not a classical redundant system, the proposed solution allows the system to work correctly in case of a failure of a communication bus, even if in some situations the periodicity of a microcycle can be overrun. However, it does not mean that the tasks have been finished. It only makes the system to reconfigure automatically in order to execute the exchange as soon as possible in the other efficient communication bus.

Acknowledgments. This work was sponsored from BK-263/2015.

References

1. Kirrmann, H., Hansson, M., Muri, P.: Bumpless recovery for highly available, hard real-time industrial networks. In: IEEE Conference on Emerging Technologies and Factory Automation, ETFA 2007, pp. 1396–1399, September 2007
2. Kirrmann, H., Weber, K., Kleineberg, O., Weibel, H.: Seamless and low-cost redundancy for substation automation systems (high availability seamless redundancy, HSR). In: 2011 IEEE Power and Energy Society General Meeting, pp. 1–7 (2011)
3. IEC 62439, Committee Draft for Vote (CDV): parallel redundancy protocol. In: Industrial Communication Networks: High Availability Automation Networks, chap. 6 (2007)
4. Kwiecień, A., Sidzina, M.: Dual bus as a method for data interchange transaction acceleration in distributed real time systems. In: Kwiecień, A., Gaj, P., Stera, P. (eds.) CN 2009. CCIS, vol. 39, pp. 252–263. Springer, Heidelberg (2009)
5. Wei, L., Xiao, Q., Xian-Chun, T., et al.: Exploiting redundancies to enhance schedulability in fault-tolerant and real-time distributed systems. IEEE Trans. Syst. Man Cybern. Part A Syst. Hum. **39**(3), 626–639 (2009)
6. Sidzina, M., Kwiecień, B.: The algorithms of transmission failure detection in master-slave networks. In: Kwiecień, A., Gaj, P., Stera, P. (eds.) CN 2012. CCIS, vol. 291, pp. 289–298. Springer, Heidelberg (2012)

7. Neves, F.G.R., Saotome, O.: Comparison between redundancy techniques for real time applications. In: Fifth International Conference on Information Technology: New Generations, ITNG 2008, pp. 1299–1300 (2008)
8. Kwiecień, A., Stój, J.: The cost of redundancy in distributed real-time systems in steady state. In: Kwiecień, A., Gaj, P., Stera, P. (eds.) CN 2010. CCIS, vol. 79, pp. 106–120. Springer, Heidelberg (2010)
9. Gaj, P., Jasperneite, J., Felser, M.: Computer communication within industrial distributed environment - a survey. IEEE Trans. Industr. Inf. 9(1), 182–189 (2013)
10. Kwiecień, A., Maćkowski, M., Stój, J., Sidzina, M.: Influence of electromagnetic disturbances on multi-network interface node. In: Kwiecień, A., Gaj, P., Stera, P. (eds.) CN 2014. CCIS, vol. 431, pp. 298–307. Springer, Heidelberg (2014)
11. Kwiecień, A., Kwiecień, B., Maćkowski, M.: Algorithms for transmission failure detection in a communication system with two buses. In: Gaj, P., Kwiecień, A., Stera, P. (eds.) 23th Conference on Computer Networks, CN 2016, Brunów, Poland. CCIS, vol. 608, pp. 141–153. Springer International Publishing Switzerland (2016)
12. Kwiecień, A., Kwiecień, B., Maćkowski, M.: Automatic scenario selection of cyclic exchanges in transmission via two buses. In: Gaj, P., Kwiecień, A., Stera, P. (eds.) CN 2015. CCIS, vol. 522, pp. 150–161. Springer, Heidelberg (2015)
13. Pereira, C.E., Neumann, P.: Industrial communication protocols. In: Nof, S.Y. (ed.) Springer Handbook of Automation, pp. 981–999. Springer, Heidelberg (2009)
14. Miorandi, D., Vitturi, S.: Analysis of master-slave protocols for real-time-industrial communications over IEEE802.11 WLAN. In: 2nd IEEE International Conference on Industrial Informatics 2004, pp. 143–148 (2004)
15. Kwiecień, A., Sidzina, M., Maćkowski, M.: The concept of using multi-protocol nodes in real-time distributed systems for increasing communication reliability. In: Kwiecień, A., Gaj, P., Stera, P. (eds.) CN 2013. CCIS, vol. 370, pp. 177–188. Springer, Heidelberg (2013)

Event-Driven Approach to Modeling and Performance Estimation of a Distributed Control System

Wojciech Rząsa and Dariusz Rzonca$^{(\boxtimes)}$

Department of Computer and Control Engineering,
Rzeszow University of Technology, al. Powstancow Warszawy 12,
35-959 Rzeszow, Poland
{wrzasa,drzonca}@kia.prz.edu.pl
http://kia.prz.edu.pl

Abstract. Currently Distributed Control Systems are commonly used in the industry. Proper operation of such systems depends on reliable and efficient communication. Simulation of the system operation in different conditions, performed during early development stages, allows to estimate performance and predict behavior of final implementation. In this paper new event-driven model of a DCS has been introduced and compared with the previous one. Structure of the model decreased simulation time and improved scalability of the approach, whereas precision of the results was not diminished.

Keywords: Distributed control systems · Communication · Modeling · Performance · Simulation

1 Introduction

Nowadays typical stand-alone controllers are often superseded by Distributed Control Systems (DCSs). In such systems it is essential to provide efficient and reliable data exchange method between system components. Different paradigms of access to the common communication link may be considered [1], however most of them are based on the Time Division Multiplexing (TDM) scheme. Alternative Frequency Division Multiplexing (FDM) method has been also discussed [2], but it was not adopted in the practice. The most common TDM models used in the DCSs are master-slave, token passing and producer-distributor-consumer [3]. Each of them defines a scenario of data exchanges in particular case, assigning appropriate time slots to every device sharing the common bus. Such scenario is typically fixed, however in some cases it might be changed automatically, especially in situation where numerous redundant data buses are present, to comply with external events (e.g. failure of one of the buses) [4]. Usually industrial systems use a fieldbus [5] and a field protocol, e.g. one of the defined in the IEC 61158 standard [6], however nonstandard solutions, such as using Hypertext

© Springer International Publishing Switzerland 2016
P. Gaj et al. (Eds.): CN 2016, CCIS 608, pp. 168–179, 2016.
DOI: 10.1007/978-3-319-39207-3_15

Transfer Protocol (HTTP) instead of specialized field protocol, have been also considered [7,8].

In this paper a simple DCS structure is analyzed by means of simulation. The goal of this analysis is to observe behavior and provide information about efficiency of the whole control system. As performance in distributed system may depend on different factors, simulation can provide information about the ones that should be considered as bottlenecks. This research considers basic structure of the DCS and estimates its efficiency depending on network reliability and run time of master's control program as discussed in [9]. This research however uses new approach to modeling and simulation that was used for the other distributed system in [10].

Similarly as in [9] Timed Colored Petri Nets (TCPN) [11] are used to ensure reliability of simulation. However, the DCS model is created using Domain Specific Language (DSL) designed to describe models of distributed systems. Thereafter the model is automatically translated to TCPN according to principles described in [12] and refined in further research. The DSL can be used to conveniently model wide variety of distributed systems and is designed to enable convenient scaling of the models. Additionally, simulator used in this research is significantly more efficient then the previous one.

Consequently, unlike the previous solution, the approach used in this paper can be easily extended to analyze more complex systems and longer periods of time. The goal of this research is to compare results obtained using the new approach with the previous ones and verify that while enhancing convenience and scalability of the method its accuracy was not diminished.

2   Structure of the DCS

The DCS analyzed in this paper consists of one controller and two slave modules. They are connected to the common serial bus (e.g. RS-485) and communicate in master-slave field protocol (e.g. Modbus RTU). The controller as a master initiates a data exchange sending a message to a slave, and the addressed slave responds. If the valid response is not received in the assumed time (e.g. due to noise or bus/slave failure) the timeout occurs. Scenario of data exchanges is determined by communication tasks. Each task defines data exchange (command-response) with particular slave, repeated periodically. In described case two communication tasks are present, one for each slave module. Simple controller without dedicated communication coprocessor is considered, so the communication tasks are served by the controller in the time slots after execution of each loop of the control program, and before defined cycle time pass. It is assumed that cycle time is kept constant, thus duration of the communication time slot is determined by current program execution time t_{prog}. Of course communication takes place asynchronously. The transmission is handled in the interrupts, so messages may be sent or response received by the controller also during execution of the control program, however processing of the response and starting of the new transmission will be delayed till new communication time slot. In such system it is interesting to measure obtained communication cycle time t_{com} (i.e. time necessary to handle all the communication tasks) as the function of parameters like t_{prog}, or probability of packet loss P_{loss}.

Such structure of the modeled DCS is taken from [9], however modeling method in this research is different. In [9] SysML diagrams have been used to define system structure and parameters. These values have been introduced, as initial marking, to the constant TCPN model of the communication loop. In this paper communication is modeled in an event-driven way, as described in Sect. 3. Event-driven approach is quite new for modeling of a DCS, but seems to be promising. Identical, quite simple structure of the DCS, as modeled in [9] is considered here, to allow direct comparison of the results between different applied methodologies.

3   Model

In order to enable analysis a model of the Distributed Control System was created. The DSL used to describe the model assumes event-driven description of *programs* running on computing resources. The model of master controller used in this research uses four events described below.

Event :cycle – Listing 1.1 – is used to model control cycle of the controller. It is registered at the beginning of simulation and its handler registers it again with the delay equal to controllers cycle time. The controller first executes iteration of its controlling loop and for the rest of time performs communication. Thus the :communication event is registered with the delay equal to controllers work_time (i.e. sum of reading inputs, writing outputs and running control program). The flag @communication unset at the beginning of the :cycle event handler ensures that no communication tasks are performed during the control program execution.

```
1  on_event :cycle do
2    @communication = false
3    work_time = params[:read_time] + params[:exec_time] +
4                params[:write_time]
5    unless @tasks.empty? && @running_task.nil?
6      register_event :cycle, delay: params[:cycle_time]
7    end
8    register_event :communication, delay: work_time
9  end
```

Listing 1.1. Model of master controller's cycle

Communication is enabled when :communication event occurs and its handler (Listing 1.2) sets @communication flag at the beginning. If there is no communication task running yet, new one is started. Otherwise, it is checked if a response to running communication task was received (using interrupts) and buffered while control program was running (call to check_response). If so, a new communication task is started as soon as the response is received.

```
1   on_event  :communication do
2       @communication = true
3       if @running_task.nil?
4           start_task
5       else
6           start_task if check_response
7       end
8   end
```

Listing 1.2. Model of master controller's communication-related actions

The `data_received` event is registered by the simulator itself whenever a data package is delivered to modeled process and is handled by code from Listing 1.3. If received data matches currently running communication task it is saved in a buffer. Otherwise, the data must be a late response to a previous communication task that was timed out and thus should be dropped. The data can be received by the DCS master in two distinct situations. If a data package (i.e. response to a running communication task) was received while master was running its control program and communication was disabled (`@communication` flag unset), then the received data is left in the buffer for later use. This data package is served later, when `:communication` event occurs. Otherwise, if communication is enabled, the response is received immediately by `check_response` function and new communication task is started. If no data package was received during the period of running control program, the `check_response` function returns false and no new communication task is started.

```
1   on_event  :data_received do |data|
2       if data.content.instance_equals? @running_task
3           @receive_buffer = data
4       end
5       if @communication
6           start_task if check_response
7       end
8   end
```

Listing 1.3. Model of master controller's data received actions

Event `:comm_timeout` (Listing 1.4) is registered with appropriate delay for each new communication task. If the event occurs while this task is still running (before response was received), the task is considered timed-out. Then, the running task buffer is emptied and a new task is started without waiting for a response for the old one.

```
1   on_event  :comm_timeout do |task|
2       if task.instance_equals? @running_task
3           @running_task = nil
4           start_task if @communication
5       end
6   end
```

Listing 1.4. Model of master controller's communication timeout actions

Event handlers use two functions to control communication tasks: start_task from Listing 1.5 and check_response from Listing 1.6. New communication tasks are taken from priority queue according to its discipline. In this research the queue is cyclic and the task cycle can be restarted. If the queue was exhausted and must be restarted, this fact is saved in simulator statistics. Thereafter, a new task instance is taken from the queue, request is sent to this task's destination and the task is stored in master controller's buffer to verify correctness of the response. Finally, new :comm_timeout event is registered with appropriate delay for this communication task.

```
1  function :start_task do
2    if @tasks.restart?
3      stats_stop :task_cycle unless @first_task
4      stats_start :task_cycle
5      @first_task = false
6    end
7    task = @tasks.shift
8    return false if task.nil?
9    send_data to: task.dst, size: task.request_size, content: task
10   @running_task = task
11   register_event :comm_timeout, args: task, delay: task.timeout
12   true
13 end
```

Listing 1.5. Starting communication task by master controller

The check_response function is used to handle responses for communication tasks. If there is no response to be served, it returns false. If there is a response, the buffers for incoming data and for currently running tasks are reset and the function returns true.

```
1  function :check_response do
2    return false if @receive_buffer.nil?
3    @receive_buffer = nil
4    @running_task = nil
5    true
6  end
```

Listing 1.6. Handling response to communication task by master controller

The model of slave controllers is implemented as :slave program (Listing 1.7). It has only one event handler for data_received event. When a data package carrying master's request is received, slave's CPU is loaded for time specified in the communication task's parameters and thereafter a response is sent.

The programs of master and slave controllers are mapped to run as processes on devices modeled as nodes with single CPUs and connected with a single bus-like network segment. Each of the controllers is represented by separate node, thus three nodes are used – one for master and two for slave controllers.

```
1  program :slave do
2    on_event :data_received do |data|
3      cpu do |c|
4        data.content.slave_delay
5      end
6      send_data to: data.src, size: data.content.response_size,
7              content: data.content
8    end
9  end
```

Listing 1.7. Model of slave controller action

The model accepts the following parameters. For master controller read_time, write_time, exec_time and cycle_time can be set denoting read time from controller inputs, write time to controller outputs, execution time of control program and cycle time of the master. Additionally, queue of communication tasks must be passed to the model of master controller. Each task also carries its parameters including task destination (dst), size of request sent to slave (request_size), size of response sent to master (response_size), time between receiving request and sending response by slave (slave_delay) and task timeout (timeout). Two additional parameters can be set for the model. First of them is drop which models network reliability and denotes probability that a data packet is dropped by the network. Second is the number of slave controllers connected to the bus (this one is always set to 2 in this research).

4   DCS Simulation

Before the model from previous section can be used for simulation, its parameters must be set. The parameters are set accordingly to the ones used in [9]. Thus, write_time and read_time are set to 0 since considered controller does not use its inputs and outputs, cycle_time is set to 200 ms. For each communication task request_size and response_size as well as bandwidth of the network segment are set to obtain required total network transmission time for each task equal to 40 ms (sum of transmission time in both directions) corresponding to the value used in [9]. Delay of slave's response (slave_delay) is random value with normal distribution for mean value 10 ms and standard deviation of 5 ms. Timeout of each communication task is 500 ms. Task queue in simulation contains two communication tasks, one for each slave module, with its destination parameter (dst) set accordingly.

During simulations some parameters are automatically changed and experiment repeated for different configurations. Simulated values of program execution time t_{prog} (denoted in the model as exec_time) are between 10 ms and 190 ms with 10 ms steps. Tested values of probability of packet loss (denoted in the model as drop) are $P_{loss} \in \{0, 0.05, 0.2\}$. The values were used to facilitate direct comparison of the results with [9].

The model is automatically translated to the TCPN and simulations are performed. As a result communication cycle time t_{com} (i.e. time necessary to handle all the communication tasks) has been logged (numerous values for each set of parameters). Petri net simulator used in this research was significantly optimized in comparison to the version used in [9]. Detailed description of the modifications are a wide topic and thus do not fit into scope of this paper. As a consequence of the optimizations, the more efficient Petri net simulator allowed to observe DCS behavior in longer periods of time. Thus, for each set of parameters period of 200 s of DCS operations (model time) was simulated instead of 20 s period used in [9]. Consequently, credibility of the obtained results is improved.

5   Results

To compare both models the same DCS system as described in [9] has been analyzed two times, once in every model. The results have been analyzed in R statistical package [13]. Graphical comparison of the results, for different probability of packet loss P_{loss} is shown in Fig. 1. The results obtained in the previous models are in the left columns, while the current results are shown on the right. Typical symbols in the statistical boxplot have been used. Line inside the box indicate the median value of the data set in particular case, while top and bottom box hinges mark the upper and lower quartile, respectively. Vertical whiskers indicate the minimum and maximum of the obtained values, which are located in one and a half interquartile range from the box. Outliers located outside the whiskers are marked as circles.

Shape of the plots in Fig. 1a, b may be easily explained. As mentioned in Sect. 2 communication tasks are handled during time slots determined by end of the execution of each loop of the control program, before defined cycle time pass. Thus if the program execution time t_{prog} is small, resulting time slot is large enough for handling all communication tasks, and obtained communication cycle time t_{com} is also small. If t_{prog} is increased, time slot will be large enough for handling only one of the tasks. In the second task the message will be sent, but response from the slave will be processed in the next program cycle. Thus stabilization of the t_{com} near 200 ms (cycle time) is observed. Further incrementation of t_{prog} results in stabilization of the t_{com} near 400 ms (twofold cycle time), because in every time slot one message is sent, but response is always analyzed in the subsequent cycle.

In Fig. 1c–f an influence of packet loss on the t_{com} is presented. Additional probability of packet loss equal to 0.05 and 0.2, respectively, has been introduced. Such huge values, unrealistic for real network, have been intentionally chosen in [9] to show behavior of the system in extreme conditions. Both models present similar results with significant degradation of network performance.

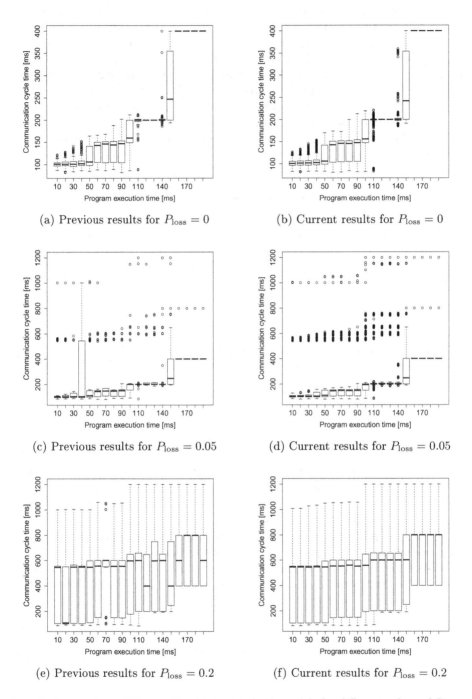

Fig. 1. Comparison of the results obtained in both models for different values of P_{loss}

As seen in Fig. 1 the results obtained in both models seem to be consistent. For some sets of parameters the results are constant and identical in both models (for $P_{\text{loss}} = 0$, $t_{\text{prog}} = 120\,\text{ms}$ and $t_{\text{prog}} = 130\,\text{ms}$ obtained $t_{\text{com}} = 200\,\text{ms}$; for $t_{\text{prog}} \in \{160, 170, 180, 190\}$ [ms] resulted $t_{\text{com}} = 400\,\text{ms}$), thus statistical tests are not applicable in such cases. The p-value of the Wilcoxon rank-sum test has been calculated for remaining populations received in both models for every set of the parameters (see Table 1). Null hypothesis for this test assumes that both samples come from populations with identical distributions. Assuming significance level $\alpha = 0.01$, test results do not support rejecting this hypothesis for any analyzed case.

Table 1. Wilcoxon rank sum test

$P_{\text{loss}} = 0$									
Cycle time [ms]	10	20	30	40	50	60	70	80	90	100
p-value	0.668	0.6144	0.9945	0.4695	0.6854	0.8942	0.9927	0.8121	0.5784	0.1321
Cycle time [ms]	110	120	130	140	150	160	170	180	190	
p-value	0.8749	NA	NA	0.2745	0.07882	NA	NA	NA	NA	
$P_{\text{loss}} = 0.05$									
Cycle time [ms]	10	20	30	40	50	60	70	80	90	100
p-value	0.7252	0.944	0.3752	0.7432	0.6533	0.4814	0.5311	0.9959	0.9916	0.2203
Cycle time [ms]	110	120	130	140	150	160	170	180	190	
p-value	0.6393	0.2288	0.3129	0.3569	0.8267	0.2553	0.4815	0.9942	0.6371	
$P_{\text{loss}} = 0.2$									
Cycle time [ms]	10	20	30	40	50	60	70	80	90	100
p-value	0.3232	0.108	0.1242	0.8753	0.9119	0.4571	0.0184	0.3798	0.721	0.9813
Cycle time [ms]	110	120	130	140	150	160	170	180	190	
p-value	0.4726	0.2931	0.9699	0.7364	0.1521	0.7914	0.9264	0.808	0.1695	

To illustrate that the results obtained for both models give populations with a similar distribution, they have been also compared using quantile-quantile plots (see Fig. 2). Due to limited space only selected plots have been shown, i.e. for $P_{\text{loss}} = 0$, $t_{\text{prog}} \in \{30, 60, 90, 120, 150, 180\}$ [ms].

As seen in Fig. 2 in numerous cases the points are located approximately along reference line representing equal quantile distribution, which is an evidence, that the distributions being compared are similar. Different distortions, which may be observed, might be related to small numbers of results from the previous model.

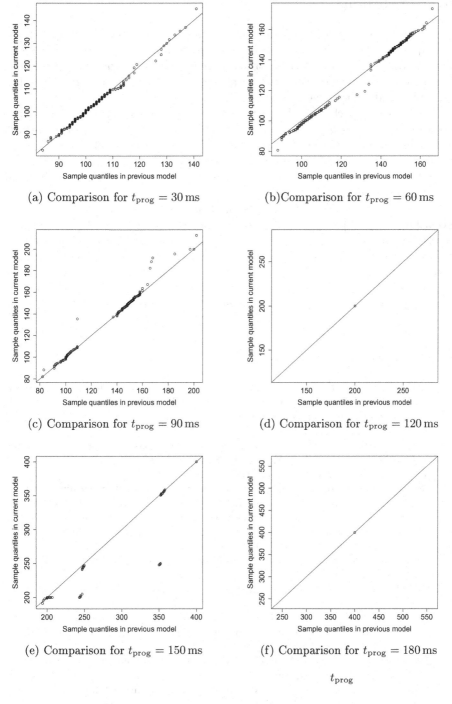

(a) Comparison for $t_{\mathrm{prog}} = 30\,\mathrm{ms}$

(b) Comparison for $t_{\mathrm{prog}} = 60\,\mathrm{ms}$

(c) Comparison for $t_{\mathrm{prog}} = 90\,\mathrm{ms}$

(d) Comparison for $t_{\mathrm{prog}} = 120\,\mathrm{ms}$

(e) Comparison for $t_{\mathrm{prog}} = 150\,\mathrm{ms}$

(f) Comparison for $t_{\mathrm{prog}} = 180\,\mathrm{ms}$

t_{prog}

Fig. 2. Comparison for different values of t_{prog}

6 Conclusion

In this paper new, event-based approach to modeling Distributed Control System was applied. The model was used to assess performance of the DCS depending on execution time of master controller's main loop and reliability of network connection. The results were statistically compared to the ones from [9] that were obtained using older approach.

The new approach to modeling DCS allows to conveniently create large models. Consequently, it is possible to analyze more complex examples in the future. Additionally, more efficient simulator allows to obtain more reliable simulation results that were already presented in this paper. The results proved to be consistent with the ones presented in [9], thus the change in modeling method did not deteriorate precision of the results.

The research also showed that the modeling and simulation method created according to principles from [12] and used in [10] can be successfully applied to model and analyze distributed aspects of control systems. This conclusion allows to use the method in future research concerning performance of more complex structures of DCS. Concurrently, flexibility of the method was confirmed as it was successfully used for control system as well as for Platform as a Service deployed web applications [10] that exemplify significantly different distributed systems.

References

1. Silva, M., Pereira, F., Soares, F., Leao, C., Machado, J., Carvalho, V.: An overview of industrial communication networks. In: Flores, P., Viadero, F. (eds.) New Trends in Mechanism and Machine Science. Mechanisms and Machine Science, vol. 24, pp. 933–940. Springer, Switzerland (2015)
2. Stój, J.: Real-time communication network concept based on frequency division multiplexing. In: Kwiecień, A., Gaj, P., Stera, P. (eds.) CN 2012. CCIS, vol. 291, pp. 247–260. Springer, Heidelberg (2012)
3. Gaj, P., Jasperneite, J., Felser, M.: Computer communication within industrial distributed environment - a survey. IEEE Trans. Ind. Inf. **9**(1), 182–189 (2013)
4. Kwiecień, A., Kwiecień, B., Maćkowski, M.: Automatic scenario selection of cyclic exchanges in transmission via two buses. In: Gaj, P., Kwiecień, A., Stera, P. (eds.) CN 2015. CCIS, vol. 522, pp. 150–161. Springer, Heidelberg (2015)
5. Thomesse, J.P.: Fieldbus technology in industrial automation. Proc. IEEE **93**(6), 1073–1101 (2005)
6. IEC 61158 Standard: Industrial Communication Networks – Fieldbus Specifications (2007)
7. Jestratjew, A., Kwiecień, A.: Using HTTP as field network transfer protocol. In: Kwiecień, A., Gaj, P., Stera, P. (eds.) CN 2011. CCIS, vol. 160, pp. 306–313. Springer, Heidelberg (2011)
8. Jestratjew, A., Kwiecien, A.: Performance of HTTP protocol in networked control systems. IEEE Trans. Ind. Inf. **9**(1), 271–276 (2013)
9. Jamro, M., Rzonca, D., Rząsa, W.: Testing communication tasks in distributed control systems with SysML and timed colored petri nets model. Comput. Ind. **71**, 77–87 (2015)

10. Rząsa, W.: Simulation-based analysis of a platform as a service infrastructure performance from a user perspective. In: Gaj, P., Kwiecień, A., Stera, P. (eds.) CN 2015. CCIS, vol. 522, pp. 182–192. Springer, Heidelberg (2015)
11. Jensen, K., Kristensen, L.: Coloured Petri Nets. Modeling and Validation of Concurrent Systems. Springer, Heidelberg (2009)
12. Rząsa, W.: Timed colored petri net based estimation of efficiency of the grid applications. Ph.D thesis, AGH University of Science and Technology, Kraków, Poland (2011)
13. R Core Team R: A language and environment for statistical computing. R Foundation for Statistical Computing, Vienna, Austria (2013)

QoE-Oriented Fairness Control for DASH Systems Based on the Hierarchical Structure of SVC Streams

Slawomir Przylucki[1(✉)], Dariusz Czerwinski[1], and Artur Sierszen[2]

[1] Lublin University of Technology, 38A Nadbystrzycka Street, 20-618 Lublin, Poland
{s.przylucki,d.czerwinski}@pollub.pl
[2] Lodz University of Technology, 116 Zeromskiego Street, 90-924 Lodz, Poland
artur.sierszen@p.lodz.pl
http://www.pollub.pl
http://www.p.lodz.pl

Abstract. Nowadays, multimedia content is the basis of most popular network services. At the same time, adaptive video streaming is becoming more widely used method for delivering such content to end users. The main purpose of the mechanisms of adaptation is to maximize the usage of network resources while ensuring the best quality of transmitted video. The implementation of adaptation, however, has a serious drawback of not considering the mutual influence of simultaneous transmission of multiple video streams. In this article, we propose a method of the QoE-oriented fairness control for DASH systems which uses a new utility function. This function based on a hierarchical structure of a SVC video stream. The properties of this solution were evaluated in a test environment that uses SDN principles. Functional features of the proposed control method were compared with standard algorithm of the adaptation used by DASH supported devices.

Keywords: Video adaptation · Resource allocation · Utility fairness · SVC video · DASH streaming · Quality of Experience (QoE) · Software Defined Network (SDN)

1 Introduction

The beginning of the twenty-first century has brought convergence of telephone systems to computer networks. Currently, a similar process is observed in relation to the market for TV and video services. Moreover, there is a new, increasingly widespread trend according to which it is expected that high-resolution video material would has combined with high and stable parameters of its transmission and reception. For this reason, modern telecommunications infrastructure must support diverse network traffic with limited capacity of individual links. This in turn inevitably leads to the possibility of congestion. Currently the TCP (Transmission Control Protocol) is able to provide efficient services for typical Internet applications. Unfortunately, there is a significant difference between

© Springer International Publishing Switzerland 2016
P. Gaj et al. (Eds.): CN 2016, CCIS 608, pp. 180–191, 2016.
DOI: 10.1007/978-3-319-39207-3_16

services such as web browsing or e-mail and video streaming services. The conventional applications are classified as the elastic services in contrast to transmission of video which represents the real time services. Streaming video and generally video broadcasting are delay sensitive and require bandwidth guarantees. The failure to comply with strict requirements of a quality of service (QoS) may affect video quality perceived by users [1]. Although bandwidth provision in computer networks has been improved in the last few years, mechanisms currently implemented only partially solve the problems arising from the transmission of video streams. A proposal aimed at improving the situation was adaptive streaming. The fundamental idea of this approach is the introduction of monitoring mechanism of a network status at the client side. The gathered data allow for adaptations of QoS requirements to changing conditions inside a computer network. The standardised implementation of this solution is DASH (Dynamic Adaptive Streaming over HTTP) [2] elaborated by the Moving Pictures Experts Group. Another factor influencing the popularity of adaptive video streaming is the layered video coding. The standard H.264/SVC (Scalable Video Coding) [3] was created by JVT (Joint Video Team) as an extension of the H.264 codec. The idea of a SVC is to enable the removal of parts of the coded data stream while ensuring proper decoding video. The resulting sub-stream, with a reduced quality has lower QoS requirements than the original video stream.

A combination of both techniques, the adaptive streaming and the scalable video coding, allows for a significant reduction of such phenomena as video pauses or too long buffering and thus greatly improved the overall user experience [4,5]. Unfortunately, even in this case, a few problems remain unsolved. The network resource adaptation has to be correlated to the perceptual quality of a video stream. In other words, it is necessary to map between an available bandwidth and the parameters QoE (Quality of Experience). This mapping could serve as a utility function. Unfortunately, the research work in this field shows that there is no simple, linear correlation between the bitrate of a video stream and an assessment of the parameters QoE. Also all relationships strongly depend on the characteristics of the source video material [6,7]. The second unresolved issue is the implementation of the process of the adaptation, which recommends the DASH standard [2]. The DASH adaptation algorithms do not define mechanisms that take into account mutual influence of the individual transmission. Each video receiver, realizes the process of adaptation in isolation from other recipients [8]. This leads to a situation in which every recipient of the service DASH selfishly trying to secure the best quality of the received video. This leads to the degradation of the parameters of the received video at all other competing clients in a network segment [9]. There is therefore a need to ensure a fair share of network resources. The QoE fairness in this context means that the perceived quality of a video content should be maximized for all users of the DASH service. Over the last several years it has been proposed many solutions to achieve a user-centric fair-share in streaming applications. In this context, particular interest focused research concerning the specificity of cooperation between the mechanisms of adaptation inside the DASH systems and the protocols: TCP and HTTP [9].

The presentation of the principles of this approach is published in [10]. The authors described the "downward spiral effect", and the way to measure it on the adaptive streaming services. They also validated the initial solutions to prevent this phenomenon. At the same time, Tian et al. studied the responsiveness and smoothness trade-off in DASH systems [11]. They presented the algorithm for the throughput prediction to attenuate video rate fluctuations. The solution that integrates both network measurement and the QoE-aware adaptation is presented in [12]. Some more detailed research has focused on the delayed bitrate update mechanism [13] and controlling the buffering behaviour to reduce the size of the receiver window at a DASH client [14].

An entirely different approach to ensure an fair distribution of network resources in the DASH system is the method called OFC (Optimal Flow Control). OFC provides an efficient congestion control mechanism for the network, but it also opens a way to develop a fair bandwidth allocation among competing DASH client. The principles of the method OFC described Kelly [15]. According to the proposal, the process of the flow control can be formulated as an optimization issue. In such a case, the rate flow control algorithm can be developed based on the optimal pricing policy for a link congestion. Moreover, Kelly showed the importance of the utility function (he used a logarithmic function) and presented that the OFC approach leads to the so-called proportional fairness in a network resource (bandwidth) allocation. Shortly later, Marbach [16] described the max-min fair allocation in a network controlled in accordance with the OFC principles. All of these solutions have been developed to control an elastic traffic. Wang et al. proposed the method OFC for real-time network traffic [17]. Authors defined the application-oriented utility function and defined a new distributed algorithm for the flow control in multiservice networks. Recent research go towards a combination of both approaches discussed above. For example, in [18] authors analyze the theoretic approach for the control of dynamic, adaptive video streaming over HTTP. They proposed the predictive control algorithm that combines the video bitrate and buffer occupancy information. This paper introduces the method for the QoE-oriented fairness control for DASH systems. It based on a new utility function. This function is derived from the analysis of a hierarchical structure of a SVC video stream.

The remainder of the paper is organised as follows. Section 2 contains presentation of the method for creating the proposed utility function. Section 3 describes the proposed method of QoE-oriented fairness control. The test network and the test results are in Sect. 4. Finally, a brief conclusion is made in Sect. 5.

2   QoE-Aware Utility Function for SVC/DASH Systems

For many recent years, the allocation of bandwidth in computer networks has been the subject of numerous studies. Recently, however, more attention is paid to the QoS performance which is described by the utility function for an network application. The following sections will be devoted to the presentation a method for the construction of a new utility function, which is based on an analysis of the hierarchical structure of SVC video streams.

2.1   Gradation of Video Quality in SVC Video Stream

Any SVC video stream consists of three types of scalability: temporal, spatial and quality (SNR), respectively. Also, each scalability is represented by the base and the enhancement layers. Removing the given enhancement layers lead to creation a new structure of the SVC stream. Temporal scalability allows for a change of a bitrate by changing the number of frames per second. Removing single spatial enhancement layer leads to a reduction in resolution of the received video. In the case of SNR layers, the situation is more complex. There are two types of the SNR scalability: the Coarse Grain Scalability (CGS) and the Medium Grain Scalability (MGS), respectively [3,19]. Considering cooperation between standard DASH and SVC coding, it has been proved that the MGS approach seems to be particularly promising [20]. This implementation allows to obtain finer granularity of the video quality by dividing each SNR enhancement layer on several sublayers (MGS layers).

The process of defining the quality gradations inside the SVC stream is therefore directly linked to the analysis of the hierarchical structure of that video stream. The process of elimination of certain enhancement layers leads to the formation of a new video stream with a lower bitrate. Any such a new structure of the SVC stream is often called as a OP (Operation Point). Typically, OPs are defined at the level of single GOP (Group of Pictures) [7]. Using this nomenclature, the OP1 is a basic OPs i.e. it contains all the SVC layers (base and enhancement). Removal of the individual enhancement layers leads to the definition of the next OPs.

2.2   Construction of Utility Function

For each video content encoded using the codec H264SVC, it is possible to define multiple sets of OPs. The method of the selection of the best hierarchical structure of the SVC stream should take into account, on the one hand the ability to provide uniform gradation of the video quality and on the other, the best perceived quality of the decoded video content. This issue was the subject of our more detailed research. Their results were presented in [21] along with a description of the algorithm allowing for the selection of the best set of OPs. This algorithm was called as the T-P (Temporal Preservation) method and it has been used during the process of defining the utility function. The main assumptions of this method are as follows:

- A stream of the SVC video is encoded using temporal and quality scalability. It was also assumed that separate representations of video content with different resolution are available on the DASH server.
- For each temporal layer (T0, T1, T2, ..., Tn) at least a basic quality layer (SNR0) is available.

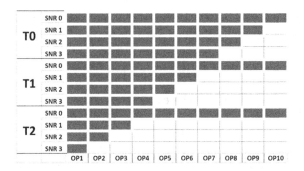

Fig. 1. Definition of OPs according to the T-P method

– A video stream created by the OP with the highest number ensures that the decoder has at least SNR0 data for each temporal layer.

To illustrate the principles of the method T-P, the Fig. 1 presents a set of OPs obtained for the case of the video stream containing three temporary layers and four SNR layers.

The video stream that was created based on the T-P method, has a finite number of OPs (in the example illustrated in the Fig. 2 is equal to 10). With the change to a higher OP decreases perceived video quality due to the removal of more and more enhancement layers. At the same time decreasing the requirement for the necessary bandwidth, which should be guaranteed during the transfer of a video content. Such a relationship can serve as a utility function since it provides a way to map bitrate of video (at particular resolution) and value of the QoE metric which is a measure of video quality perceived by the users of a video service. Figure 2 shows this type of the relationship for two video resolutions, 1080p and 720p, respectively. The data were obtained for the "Elephants Dreams" video sequence (1080p, 24 f/s) [22]. The video was encoded using the reference software JSVM (Joint Scalable Video Model) [23].

Unfortunately, direct use of the presented relationships as the utility function has one very significant drawback. For each particular video, the relationship between bitrate and perceived quality is different. There are also differences for each resolution of the same video content. Therefore, the practical use of such utility functions would require to build a database containing the utility function per video at each resolution. For this reason, we propose a different approach to the issue of construction of the utility function.

Each client of the service DASH is interested in getting the best quality of the received video. For SVC video, this is equivalent to the possibility of receiving a video stream containing the highest possible number of layers (the lower OP). In other words, a measure of "success" of the receiver DASH/SVC is to use a video representation of the highest bitrate (representation ensuring the highest perceived quality). Any available video representation is an implementation of a particular OP. Therefore we defined the RU indicator describing the ratio of

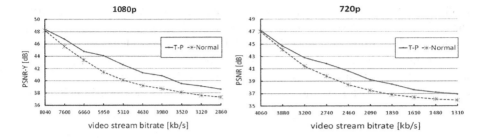

Fig. 2. Bit rate vs. PSNR-Y for the analyzed video sequence

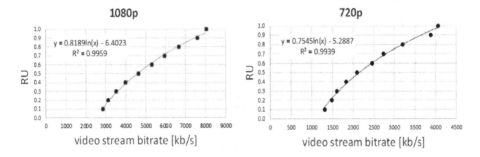

Fig. 3. Bit rate vs. RU for the analyzed video sequence

the number of the OPs, a client can receive in a particular state of the network to all the available OPs for the received video content. RU (Relative Utilization) indicator is therefore a measure of the relative use of video data in the context of maximizing the quality of the received content. For a video containing 10 OPs (case shown in Fig. 2), the relationship between RU and bit rate of the video stream is shown in Fig. 3.

The relationship between bitrate and RU is nonlinear. As it is shown in Fig. 3, with increasing bandwidth also increases the value of the indicator RU. Also, the gain in video quality (represented by increasing RU values) is gradually saturated. This relationship defines a new utility function whose usefulness will be reviewed later in this article. In order to use this utility function for the QoE-oriented fairness control (presented in the next section), it is necessary to conduct the curve fitting. For video content used as the example, the best results were obtained by the method of logarithmic regression. The final parameters of this fitting procedure are also shown in Fig. 3. The model that describes the utility function is not critical in our approach. It is only necessary for deriving a set of parameters required by proposed control method. On the other hand, it should be strongly emphasized another feature of the proposed utility function. The parameters (shape) of the proposed utility function obviously depends on the characteristics of the video. Unlike other solutions [24], all the data needed to define it can be obtained on the basis of the information contained in the MDF file. This file contains descriptions of all video representations (the number of available OPs) and the stream bitrate for each of them. This means that after the MDF file is downloaded, it is possible to determine the RU and to create proposed utility function. There is no need for a separate, dedicated database, which contains the parameters of all the offered movies and also there is no need to create custom methods for transmitting this information to the DASH clients. This allows for simple implementation of the proposed utility function in existing video streaming services.

3   QoE-Oriented Fairness Control for Optimization of Bandwidth Allocation in the System DASH

The utility function is the primary but not the only element necessary to ensure the QoE-oriented fairness control. It can be said that the utility function is only the model required to obtain the set of bitrates that ensures QoE fairness across all DASH clients. The problem to be solved, can be defined as follows. Let UF_i be a utility function and x_i the bitrate for i-th video stream. Also, let BT represents the total amount of available bandwidth for all streams. We need to find the optimum set of bitrates for all video streams (all DASH receivers). This means the need to calculate the value of Y_{optimal} satisfying the Formula (1).

$$Y_{\text{optimal}} = \max[\min_i(UF_i(x_i))]   \Sigma(x_i) \leq BT. \tag{1}$$

In the case of continuous and strictly increasing function UF_i, the solution is relatively easy to find [16,17]. Unfortunatelly, in the DASH system, video content is available only at a finite set of OPs. Therefore, it is necessary to apply the algorithm for finding solutions to optimization problem in which the variables must be integers. To solve our problem we chose an integer-programming algorithm which belongs to the branch and bound family of algorithms. It was proposed by Dakin in [25]. It assumes that an algorithm exists for finding solution to problem in a continuous domain with the addition of upper or lower bounds on any of the integer variables. Based on the information provided in the previous parts of the article, we can say that our problem meets these conditions. Unfortunately, the algorithm provides the optimum bitrate values which do not correspond with the bitrates for the available OPs. For this reason, each calculated theoretical values of bitrate must be changed to the nearest, lower value of the bitrate associated with the particular OP. The proposed control method consists of four steps. First three of them are mandatory, the last one is optional (in this regard, we plan separate tests). These steps are as follows:

- MDF file downloads, construction of utility function,
- finding solutions to optimization problem (Formula (1)),
- downgrading the optimal, theoretical bitrates to the bitrate (lower) of the closest OP,
- redistribution of the available bandwidth that remains after reducing the bitrate.

4   Test Scenario and Evaluation

Video material that was used in all test scenarios was the animated movie "Elephants Dreams" [22]. Three representations of the video were available on the server DASH, with resolutions 1080p, 720p and 360p, respectively. All of them have been encoded by the codec H264SVC and divided into chunks with a duration of 20 s each. The structure of the SVC GOP was created according to T-P method and consists of 10 OPs. The MDF file also was placed on the server DASH. It described the parameters of the prepared video. The values of the bitrate for OP1 and for each resolution were as follows: 8140 kb/s for 1080p, 4060 kb/s for 720p and 1620 kb/s for 360p.

4.1   Testbed Configuration

In order to assess the functional characteristics of the proposed control method, we implemented a simple network environment. Its fundamental element is a network controller. The controller should have the following characteristics:

– It should has knowledge of all video streams, which are ordered or received in the network (MDF file contents, the maximum resolution and bitrate of the video stream supported by each of the receivers).
– It should be able to monitor the amount of free bandwidth and should know the limits imposed for a given service or a particular receiver.
– It should be able to delegate to individual clients information about the parameters of video representation, what should be downloaded in the next step of the adaptation process (DASH receivers do not make an independent evaluation of the network status and do not perform individually the adaptation process).

These features offer a new approach to the control of the packet forwarding, which is known as SDN (Software Defined Networks) [26]. SDN networks allow for separation of network resource management plane from the packet forwarding plane. The management plane is implemented as a dedicated software, called controller SDN. It has two interfaces for communication with outer systems. The northbound interface is used to exchange data with other controllers or external applications (in our case, receivers DASH). On the other hand, the transmission of packet-forwarding rules from the controller to physical switching devices passes through the southbound interface. This interface is also used to collect data on the status of individual devices and to monitor network parameters. A more detailed discussion of the use of SDN in conjunction with the DASH we presented in [27].

Considering the above facts, we decided that he fundamental part of the testbed was SDN controller. It has been implemented on the basis of Open-Daylight [28]. The SDN controller supervised a network switch that supports the OpenFlow protocol (in our case it was Linksys WRTG54L with Pantou installed). DASH Server was implemented based on DASH-JS library [22]. The

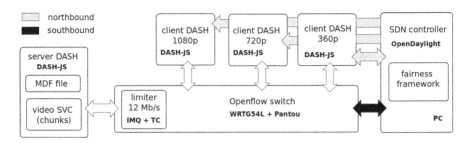

Fig. 4. The structure of the testbed

network included three receiver DASH with a maximum playback resolution, respectively: 1080p (DASH1), 720p (DASH2), and 360p (DASH3). All DASH clients have been also implemented based on DASH-JS. Throughput on the link to the receiving network was limited to 12 Mb/s. The structure of the testbed is illustrated in Fig. 4.

4.2 Test Results

Three tests were planned for the functional assessment of the proposed control method. Each of the test scenarios assumed to stream video for 4 min. DASH1 has been receiving video data during the whole test period. The DASH2 began receiving after 60 s and finally, the DASH3 after 120 s. In the first scenario (Test A), we used unmodified DASH clients and the standard network switch. Next, we configured the SDN controller to manage the allocation of the available bandwidth equally between all three receivers DASH (Test B). The last test scenario consisted of the use of utility function based on the hierarchical structure of SVC streams and proposed control method (Test C). The values of the video stream bitrate which was received by each client DASH, collected during subsequent tests, are shown in Fig. 5. In addition, this figure includes information about the transitions between different OPs that have taken place during each test.

Standard adaptation algorithm (Test A), which was implemented in the DASH clients led to the highest bitrate instability of the transmitted video. All three clients receive a video stream, whose internal structure has changed almost every 20 s (with each new decision about getting another chunk). On the basis of our subjective assessment of the quality of the received video we can conclude that it significantly affected the perceived video quality. Deterioration has been particularly noticeable for DASH1 (1080p) client. Frequent skipping between the OPs is not the only registered disadvantage of this process of adaptation. A definite winner of the competition for the available bandwidth turned out to be DASH3, the receiver with the lowest performance. In turn, HD receiver (DASH1) had to settle for video stream OP4 (5950 kb/s). This is two thirds of the maximum bit rate, which was 8040 kb/s. An even greater reduction in the bitrate for this receiver took place during the test B. In this case, after 4 min of receiving the video stream, the bitrate has the value for OP7 (3980 kb/s). The

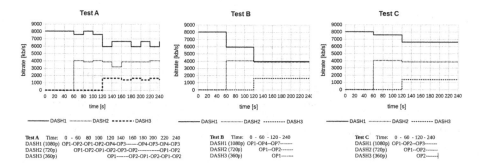

Fig. 5. The values of the video stream bitrate, for each client DASH and the transitions between different OPs that have taken place during each test

parameters of the video stream for the other receivers "happily" fit in rigid limits of the available bandwidth and are therefore not (or almost not) affected. The allocation of the available bandwidth equally between all three DASH clients greatly increases the stability of the parameters of the received video stream. Unfortunately, due to the fact that in this method the receivers with the highest requirements are most strongly affected, it is not suitable for practical use. Comparing the results of the test A and test B with our solution of QoE-oriented fairness control (test C), we can conclude that our proposal is the best one among all three. The final values of the bitrate for all three video streams are comparable with test A. However, it is clearly noticeable improvement in the stability of transmission parameters. In addition, all three receivers fairly shared the quality loss caused by the lack of sufficient bandwidth.

5   Conclusion and Future Work

The main element of the proposed method of the QoE-oriented fairness control for DASH systems is the utility function. It is based on a hierarchical structure of a SVC video stream. This function is described in Sect. 2 and its usage does not require any changes to the rules of the DASH service. The implementation of the test environment described in Sect. 4, shows the usefulness of SDN and OpenFlow protocol for the implementation of complex algorithms for the management of real-time network traffic. The conducted tests confirmed that the proposed method of the fairness control is superior to the standard rules of adaptation DASH and it offers increased stability of parameters of the received video. In subsequent studies, we plan to evaluate the behavior of the proposed solution in the presence of a background traffic of a different characteristics (generated by other real-time applications as well as elastic applications) and on the basis of a larger number of video sequences with different motion characteristics (slow motion, action scenes and high dynamic sequences). Also, we are going to conduct separate tests of algorithms for the step 4 in the proposed QoE-oriented control method.

References

1. Shenker, S.: Fundamental design issues for the future Internet. IEEE J. Sel. Areas Commun. **13**(7), 1176–1188 (1995)
2. ISO/IEC 23009–1:2014, Information technology – Dynamic adaptive streaming over HTTP (DASH) – Part 1: Media presentation description and segment formats. http://standards.iso.org/ittf/PubliclyAvailableStandards/c065274_ISO_IEC_23009-1_2014.zip
3. Reichel, J., Schwarz, H., Wien, M., Vieron, J.: Joint Scalable Video Model 9 of ISO/IEC 14496–10:2005/AMD3 Scalable Video Coding. Joint Video Team (JVT), Docs. JVT-X202 (2007)
4. Sieber, C., Hossfeld, T., Zinner, T., Tran-Gia, P., Timmerer, C.: Implementation and user-centric comparison of a novel adaptation logic for DASH with SVC. In: IFIP/IEEE International Symposium on Integrated Network Management, pp. 1318–1323. IEEE Press, Ghent (2013)
5. Kalva, H., Adzic, V., Furht, B.: Comparing MPEG AVC and SVC for adaptive HTTP streaming. In: IEEE International Conference on Consumer Electronics, pp. 158–159. IEEE Press, Las Vegas (2012)
6. Cermak, G., Pinson, M., Wolf, S.: The relationship among video quality, screen resolution, and bit rate. IEEE Trans. Broadcast. **57**, 258–262 (2011)
7. Daronco, L.C., Roesler, V., Lima, J.V.: Subjective video quality assessment applied to scalable video coding and transmission instability. In: ACM Symposium on Applied Computing, pp. 1898–1904, ACM, New York (2010)
8. Stockhammer, T.: Dynamic adaptive streaming over HTTP - Standards and design principles. In: Second Annual ACM Conference on Multimedia Systems, pp. 133–144. ACM, San Jose (2011)
9. Akhshabi, S., Begen, A., Dovrolis, C.: An experimental evaluation of rate-adaptation algorithms in adaptive streaming over HTTP. In: Second Annual ACM Conference on Multimedia Systems, pp. 157–168. ACM, San Jose (2011)
10. Huang, T.-Y., Handigol, N., Heller, B., McKeown, N., Johari, R.: Confused, timid, and unstable: picking a video streaming rate is hard. In: ACM Conference on Internet Measurement, pp. 225–238. ACM, Boston (2012)
11. Tian, G., Liu, Y.: Towards agile and smooth video adaptation in dynamic HTTP streaming. In: 8th International Conference on Emerging Networking Experiments and Technologies, pp. 109–120. ACM, Nice (2012)
12. Mok, R.K.P., Luo, X., Chan, E.W.W., Chang, R.K.C.: QDASH: a QoE-aware DASH system. In: 3rd Multimedia Systems Conference, pp. 11–22. ACM, Chapel Hill (2012)
13. Jiang, J., Sekar, V., Zhang, H.: Improving fairness, efficiency, and stability in http-based adaptive video streaming with festive. IEEE/ACM Trans. Network. **22**(1), 326–340 (2014)
14. Mansy, A., Ver Steeg, B., Ammar, M.: SABRE: a client based technique for mitigating the buffer bloat effect of adaptive video flows. In: 4th ACM Multimedia Systems Conference, pp. 214–225. ACM, Oslo (2013)
15. Kelly, F.P.: Charging and rate control for elastic traffic. Eur. Trans. Telecommun. **8**, 33–37 (1997)
16. Marbach, P.: Priority service and max-min fairness. IEEE/ACM Trans. Network. **11**(10), 733–746 (2003)
17. Wang, W.-H., Palaniswami, M., Low, S.H.: Application-oriented flow control: fundamentals, algorithms and fairness. IEEE/ACM Trans. Network. **14**(6), 1282–1291 (2006)

18. Yin, X., Jindal, A., Sekar, V., Sinopoli, B.: A control-theoretic approach for dynamic adaptive video streaming over HTTP. SIGCOMM Comput. Commun. Rev. **45**(5), 325–338 (2015)
19. Unanue, I., Urteaga, I., Husemann, R., Del Ser, J., Roesler, V., Rodriguez, A., Sanchez, P.: A tutorial on H264/SVC scalable video conding and its tradeoof between quality, coding efficiency and performance. In: Lorente, J.D.S. (ed.) Recent Advances on Video Coding, Chap. 1. InTech, Rijeka (2011)
20. Gupta, R., Pulipaka, A., Seeling, P., Karam, L.-J., Reisslein, M.: H.264 coarse grain scalable (CGS) and medium grain scalable (MGS) encoded video: a trace based traffic and quality evaluation. IEEE Trans. Broadcast. **58**(3), 428–439 (2012)
21. Przylucki, S., Sierszeń, A.: Gradation of video quality in DASH services using the H264/SVC coding. Image Process. Commun. **20**, 15–24 (2016)
22. ITEC – Dynamic Adaptive Streaming over HTTP. http://www-itec.uni-klu.ac.at/dash/
23. Joint Scalable Video Model – reference software. http://www.hhi.fraunhofer.de/departments/video-coding-analytics/research-groups/image-video-coding/research-topics/svc-extension-of-h264avc/jsvm-reference-software.html
24. Georgopoulos, P., Elkhatib, Y., Broadbent, M., Mu, M., Race, N.: Towards network-wide QoE fairness using openflow-assisted adaptive video streaming. In: ACM SIGCOMM Workshop on Future Human-Centric Multimedia Networking, pp. 15–20. ACM, Hong Kong (2013)
25. Dakin, R.J.: A tree-search algorithm for mixed integer programming problems. Comput. J. **8**(3), 250–255 (1965)
26. ONF: Open Networking Foundation Software-Defined Networking (SDN) Definition. https://www.opennetworking.org/sdn-resources/sdn-definition
27. Przylucki, S., Czerwinski, D.: The ability to ensure a fair distribution of network resources for video streaming services based on DASH standard. Studia Informatica **36**(2), 73–82 (2015)
28. Linux Foundation, OpenDaylight. https://www.opendaylight.org/lithium

The Impact of the Degree of Self-Similarity on the NLREDwM Mechanism with Drop from Front Strategy

Adam Domański[1]([✉]), Joanna Domańska[2], and Tadeusz Czachórski[2]

[1] Institute of Informatics, Silesian Technical University,
Akademicka 16, 44-100 Gliwice, Poland
adamd@polsl.pl
[2] Institute of Theoretical and Applied Informatics,
Polish Academy of Sciences, Baltycka 5, 44-100 Gliwice, Poland
{joanna,tadek}@iitis.gliwice.pl

Abstract. This paper examines the impact of the degree of self-similarity on the selected AQM mechanisms. During the tests we analyzed the length of the queue, the number of rejected packets and waiting times in queues. We use fractional Gaussian noise as a self-similar traffic source. The quantitative analysis is based on simulation.

Keywords: Self-similarity · Active queue management · Non-linear RED · Dropping packets

1 Introduction

The development of the Internet is partially based on new solutions for traffic control to improve the Quality of Service (QoS) provided at the network layer. Among others, the studies are related to real-time applications such as Voice over IP or Video on Demand. To ensure QoS, the Internet Engineering Task Force (IETF) has proposed *Integrated Services* (IntServ) and *Differentiated Services* (DiffServ) architectures. They include a number of mechanisms, in particular for queue management in routers and the efficiency of the TCP protocol depends largely on them. Queue management may be passive or active. In passive solutions, packets coming to a buffer are rejected only if there is no space in the buffer to store them, hence the senders have no earlier warning on the danger of the increasing congestion and all packets coming during saturation of the buffer are lost.

To enhance the throughput and fairness of a link sharing, also to eliminate the synchronization, the IETF recommends active algorithms of buffer management (Active Queue Management, AQM) [1]. They incorporate mechanisms of preventive packet dropping already when there is still place to store packets, this way advertising that the queue is increasing and the danger of congestion is ahead. The packets are dropped randomly, hence only certain users are notified and the global synchronization of connections is avoided. The probability of packet rejection is increasing with the level of congestion.

© Springer International Publishing Switzerland 2016
P. Gaj et al. (Eds.): CN 2016, CCIS 608, pp. 192–203, 2016.
DOI: 10.1007/978-3-319-39207-3_17

The basic active queue management algorithm is Random Early Detection (RED) algorithm. It was primarily proposed in 1993 by Sally Floyd and Van Jacobson [2]. Its performance is based on a drop function giving probability that a packet is rejected. The argument avg of this function is a weighted moving average queue length determined at arrival of a packet:

$$avg = (1 - w_q)avg' + w_q q$$

where q is the current queue length, avg' is the previous value of avg and w_q is a weight parameter, typically $w_q \ll 1$, thus avg varies much more slowly than q. Therefore avg indicates long-term changes of q. If $avg < Min_{th}$, all packets are admitted. If $Min_{th} < avg < Max_{th}$, then dropping probability p is increasing linearly:

$$`p = p_{max}\frac{avg - Min_{th}}{Max_{th} - Min_{th}}.$$

The value p_{max} corresponds to a probability of packet rejecting at $avg = Max_{th}$. If $avg > Max_{th}$ then all packets are dropped. Dropping probability p is thus dependent on network load.

Efficient operation of the RED mechanism depends on the proper selection of its parameters. There were several works on the impact of various parameters on the RED performance [3] and many variants of RED mechanism were developed to improve its performance [4–6]. They may be classified according to the dropping packet function and according to the parameters of the algorithm. Section 2 briefly reviews the modifications of the RED mechanism studied in this article.

Research related to the Internet traffic aims to provide a better understanding of the modern Internet, inter alia, by presenting the current characteristics of Internet traffic based on a large number of experimental data and introducing the internet traffic models. The understanding of the traffic nature of the modern Internet is important to the Internet community. It supports optimization and development of protocols and network devices, improves the network applications security and the protection of network users.

Measurements and statistical analysis (performed already in the 90s) of packet network traffic show that this traffic displays a complex statistical nature including self-similarity, long-range dependence and burstiness [7–10].

Self-similarity of a process means that the change of time scale does not influence the statistical characteristics of the process. It results in long-distance auto-correlation and makes possible the occurrence of very long periods of high (or low) traffic intensity. These features have a great impact on a network performance [11]. They enlarge mean queue lengths at buffers and increase the probability of packet loss, deteriorating this way the quality of services provided by a network.

In consequence, it is needed to propose new or to adapt known types of stochastic processes able to model these negative phenomena in network traffic. Several models have been introduced to imitate self-similar processes in the network traffic. They use fractional Brownian Motion, chaotic maps, fractional Autoregressive Integrated Moving Average (fARIMA), wavelets and multifractals, and processes based on Markov chains: SSMP (Special Semi-Markov

Process), MMPP (Markov-Modulated Poisson Process) [12], HMM (Hidden Markov Model) [13]. Section 3 briefly describes the self-similar traffic source used in this article.

2   Our Modifications of the RED Mechanism

Our previous works [14–16] presented a study of the influence of RED modifications on its performance in the presence of self-similar traffic.

In classic RED and its variations described in the literature a packet to be dropped is taken usually from the end of the queue. As Sally Floyd wrote: "when RED is working right, the average queue size should be small, and it shouldn't make too much difference one way or another whether you drop a packet at the front of the queue or at the tail". Our article [14] reconsiders the problem of choosing tail or front packets in presence of self-similar traffic.

It was shown that in the case of light non-self-similar traffic the obtained results confirmed the opinion of S. Floyd. If the mean queue length is relatively low, the influence of dropping scheme on queueing time is negligible: the introduction of drop-front strategy gives less then 1 % shorter mean queueing time. In the case of heavy traffic, drop from front strategy gives two times shorter mean queueing times. However, when the Poisson traffic is replaced by self-similar one with the same intensity and the same parameters of RED are preserved, the length of the queue increases and the influence of the dropping scheme is more visible: drop from front strategy reduces mean queueing time by about 16 % even in the case of light load. This fact confirms the advantage of drop from front strategy if the traffic exhibits the self-similarity.

In [15] we investigated the influence of the self-similarity on the non-linear packet dropping function in a special case of NLRED queues. In the NLRED mechanism the linear packet dropping function is usually replaced by a quadratic function. We introduced a linear combination of independent polynomials of 3rd degree:

$$p(x, a_1, a_2, p_{\max}) = \begin{cases} 0 & \text{for} \quad x < Min_{\text{th}} \\ \varphi_0(x) + a_1\varphi_1(x) + a_2\varphi_2(x) & \text{for} \quad Min_{\text{th}} \leq x \leq Max_{\text{th}} \\ 1 & \text{for} \quad x > Max_{\text{th}} \end{cases}$$

where the set of basis function is defined as follows:

$$\varphi_0(x) = p_{\max} \frac{x - Min_{\text{th}}}{Max_{\text{th}} - Min_{\text{th}}},$$
$$\varphi_1(x) = (x - Min_{\text{th}})(Max_{\text{th}} - x),$$
$$\varphi_2(x) = (x - Min_{\text{th}})^2(Max_{\text{th}} - x).$$

The process of finding the best values of $p_{\max}$, a_1 and a_2 for a given type of traffic may be considered as optimization problem in 3-dimensional space. The experimental results show the existence of one optimal set of values of parameters; self-similarity of network traffic and traffic load have no influence on the choice of the optimal dropping packet function. The results obtained for this optimal set of parameter values demonstrate that the mean waiting time is two and half times shorter compared to the RED mechanism in the case of non-self-similar traffic and it is four times shorter in the case of self-similar traffic.

Then in [16] we investigated the impact of the way how the weighted moving average is defined on the performance of the RED mechanism in the presence of self-similar traffic. The proposed approach, named REDwM, is an extension of RED where the average queue length $A(n)$ at a moment n is given by the first order difference equation

$$A(n) = a_1 A(n-1) + a_2 A(n-2) + \ldots + a_k A(n-k)$$
$$+ b_0 Q(n) + b_1 Q(n-1) + \ldots + b_m Q(n-m)$$

where a_j $(j = 1, \ldots, k)$ and b_i $(i = 0, \ldots, m)$ are constant, $A(l)$ is the average queue length at the l-th moment of time, and $Q(l)$ is the current length of the packet queue at the l-th moment; a_j and b_i are subject to constraints:

$$\sum_{j=1}^{k} a_j + \sum_{i=0}^{m} b_i = 1 \ \land \ a_j \geq 0 \ \land \ b_i \geq 0.$$

The optimal values of equation coefficients were found during minimization of the score function. The improvements, following numerical experiments, are over 5 % if we refer to results given by the classic RED approach (for the assumed score function based on the mean waiting time).

The improvements of the RED mechanism described above may be combined making NLREDwM mechanism. The primary goal of this article is to study its performance.

3   Self-Similar Traffic Source

Previously in [14–16] we used the SSMP markovian traffic source to represent the self-similar traffic. Such Markov based model can generate a self-similar traffic over a finite number of time scales. Here we use fractional Gaussian noise which is an exactly self-similar traffic source.

Fractional Gaussian noise (fGn) has been proposed in [17] as a model for the long-range dependence postulated to occur in a variety of hydrological and geophysical time series. Nowadays, fGn is one of the most commonly used self-similar processes in network performance evaluation [18] and the only stationary Gaussian process being exactly self-similar.

The autocorrelation function of fGn process [7]

$$\rho^{(m)}(k) = \rho(k) = \frac{1}{2} \left[(k+1)^{2H} - 2k^{2H} + (k-1)^{2H} \right]$$

assures second-order self-similarity.

Deviation

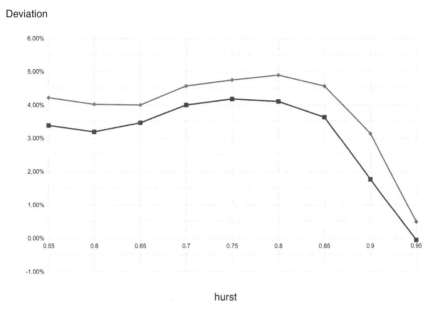

hurst

Fig. 1. Maximum and minimum difference between assumed and estimated Hurst parameter

The synthetic generation of sample paths (traces) of self-similar processes is an important problem [18]. In this paper we use a fast algorithm for generating approximate sample paths for a fGn process, first introduced in [19].

The Hurst parameter H characterizes a process in terms of the degree of self-similarity, the degree increases with the increase of H. Thge value $H \leq 0.5$ denotes the lack of long-range dependence, but the process is still self-similar, [20]. We have generated the sample traces with the Hurst parameter with the range of 0.5 to 0.95. After each trace generation, the parameter was estimated with the use of aggregated variance method [21]. Table 1 presents results of this estimation for 10 generated traces with the Hurst parameter assumed to be equal to 0.7. These results show that the assumed and estimated Hurst parameters are not the same. This situation repeated for each value of Hurst parameter (see Fig. 1).

Table 1. Estimated Hurst parameters obtained for sample fGn traces generated for assumed Hurst parameter $H = 0.7$

Trace number	Estimated Hurst parameter	Trace number	Estimated Hurst parameter
1	0.7279822	6	0.73197
2	0.7299411	7	0.7311628
3	0.7288566	8	0.7291909
4	0.731594	9	0.7290085
5	0.7313482	10	0.7284157

Table 2. FIFO queue

	Hurst parameter	Mean queue length	Mean waiting time	Rejected packets	
Tail drop	0.50	299.099	119.380	249520	49.90 %
Front drop	0.50	299.089	119.223	249494	49.89 %
Tail drop	0.70	298.118	119.158	249879	49.97 %
Front drop	0.70	298.132	119.118	250034	50.01 %
Tail drop	0.80	296.878	118.883	250354	50.07 %
Front drop	0.80	296.870	118.772	250383	50.08 %
Tail drop	0.90	248.553	102.061	256587	51.32 %
Front drop	0.90	247.848	101.399	255954	51.19 %

fGn generator usually generates the traffic with a slightly greater Hurst parameter. The difference between assumed and estimated Hurst parameter decreases with the increase of the value of Hurst parameter. For our purpose we chose the samples with the smallest difference.

4   Obtained Results

The simulation model of an appropriate AQM mechanism was prepared with the use of SimPy. SimPy is a process-based discrete-event simulation framework based on the language Python. Its event dispatcher is based on Python's generators [22]. SimPy is released under the MIT License (free software license originating at the Massachusetts Institute of Technology).

We investigated the influence of combination of modifications described in Sect. 2 on RED performance. We also studied the impact of the degree of self-similarity on the examined AQM mechanisms. During the tests we analyzed the following parameters of the transmission with AQM: the length of the queue, queue waiting times and the number of rejected packets. The service time represented the time of a packet treatment and dispatching. The input process, following fGn was based on descrete time slots (1 or 0 arrivals in a time slot), the average interarrival time was 2 time slots. The size of this time slot was our symbolic time unit presented in figures. The service time was geometrically distributed with the average of 4 time slots. That means a heavy charge, leading to the link saturation, as our goal was to study the mechanisms performance at high load intervals. The type of the distributions and their mean values were the same for all Hurst parameters.

Table 2 shows the results obtained for the FIFO queue without any AQM mechanism. The introduction of *drop from front strategy* gives shorter mean waiting time compared to *drop tail strategy*. Additionally, an increase in the degree of self-similarity causes an increase in number of rejected packets. The

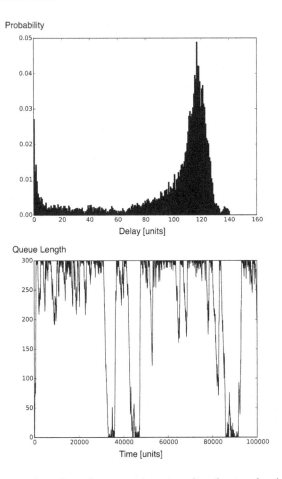

Fig. 2. FIFO queue – drop from front, waiting time distribution (top), fluctuations of queue length (bottom), Hurst parameter $H = 0.90$

reason of it is bursty nature of self-similar traffic. Figure 2 shows fluctuations of queue length in the case of $H = 0.9$.

The same results were obtained in case of the RED queue (see Table 3). The RED parameters were: buffer size 300 packets, threshold values $Min_{th} = 100$ and $Max_{th} = 200$, $p_{max} = 0.1$, $w = 0.02$. We distinguish packets rejected when RED starts dropping packets and packets rejected when the walking average of the queue is at the maximum threshold. Table 3 shows that majority of packets were rejected when the average is at the maximum threshold. The number of packets rejected because of reaching the maximum threshold increased with the increase of Hurst parameter.

Tables 4 and 5 show the results obtained for two sets of parameter values of our NLRED mechanism (described in Sect. 2). The first set of values of parameters (Table 4) refers to the case with the minimal value of average waiting time

Table 3. RED queue

	Hurst parameter	Mean queue length	Mean waiting time	No. of rejected packets			
				$\leq Max_{th}$		$> Max_{th}$	
Tail drop	050	199.801	79.865	27616	5.52 %	222436	44.48 %
Front drop	050	199.841	79.477	27083	5.41 %	222318	44.46 %
Tail drop	070	198.549	79.485	26948	5.38 %	223461	44.69 %
Front drop	070	198.524	79.079	27230	5.44 %	222519	44.50 %
Tail drop	080	196.941	78.473	26805	5.36 %	222453	44.49 %
Front drop	080	196.864	78.604	26491	5.29 %	223819	44.76 %
Tail drop	090	158.818	66.119	19108	3.82 %	240991	48.19 %
Front drop	090	158.541	65.797	19487	3.89 %	240405	48.08 %

at the expense of the number of rejected packets (the best case). The second set (Table 5) refers to the case with the maximal value of average waiting time (the worst case). The results obtained for the worst case resemble those obtained for the classical RED mechanism. In the best case all packets are rejected when RED starts dropping packets. In this case the mean waiting time is 1.77 times shorter compared to the RED mechanism (in the case of $H = 0.9$). The results presented in both tables show the impact of degree of self-similarity on mean queue length, mean waiting time and number of rejected packets. Figures 3 and 4 compare the waiting time distributions and queue length fluctuations (the best case of NLRED) for non-self-similar traffic to the case of self-similar traffic with $H = 0.9$.

Table 4. NLRED queue; $a1 = 0.00042$, $a2 = -0.0000038$, $p_{\max} = 0.855$

	Hurst parameter	Mean queue length	Mean waiting time	No. of rejected packets		
				$\leq Max_{th}$		$> Max_{th}$
Tail drop	050	112.3725	44.8316	249846	49.96 %	0
Front drop	050	112.4435	44.6774	249943	49.98 %	0
Tail drop	070	111.3998	44.6255	250865	50.17 %	0
Front drop	070	111.3741	44.2123	249685	49.93 %	0
Tail drop	080	109.9958	43.8370	249588	49.91 %	0
Front drop	080	109.9371	43.6606	249788	49.95 %	0
Tail drop	090	87.7811	36.8922	264244	52.84 %	0
Front drop	090	87.7484	37.0343	264662	52.93 %	0

Table 5. NLRED queue; $a1 = -0.00008$, $a2 = -0.0000008$, $p_{\max} = 0.6$

	Hurst parameter	Mean queue length	Mean waiting time	No. of rejected packets			
				$\leq Max_{th}$		$> Max_{th}$	
Tail drop	050	194.1852	77.8825	227407	45.48 %	23477	4.69 %
Front drop	050	194.1163	77.3081	227516	45.50 %	22249	4.44 %
Tail drop	070	191.3899	76.4370	201347	40.26 %	48478	9.70 %
Front drop	070	191.3007	76.2139	200608	40.12 %	49243	9.85 %
Tail drop	080	188.8292	75.4878	183403	36.68 %	66689	13.34 %
Front drop	080	188.8145	75.1376	184170	36.83 %	65392	13.08 %
Tail drop	090	152.5569	63.7271	143043	28.61 %	117863	23.57 %
Front drop	090	152.4339	63.4312	142710	28.54 %	118333	23.67 %

Table 6. NLREDwM mechanism; $a1 = 0.00042$, $a2 = -0.0000038$, $p_{\max} = 0.855$

	Hurst parameter	Mean queue length	Mean waiting time	No. of rejected packets			
				$\leq Max_{th}$		$> Max_{th}$	
Tail drop	050	112.4056	44.8221	249713	49.94 %	0	
Front drop	050	112.4650	44.8019	250599	50.12 %	0	
Tail drop	070	111.3513	44.3102	249209	49.84 %	0	
Front drop	070	111.3267	44.1467	249418	49.88 %	0	
Tail drop	080	109.9675	43.7743	249306	49.86 %	0	
Front drop	080	110.1168	43.7970	250160	50.03 %	0	
Tail drop	090	87.5870	37.1751	264962	52.99 %	0	
Front drop	090	87.4502	36.8666	264395	52.88 %	0	

Table 7. NLREDwM mechanism; $a1 = -0.00008$, $a2 = -0.0000008$, $p_{\max} = 0.6$

	Hurst parameter	Mean queue length	Mean waiting time	No. of rejected packets			
				$\leq Max_{th}$		$> Max_{th}$	
Tail drop	050	194.2196	77.5939	228135	45.63 %	21802	4.36 %
Front drop	050	194.1596	77.3485	226814	45.36 %	23049	4.61 %
Tail drop	070	191.4063	76.5439	201466	40.29 %	48687	9.74 %
Front drop	070	191.4217	76.4816	201918	40.38 %	48654	9.73 %
Tail drop	080	188.8828	75.3979	184848	36.97 %	64876	12.98 %
Front drop	080	188.8172	75.1325	184379	36.88 %	65160	13.03 %
Tail drop	090	152.4997	63.7525	141616	28.32 %	119465	23.89 %
Front drop	090	152.3294	63.4188	142388	28.48 %	118311	23.66 %

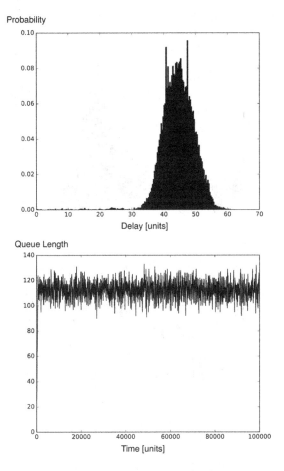

Fig. 3. NLRED – tail drop, waiting time distribution (top), fluctuations of queue length (bottom), $H = 0.50$, $a1 = 0.00042$, $a2 = -0.0000038$, $p_{max} = 0.855$

Tables 6 and 7 show the results obtained for NLREDwM mechanism (best and worse case), which is a combination of our NLRED and REDwM mechanisms (described in Sect. 2). The introduction of modified weighted moving average function gives about 0.1 % of changes compared to NLRED. This improvement increases with the increase of Hurst parameter.

5   Conclusions

The article confirms the important impact of the degree of self-similarity (expressed in terms of Hurst parameter) on the following parameters of the transmission with AQM: the length of the queue, queue waiting times and the number of rejected packets. We discuss the problem of choosing the optimal shape of dropping packet function for NLRED algorithm and at the same time investigate the influence of the weighted moving average on packet waiting time reduction for this NLRED mechanism.

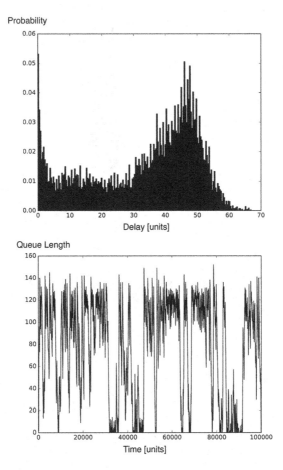

Fig. 4. NLRED – tail drop, waiting time distribution (top), fluctuations of queue length (bottom), $H = 0.90$, $a1 = 0.00042$, $a2 = -0.0000038$, $p_{max} = 0.855$

Drop from front strategy, when applied in place of tail drop one, results in reduction of packet waiting time in examined AQM mechanism. Obtained results are closely related to the level of self-similarity. Hence the application of presented AQM mechanism may be recommended for bursty traffic connections with real-time requirements.

References

1. Braden, B., et al.: Recommendations on queue management and congestion avoidance in the internet. RFC 2309, IETF (1998)
2. Floyd, S., Jacobson, V.: Random early detection gateways for congestion avoidance. IEEE/ACM Trans. Network. **1**(4), 397–413 (1993)
3. May, M., Diot, C., Lyles, B., Bolot, J.: Influence of active queue management parameters on aggregate traffic performance. Research Report, INRIA (2000)
4. Ho, H.-J., Lin, W.-M.: AURED - Autonomous random early detection for TCP congestion control. In: 3rd International Conference on Systems and Networks Communications, Malta (2008)

5. Bhatnagar, S., Patro, R.: A proof of convergence of the B-RED and P-RED algorithms for random early detection. IEEE Commun. Lett. **13**(10), 809–811 (2009)
6. Domański, A., Domańska, J., Czachórski, T.: Comparison of CHOKe and gCHOKe active queues management algorithms with the use of fluid flow approximation. In: Kwiecień, A., Gaj, P., Stera, P. (eds.) CN 2013. CCIS, vol. 370, pp. 363–371. Springer, Heidelberg (2013)
7. Karagiannis, T., Molle, M., Faloutsos, M.: Long-range dependence: ten years of internet traffic modeling. IEEE Internet Comput. **8**(5), 57–64 (2004)
8. Domański, A., Domańska, J., Czachórski, T.: The impact of self-similarity on traffic shaping in wireless LAN. In: Balandin, S., Moltchanov, D., Koucheryavy, Y. (eds.) NEW2AN 2008. LNCS, vol. 5174, pp. 156–168. Springer, Heidelberg (2008)
9. Domańska, J., Domański, A., Czachórski, T.: A few investigation of long-range dependence in network traffic. In: Czachórski, T., Gelenbe, E., Lent, R. (eds.) Information Science and Systems 2014, pp. 137–144. Springer, Heidelberg (2014)
10. Domańska, J., Domański, A., Czachórski, T.: Estimating the intensity of long-range dependence in real and synthetic traffic traces. In: Gaj, P., Kwiecień, A., Stera, P. (eds.) CN 2015. CCIS, vol. 522, pp. 11–22. Springer, Heidelberg (2015)
11. Domańska, J., Domański, A.: The influence of traffic self-similarity on QoS mechanism. In: International Symposium on Applications and the Internet, SAINT, Trento, Italy (2005)
12. Domańska, J., Domański, A., Czachórski, T.: Modeling packet traffic with the use of superpositions of two-state MMPPs. In: Kwiecień, A., Gaj, P., Stera, P. (eds.) CN 2014. CCIS, vol. 431, pp. 24–36. Springer, Heidelberg (2014)
13. Domańska, J., Domański, A., Czachórski, T.: Internet traffic source based on hidden markov model. In: Balandin, S., Koucheryavy, Y., Hu, H. (eds.) NEW2AN 2011 and ruSMART 2011. LNCS, vol. 6869, pp. 395–404. Springer, Heidelberg (2011)
14. Domańska, J., Domański, A., Czachórski, T.: The drop-from-front strategy in AQM. In: Koucheryavy, Y., Harju, J., Sayenko, A. (eds.) NEW2AN 2007. LNCS, vol. 4712, pp. 61–72. Springer, Heidelberg (2007)
15. Domańska, J., Augustyn, D.R., Domański, A.: The choice of optimal 3rd order polynomial packet dropping function for NLRED in the presence of self-similar traffic. Bull. Polish Acad. Sci., Tech. Sci. **60**(4), 779–786 (2012)
16. Domańska, J., Domański, A., Augustyn, D.R., Klamka, J.: A RED modified weighted moving average for soft real-time application. Int. J. Appl. Math. Comput. Sci. **24**(3), 697–707 (2014)
17. Mandelbrot, B.B., Ness, J.V.: Fractional brownian motions, fractional noises and applications. SIAM Rev. **10**, 422–437 (1968)
18. Lopez-Ardao, J.C., Lopez-Garcia, C., Suarez-Gonzalez, A., Fernandez-Veiga, M., Rodriguez-Rubio, R.: On the use of self-similar processes in network simulation. ACM Trans. Model. Comput. Simul. **10**(2), 125–151 (2000)
19. Paxson, V.: Fast, approximate synthesis of fractional Gaussian noise for generating self-similar network traffic. ACM SIGCOMM Comput. Commun. Rev. **27**(5), 5–18 (1997)
20. Samorodnitsky, G., Taqqu, M.S.: Stable Non-Gaussian Random Processes: Stochastic Models with Infinite Variance. Chapman and Hall, New York (1994)
21. Clegg, R.G.: A practical guide to measuring the Hurst parameter. Int. J. Simul. **7**(2), 3–14 (2006)
22. http://simpy.readthedocs.org

Teleinformatics and Telecommunications

Influence of Noise and Voice Activity Detection on Speaker Verification

Adam Dustor[✉]

Institute of Electronics, Silesian University of Technology,
Akademicka 16, 44-100 Gliwice, Poland
adam.dustor@polsl.pl

Abstract. The scope of this paper is to check influence of voice activity detection VAD procedure and its accuracy on speaker verification error rates. It is shown that for speech of high quality, it is absolutely necessary to remove silence from the signal as the errors increase radically. It is better to remove more than less from the signal as the equal error rate EER is the worst for the original speech with silence. Additionally influence of white noise, which was added to speech utterances, was examined. Presented results show that in order to achieve highly reliable speaker verification system it must be insensitive to low quality of speech, since noise is the most important factor responsible for high error rates.

Keywords: Biometrics · Security · Speaker verification · Voice activity detection

1 Introduction

Division of Telecommunication, a part of the Institute of Electronics and Faculty of Automatic Control, Electronics and Computer Science of the Silesian University of Technology, for many years specializes in speech and speaker recognition [1–11]. One of the results of conducted research is presented in this paper which is devoted to speaker verification.

Speaker recognition is the process of automatically recognizing who is speaking by analysis speaker-specific information included in spoken utterances. This process may be divided into identification and verification. The purpose of speaker identification is to determine the identity of an individual from a sample of his or her voice. The purpose of speaker verification is to decide whether a speaker is whom he claims to be.

The paper is organized in the following way. At first fundamentals of speaker verification are presented, next research procedure with obtained results are shown and discussed.

2 Speaker Verification

Typical structure of speaker verification system consists of feature extraction, similarity calculation, construction of speaker model and making an

© Springer International Publishing Switzerland 2016
P. Gaj et al. (Eds.): CN 2016, CCIS 608, pp. 207–215, 2016.
DOI: 10.1007/978-3-319-39207-3_18

accept/reject decision. Basic structure of speaker verification system is shown in Fig. 1. Speech signal is cut into short fragments, which usually last for 20–30 ms known as speech frames. Feature extraction is responsible for extracting from each frame a set of parameters known as feature vectors. Extracted sequence of vectors is then compared to speaker model by pattern matching. The purpose of pattern matching is to measure similarity between test utterance and speaker model. In verification the similarity between input test sequence and claimed model must be good enough to accept the speaker as whom he claims to be.

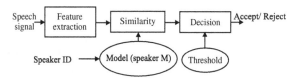

Fig. 1. Speaker verification scheme

Feature extraction is one of the most important procedures in speaker recognition. Applied parameters should be stable in time – physical and mental state of the speaker should have low impact on recognition performance which means low intraspeaker variability. Features should also posses high interspeaker variability in order to discriminate well between speakers. The most often applied features are parameters based on frequency spectrum of the speech like MFCC. These parameters are based on the nonlinear human perception of the frequency of sounds. They can be computed as follows: window the signal, take the FFT, take the magnitude, take the log, warp the frequency according to the mel scale and finally take the inverse FFT. Mel warping transforms the frequency scale to place less emphasis on high frequencies [12].

Speaker recognition is based on similarity calculation between test utterance and the reference model. As a result, the problem of construction of a good model is crucial. One of methods used to create voice model is based on vector quantization VQ. Speaker is represented as a set of vectors that possibly in the best way represent speaker. This set of vectors is called a codebook. In this case during recognition each test vector is compared with its nearest neighbour from the codebook and the overall distance for the whole test utterance is computed. Calculation of normalized distance D for M frames of speech is given by

$$D = \frac{1}{M} \sum_{i=1}^{M} \min(d(x_i, c_q)) \quad 1 \le q \le L, \tag{1}$$

where x_i is a test vector and c_q a code vector from a codebook of size L. As it can be seen for M frames and L code vectors its necessary to calculate ML distances. The most often used measure of similarity is an Euclidean distance

$$d(x_i, c_q) = \sum_{k=1}^{p} (x_i(k) - c_q(k))^2 \tag{2}$$

where p is a dimension of a vector. How to find the best codebook for speaker from a lot of training data? To solve this problem a kind of clustering technique is required, which can find a small set of the best representative vectors of a speaker. One of applied algorithms is k-means procedure.

K-means algorithm is an iterative procedure and consists of four major steps. At first arbitrarily choose L vectors from the training data, next for each training vector find its nearest neighbour from the current codebook, which corresponds to partitioning vector space into L distinct regions. The third step requires updating the code vectors using the centroid of the training vectors assigned to them and the last step – repeat steps 2 and 3 until some convergence criterion is satisfied. The convergence criterion is usually an average quantization error expressed in the same way as in (1) with an exception that x_i is a training vector.

3   Research Procedure

All research was done on Polish database ROBOT [13]. This database consists of 2 CD with 1 GB of speech data. The speech utterances were collected from 30 speakers of both sex in a several time-separated sessions to catch intraspeaker variability. Main specifications of ROBOT are the following: sampling frequency 22 kHz, language – Polish, quantization 16 bit, file format ".wav", lack of files compression, recording environment – quiet, each file is preceded and followed by the silence. Recorded utterances consist of the words belonging to three dictionaries (L1, L2, and L3). Words in L1 and L3 are numbers from 0 to 9 and 10 to 99 respectively. Dictionary L2 consists only from commands used in robot control (start, stop, left, right, up, down, drop, catch, angle). These dictionaries were used to construct seven different sets of utterances Z1...Z7. Set Z3 consists of 5 sentences – isolated numbers 04, 17, 49, 72, 93 (each sentence repeated 15 times in order to catch voice variability). Z4 is made of 11 sequences of numbers. Each sequence is a 3-element combination of numbers from Z3, e.g. 04-17-93, 49-72-93.

During training and testing of the speaker verification system the same signal processing procedure was used. Speech files, before feature extraction, were processed to remove silence. Voice activity detection was based on the energy of the signal. Next signal was preemphasized and segmented. Hamming windowing was applied. From each frame MFCC parameters were computed (18 features per segment).

There were 30 speaker models. Each model was trained by 75 utterances of the speaker. All training utterances came from set Z3 of ROBOT corpus. Text dependent speaker identification was implemented. Set Z4 was used for testing. The test utterances came from 30 speakers. Each speaker provided 11 test sequences of approximately 5 s each. There were 330 (30*11) genuine attempts (possible false rejection FR error) and 9570 (29*11*30) impostor attempts (possible false acceptance FA error, speaker has its own model in a system – seen impostor).

Verification performance was characterized in terms of the two error measures, namely the false acceptance rate FAR and false rejection rate FRR.

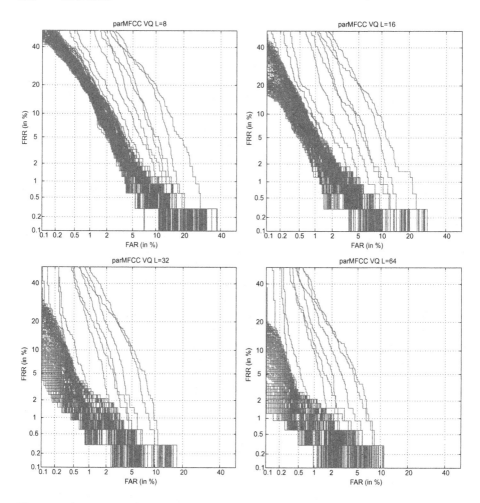

Fig. 2. Influence of VAD decision level on speaker verification for $L = 8, 16, 32, 64$ codevectors per speaker model

These measures correspond to the probability of acceptance an impostor as a valid user and the probability of rejection of a valid user. Changing the decision level, DET curves which show dependence between FRR and FAR can be plotted. Another very useful performance measure is an equal error rate EER which corresponds to error rate achieved for the decision threshold for which $FRR = FAR$. In other words EER is just given by the intersection point of the main diagonal of DET plot with DET curves.

3.1   Influence of Voice Activity Detection on Speaker Verification

For the high quality speech, influence of VAD on speaker verification was tested. Discrimination between speech and silence segments was based on the energy of

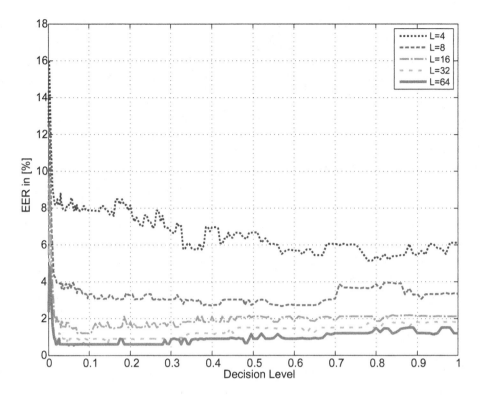

Fig. 3. Influence of VAD decision level on EER

the signal. If the energy was above some specified decision level it was assumed that this signal segment contains speech, otherwise it was removed. Decision level DL was calculated as the fraction of the mean energy of the utterance samples. There were 181 values of DL – starting from 0 (which means that all frames were classified as containing speech) to 1 (decision level was equal to mean energy of samples in utterance, approximately only 25 % segments were left as containing speech). Influence of DL was tested for different complexity of speaker models ($L = 8, 16, 32, 64$ codevectors per model). Each speaker model was obtained with the k-means procedure.

Achieved speaker verification performance was shown in Fig. 2. It can be seen that increasing complexity of the model enables to obtain lower error rates. The lowest $EER \approx 0.5 \%$ was achieved for the system with 64 codevectors per speaker model. Value of decision level has serious impact on verification performance but it is difficult to decide where is its optimum value. Figure 3 shows obtained results in a different way, namely EER as a function of DL for each complexity of the model. Obtained results are very interesting as this figure clearly shows that leaving all silence parts in utterance is responsible for drastically worse performance (strong peak near $DL = 0$). If decision level is

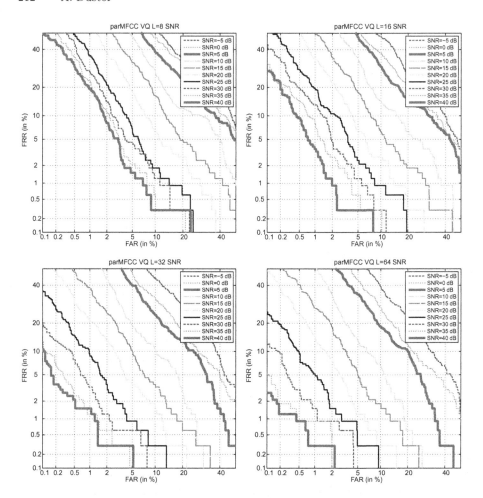

Fig. 4. Influence of signal to noise ratio on speaker verification for $L = 8, 16, 32, 64$ codevectors per speaker model and decision level 0.1

high, then EER also increases but in a moderate way. The optimum value of DL for the most complex speaker model may be estimated between 0.1 and 0.3.

Interesting conclusion may be drawn from this research that it is better to remove more than less frames from the utterance since silence is responsible for serious degradation of the speaker verification performance.

3.2   Influence of Noise on Speaker Verification

In order to test influence of speech signal quality on verification, white gaussian noise was added to each utterance. The rest of research procedure was the same as the previous one. Verification was tested for different values of signal to noise ratio, starting from the very poor quality of speech ($SNR = -5$ dB) to

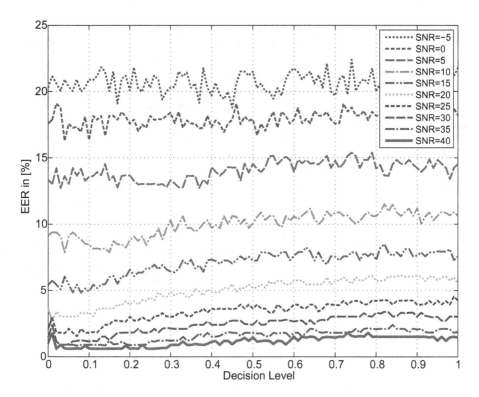

Fig. 5. Influence of SNR [dB] and VAD decision level on EER for $L = 64$ codevectors per speaker model

high quality ($SNR = 40$ dB). Obtained results for speaker models consisting of $L = 8, 16, 32, 64$ codevectors and $DL = 0.1$ are shown in Fig. 4. Dependence between EER and decision level DL in voice activity detection for the most complex speaker model was shown in the Fig. 5.

It can be seen (Fig. 4) that noise has dramatic impact on speaker verification performance. For the most complex model EER increases from 1 % for $SNR = 40$ dB to 21 % for $SNR = -5$ dB. It shows that low error rates in speaker recognition systems are possible to achieve only for the high quality speech. Figure 5 shows that optimum value of $DL \in (0.1, 0.3)$ for high quality speech ($SNR = 40$ dB), $DL \approx 0.1$ for medium quality ($SNR = 20$ dB) and for lower SNR values voice activity detection does not matter.

4   Discussion

It would be interesting to check whether the results obtained for other languages would be similar to the ones obtained for Polish language. However, it would require English speech corpus which was comparable to Polish corpus ROBOT (similar recording conditions, similar age of speakers, similar proportion between

male and female speakers, etc....). Only under these conditions results obtained
for other languages could be in some way compared to results obtained for Polish
language. Since during VAD procedure no statistical properties of language are
utilized, results obtained for other languages should be similar to the Polish ones.

Most papers devoted to speaker recognition are focused on speaker modeling
and feature extraction and VAD is only a small part of speech signal processing.
Usually authors write that silence was removed and nothing more. The aim
of this paper was not to find the best method for silence removing but rather
checking impact of silence removing on achieved recognition errors. In literature
one may find that it is better to remove more than less silence excerpts from
the speech but does it apply for only high quality of speech signal or also for
noisy speech? Presented research show that in case of high quality signal there
is an optimum value of threshold. It is true especially for more complex speaker
models. For the most complex model VQ 64 the lowest obtained $EER \approx 0.5\%$
(Fig. 3), for non-optimum threshold it can reach more than 1 %, whereas omitting
VAD procedure yields $EER \approx 5\%$. For noisy speech such optimum value is
difficult to find or even does not exist (Fig. 5). As a result VAD is definitely more
important for high quality signals where it may substantially improve recognition
results and reduce numerical complexity of recognition process.

5   Conclusion

Conducted research clearly show that for high values of SNR it is extremely
important to remove silence excerpts from the speech utterance. Such procedure
decreases EER values significantly and simplifies next steps of speaker recognition as less data are needed to proceed. Strong noise is definitely the worst
scenario for the recognition system and is responsible for bad performance of
verification process. In that case VAD is less important as EER is independent
of VAD decision level.

Acknowledgment. This work was supported by the Ministry of Science and Higher
Education funding for statutory activities.

References

1. Dustor, A.: Voice verification based on nonlinear Ho-Kashyap classifier. In: International Conference on Computational Technologies in Electrical and Electronics
Engineering SIBIRCON 2008, pp. 296–300. Novosibirsk (2008)
2. Dustor, A., Szwarc, P.: Application of GMM models to spoken language recognition. In: Napieralski, A. (ed.) MIXDES 2009: Proceedings of the 16th International Conference Mixed Design of Integrated Circuits and Systems Lodz, Poland,
pp. 603–606 (2009)
3. Dustor, A.: Speaker verification based on fuzzy classifier. In: Cyran, K.A.,
Kozielski, S., Peters, J.F., Stańczyk, U., Wakulicz-Deja, A. (eds.) Man-Machine
Interactions. AISC, vol. 59, pp. 389–397. Springer, Heidelberg (2009)

4. Dustor, A., Szwarc, P.: Spoken language identification based on GMM models. In: Pulka, A., Golonek, T. (eds.) Inetrnational Conference on Signals and Electronic Systems (ICSES 2010): Conference Proceedings, Poland, Gliwice, pp. 105–108 (2010)

5. Dustor, A., Kłosowski, P.: Biometric voice identification based on fuzzy kernel classifier. In: Kwiecień, A., Gaj, P., Stera, P. (eds.) CN 2013. CCIS, vol. 370, pp. 456–465. Springer, Heidelberg (2013)

6. Kłosowski, P., Dustor, A.: Automatic speech segmentation for automatic speech translation. In: Kwiecień, A., Gaj, P., Stera, P. (eds.) CN 2013. CCIS, vol. 370, pp. 466–475. Springer, Heidelberg (2013)

7. Dustor, A., Kłosowski, P., Izydorczyk, J.: Influence of feature dimensionality and model complexity on speaker verification performance. In: Kwiecień, A., Gaj, P., Stera, P. (eds.) CN 2014. CCIS, vol. 431, pp. 177–186. Springer, Heidelberg (2014)

8. Dustor, A., Klosowski, P., Izydorczyk, J.: Speaker recognition system with good generalization properties. In: 2014 International Conference on Multimedia Computing and Systems (ICMCS), Marrakech, Morocco, pp. 206–210 (2014)

9. Kłosowski, P., Dustor, A., Izydorczyk, J., Kotas, J., Ślimok, J.: Speech recognition based on open source speech processing software. In: Kwiecień, A., Gaj, P., Stera, P. (eds.) CN 2014. CCIS, vol. 431, pp. 308–317. Springer, Heidelberg (2014)

10. Dustor, A., Kłosowski, P., Izydorczyk, J., Kopański, R.: Influence of corpus size on speaker verification. In: Gaj, P., Kwiecień, A., Stera, P. (eds.) CN 2015. CCIS, vol. 522, pp. 242–249. Springer, Heidelberg (2015)

11. Kłosowski, P., Dustor, A., Izydorczyk, J.: Speaker verification performance evaluation based on open source speech processing software and TIMIT speech corpus. In: Gaj, P., Kwiecień, A., Stera, P. (eds.) CN 2015. CCIS, vol. 522, pp. 400–409. Springer, Heidelberg (2015)

12. Fazel, A., Chakrabartty, S.: An overview of statistical pattern recognition techniques for speaker verification. IEEE Circuits Syst. Mag. **11**(2), 62–81 (2011)

13. Adamczyk, B., Adamczyk, K., Trawiński, K.: Zasób mowy ROBOT. Biuletyn Instytutu Automatyki i Robotyki WAT **12**, 179–192 (2000)

Load Balancing in LTE by Tx Power Adjustment

Krzysztof Grochla and Konrad Połys[✉]

Institute of Theoretical and Applied Informatics, Polish Academy of Science,
Bałtycka 5, 44-100 Gliwice, Poland
{kgrochla,kpolys}@iitis.pl

Abstract. The paper describes a novel method of load balancing in cellular networks based on the management of the reference signal transmit power. The method described is easy in implementation, as it requires only reconfiguration of one of the parameters already defined in the LTE eNodeB in response to the changes in spatial distribution of the clients in the network. The proposed method is evaluated using simulation. The results prove that it allows to significantly decrease the number of unsatisfied clients (by up to 50 % for partially overloaded network), while does not decrease the total efficiency of the network in terms of the summary amount of bits per second transferred by network in time.

Keywords: LTE · LB · Load balancing · SON · HO · Handover

1 Introduction

The LTE has become a leading standard in wireless cellular communication, offering very high throughput and interoperability with exiting 3G and packet networks. As the traffic in cellular networks is rapidly growing [1], the cellular networks must improve the efficiency of radio reuse utilization. The LTE networks, also called Evolved UMTS Terrestrial Radio Access Network (E-UTRAN) improves the utilization of radio spectrum thanks to the implementation of OFDMA modulation and reuse of the full spectrum available for an operator in each of the cells. The LTE network can achieve the throughput of up to 300 MBit/s per cell. However this bandwidth is shared by all users equipments (UEs) using the same cell. The number of mobile devices using the cellular networks grows rapidly, so to limit the number of UEs sharing the cell capacity the network operators are deploying multiple low power and low range base stations.

The growing number of cells also increases the amount of intra-cell interferences. To minimize the effect of interferences the frequency allocation schemes such as Fractional Frequency Reuse (FFR) and Soft Frequency Reuse (SFR) have been proposed [2]. In FFR scheme the cell bandwidth (all available subcarriers) is divided into two or more parts, of which one part is assigned to the users near to the centre of the cell and the other part(s) to the users near to the border. The eNodeB uses lower transmission power to send the data to the users within the center of the cell (the inner part), and higher power to the rest of the

P. Gaj et al. (Eds.): CN 2016, CCIS 608, pp. 216–225, 2016.
DOI: 10.1007/978-3-319-39207-3_19

users to maximize the received signal level. In SFR each cell uses one part of the subcarriers (usually not used in the outer parts of neighbouring cells) near the border, and the remaining subcarriers in the centre.

The LTE uses different modulation and coding schemes (MCS) depending on the signal quality. The selected MCS defines the amount of bits per radio resource blocs (RRB) which can be transmitted to and by a particular UE, as it is described in Sect. 2. As the result, the maximum bitrate at which a particular client may receive data changes between approximately 4 Mbit/s and 25 Mbit/s (for a 5 MHz channel without MIMO) [3]. On the other hand the clients expect to be serviced independently of the location in which they are using the mobile device. Thus the network management layer must allocate the radio resources to minimize the amount of clients which are not serviced with the expected throughput on one hand, and maximize the amount of data being transmitted on another hand.

Typically the clients are serviced by the eNodeB which has the highest measured level of reference signals (the highest value of the RSRP – Reference Signal Received Power). This strategy is efficient as long as the amount of clients within each cell is similar. When the UE spatial distribution is not uniform, the same amount of radio resources needs to be shared between much more clients in densely occupied cell and less number of clients in another cell.

The mobile clients roam within the area covered by the network, moving from one cell to another. The communication between the clients and the network is handled by different eNodeBs (base stations). If the data or voice transmission is in progress while the client moves, the handoff procedures are initiated to provide the client with continuous transmission. Handoff is a process of transferring an ongoing call or data session from one base station to another in wireless networks. The handoffs increase the dynamics of network load distribution and shift part of the traffic served by the network from one cell to another. Thus the load balancing scheme must react dynamically to the shifts in network load distribution to constantly optimize the configuration of eNodeB.

To improve the fairness of radio resource allocation between the cells a load balancing algorithms have been proposed, shifting the load from the heavily occupied cells to the rest of the network. There are quite a few load balancing schemes proposed for LTE which try to equalize the distribution of the users among the different cells. The most popular approach is to shift the hysteresis to prefer the execution of handover to a less loaded cell. This method is described in details in [4]. Proposed solution was evaluated in simulator with different scenarios and shows quite good results in reduction of overload when users are in hotspot. Another possible method is based on Markov Decision Processes, in the paper [5] the vertical handoff procedure was described and in [6] is described a hybrid decision approach for the association problem in heterogeneous networks. Next possible approach is a game theory, described in [7] where the selection of the best radio access technology was studied and in [8] which described the dynamics of network selection in a heterogeneous wireless network. Alternative way is a Cell Range Expansion approach which adds an offset to RSRP of pico cell to offload more users from macro cell [9].

A novel approach to load balancing between the cells in LTE is to change the transmit power (TX power) of the eNodeB dynamically in response to the changing spatial distribution of the clients with consideration of Soft Frequency Reuse technique. In literature are similar solutions but widely used in other technologies, like WLAN [10] or taking into account only virtual cell breathing which do not have any Interference Mitigation methods [11].

In this paper we propose a novel load balancing scheme which regulates the load of base stations by changing not only the TX power, but also adjusting the SFR parameters, what influences very beneficial on reduction of interferences [12].

The rest of the paper is organized as follows: second section is a description of presented algorithm with pseudo code. Section 3 contain the simulation model used to analyse the performance of the proposed scheme. In Sect. 4 the results are discussed. The last section finish the paper with a short conclusion.

2 Algorithm Description

It was assumed that modifying Tx power will change the assignation of UEs to eNBs. This solution is intended to work as centralized manager of the network. It tries to offload crowded cells by reduction of TX power in outer area of those cells and reduction of the proportion of this area to the inner area because of this region has a lower number of subcarriers. It also raise a little Tx power of inner area to increase the achieved MCS by UEs what gives higher bitrates. Simultaneously to reconfiguration of crowded eNBs, the algorithm finds the best candidate for takeover of some UEs and increase its Tx power, both in inner and outer area and also reduce the area of outer part. Variables description:

lLoadIn – the power of inner area for eNB with low load,
lLoadOut – the power of outer area for eNB with low load,
lLPc – point of change between inner and outer area for eNB with low load,
hLoadIn – the power of inner area for eNB with high load,
hLoadOut – the power of outer area for eNB with high load,
hLPc – point of change between inner and outer area for eNB with high load,
defIn – the power of inner area for eNB with default settings,
defOut – the power of outer area for eNB with default settings,
defPc – point of change between inner and outer area for eNB with default settings,
eNB – single eNodeB,
eNBs – list of all eNodeBs,
crowdedeNBs – list of eNodeBs with high load,
lleNBs – list of eNodeBs with low load,
UEsTable – list of UEs connected to the current eNodeB,
alternativeeNBsTable – list of eNodeBs which are not highly loaded and UEs could connect to them.

```
setVariablesForPowers(lLoadIn, lLoadOut, lLPc,
hLoadIn, hLoadOut, hLPc, defIn, defOut, defPc)
setVariablesForLoad(overLoad, medLoad)
for each eNB in listOf(eNBs): eNB.getLoad()
   if eNB.getLoad() > overLoad: crowdedeNBs.put(eNB)
for each eNB in listOf(crowdedeNBs):
   UEsTable=eNB.getListOfConnectedUEs()
   for each UE in listOf(UEsTable):
      alternativeeNBsTable.put(UE.geteNBwith2ndBestReception)
   alternativeeNBsTable.removeeNBsWithLoadHigherThan(medLoad)
   lleNBs.put(alternativeeNBsTable.findMostPopular())
lleNBs.leaveOnlyUniqeeNBs()
for each eNB in listOf(lleNBs):
   eNB.setTxPowerInside(lLoadIn)
   eNB.setTxPowerOutside(lLoadOut)
   eNB.setPointOfChangeBetweenInsideOutside(lLPc)
for each eNB in listOf(crowdedeNBs):
   eNB.setTxPowerInside(hLoadIn)
   eNB.setTxPowerOutside(hLoadOut)
   eNB.setPointOfChangeBetweenInsideOutside(hLPc)
for each eNB in listOf(eNBS):
   if eNB is not in(crowdedeNBs and lleNBs):
      eNB.setTxPowerInside(defIn)
      eNB.setTxPowerOutside(defOut)
      eNB.setPointOfChangeBetweenInsideOutside(defPc)
```

Next, the rest of eNBs, which are not reconfigured in this round, have set their settings to default. In described scenario the following values were used: lLoadIn = 40 dBm, lLoadOut = 43 dBm, lLPc = 90 %, hLoadIn = 38 dBm, hLoadOut = 39 dBm, hLPc = 80 %, defIn = 37 dBm, defOut = 40 dBm, defPc = 60 %, overload = 1.0, medLoad = 0.85. Values for high and low load were attained experimentally. We express the network load as a percentage of used radio resource blocks (RRB) among all RRBs in all cells in the network. Load equal 100 % means that all RRBs were allocated.

3 Simulation Model

To verify the efficiency of load balancing algorithm we have implemented a simulation model. It consists of 19 LTE base stations (eNodeB) and number of UEs. Base stations were located according to the honeycomb topology. The area modelled within the simulation scenario is 14.9 × 13.3 km. The simulation model has been implemented using OMNeT++ with INET framework. UEs move according to RandomWaypoint mobility model [13]. Velocity of users was set to 50 km/h. The 24 hours time was simulated. The signal attenuation was modeled using the SUI [14] propagation model. In equal time interval the Signal To Interferences and Noise Ratio (SINR) for every UE was calculated and the next the possible bitrate, based on SINR and Modulation and Coding Scheme (MCS) table. Described in more details in subsequent paragraphs.

The cell size radius Rc is equal 1666 m. Radio resources were allocated with Soft Frequency Reuse (SFR) approach which divides the cell into two regions. For the inner area the 2/3 of all subcarriers is available and eNB transmits there with lower power and for the outer area the 1/3 of subcarriers are used with higher power. In our scenarios in the inner part of cell the eNB transmits with power of 37 dBm and in the outer part with power of 40 dBm. The point of the change between both parts is set to 60 % of Rc. Parameters described in this paragraph were taken from [15]. Subcarrier allocation was made according to [16].

Evaluation of the algorithm was made for five different densities of the devices: 100, 200, 300, 400, 500 UEs. For all instances we simulated two cases – with and without optimization of TX power. We assumed that every UE needs to be served with at least the bitrate equal 512 Kbit/s [17]. We simulate only active users, the users which are not receiving traffic are omitted as they do not influence the results. For the sake of simplification of the model only the downlink traffic is considered. Two main metrics were used to evaluate the performance of LTE network: the number of UEs unsatsfied (where CBR 512 Kbit/s cannot be guaranteed) and the total (summary) bitrate offered for all UEs. We did not simulate more than 500 users because the network load is quite high with such number of active clients and the model assumptions taken for the download ration (512 Kbit/s). Additionally, the mobility of UEs sometimes causes higher density of them in some cells.

The Physical Downlink Shared Channel (PDSCH) is used to transfer application data according to the LTE standard. The throughput achieved by the PDSCH is determined mainly by the modulation and coding scheme (MCS) and the channel width. A single UE can be assigned to only one MCS in each transmission time interval (TTI) or scheduling period. The possible modulation and coding schemes are presented in the Table 1.

In our model the maximum throughput available for a particular client is calculated basing on the selected MCS for this client. The LTE standard does not define a method of MCS selection – a comparative study of different methods is given in [18], for the sake of this work we use a simple method used in [19] on the basis on mapping tables [20], in which the MCS with highest code rate is selected, for which the SINR Threshold defined in the Table 1 is lower than the calculated SINR for a given UE. Basing on the selected MCS the efficiency η_i of the i-th transmission mode (channel quality indication index – CQI index) is derived using the equation:

$$\eta_i = r_i \cdot \log_2(M_i). \tag{1}$$

The throughput offered to a particular client can be calculated using the equation above as a function of the code rate r_i and the amount of radio Physical Resource Blocks (PRBs) allocated to this client. We assume that the Round Robin scheduler is used and the total number of PRB is divided fairly between all clients within the cell – each client is allocated with the same amount of PRBs, proportional to the channel bandwidth (see Table 2).

Table 1. MCS table

CQI Index	MCS	Modulation	Code Rate x1024, r_i	Modulation size, M_i	SINR Threshold
1	0	QPSK	78	4	-5.45
2	2	QPSK	120	4	-3.63
3	4	QPSK	193	4	-1.81
4	6	QPSK	308	4	0
5	8	QPSK	449	4	1.81
6	10	QPSK	602	4	3.63
7	12	16QAM	378	16	5.45
8	14	16QAM	490	16	7.27
9	16	16QAM	616	16	9.09
10	18	64QAM	466	64	10.90
11	20	64QAM	567	64	12.72
12	22	64QAM	666	64	14.54
13	24	64QAM	772	64	16.36
14	26	64QAM	873	64	18.18
15	28	64QAM	948	64	20

Table 2. PRBs allocated per TTI

Bandwidth (BW) [MHz]	1.5	2.5	5	10	15	20
Numer of PRBs per TTI	12	24	50	100	150	200

4 Results

The analysis shows that presented load balancing algorithm significantly improves the radio resources utilization when the whole network is not overloaded or when there are only some overcrowded spots and it is possible to takeover some load by neighbouring cells. It was assumed that UEs require the minimum download data rate equal 512 Kbit/s. The results were normalized to show the relative decrease of the number unsatisfied UEs: first the simulation has been executed without the load balancing algorithm and the number of unsatisfied client and the bitrate was measured. It was perceived as a 100 %. Next the simulation was run with the proposed load balancing algorithm and the measured value of bitrate or no of unsatisfied clients was divided by the same value for the previous simulation run. The plots represent the value measured with the load balancing divided by the same value without load balancing, multiplied by 100 %.

The analysis show that the proposed load balancing algorithm contributed to the improvement of more than 50 % in count of unsatisfied clients when there are 100 UEs on the network, by more than 30 % with 200 UEs and about 10 %

with 300 UEs, is it shown on Fig. 1. These values are compared to the scenario
without any optimization. At the same time the average network throughput
remains almost unchanged where there is up to 200 UEs. With the raise of UEs
count the average bitrate slightly falls down, but only to 1.6 % in case with
500 UEs, it is shown on Fig. 2. This is caused by the high utilization of the
radio resources when number of users is high, what makes the load balancing
ineffective, because all cells are almost fully loaded.

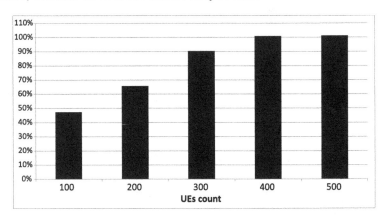

Fig. 1. Unsatisfied UEs with tx power adjustment in comparison to no optimization

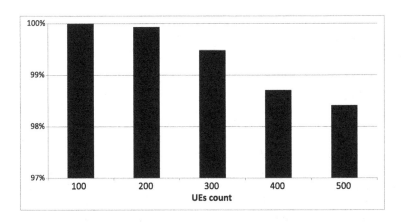

Fig. 2. Bitrate with TX power adjustment in comparison to no optimization

The Figs. 3 and 4 present the distributions (experimental probability density
functions) of number of unsatisfied clients per cell and summary throughput
of all clients in the network in time. It shows that when the load balancing
algorithm was used the number of cells which were able to serve all the clients
(with 0 unsatisfied UEs) was significantly higher. The dynamic management of
TX power allowed also to decrease the probability of significant overload in a cell,
when 4 or more client were not able to receive the data with the required speed.

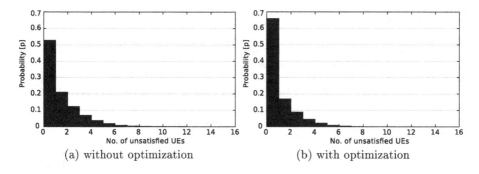

(a) without optimization   (b) with optimization

Fig. 3. Histogram of unsatisfied UEs (200 UEs)

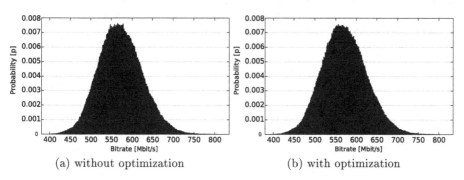

(a) without optimization   (b) with optimization

Fig. 4. Histogram of network bitrate (200 UEs)

The Fig. 4 shows that the implementation of proposed load balancing scheme did not change the variability of the total throughput offered to the clients.

5 Conclusion

The paper describes a novel method of load balancing in cellular networks based on the management of the transmit power. The method described is easy in implementation, as it requires only reconfiguration of one of the parameters already defined in the LTE eNodeB in response to the changes in spatial distribution of the clients in the network. The results prove that it allows to significantly decrease the number of unsatisfied clients, while does not decrease the total efficiency of the network in terms of the summary amount of bits per second transferred by network and its variability in time.

Acknowledgement. The work is partially supported by NCBIR Project LIDER/ 10/194/L-3/11/.

References

1. Cisco: Cisco visual networking index: global mobile data traffic forecast update, 2013–2018. Cisco Public Information (2014)
2. Boudreau, G., Panicker, J., Guo, N., Chang, R., Wang, N., Vrzic, S.: Interference coordination and cancellation for 4G networks. IEEE Commun. Mag. **47**(4), 74–81 (2009)
3. Johnson, C.W.: Long Term Evolution in Bullets. CreateSpace Independent Publishing Platform, North Charleston (2010)
4. Lobinger, A., Stefanski, S., Jansen, T., Balan, I.: Coordinating handover parameter optimization and load balancing in LTE self-optimizing networks. In: 2011 IEEE 73rd Vehicular Technology Conference (VTC Spring), pp. 1–5. IEEE (2011)
5. Stevens-Navarro, E., Lin, Y., Wong, V.W.: An MDP-based vertical handoff decision algorithm for heterogeneous wireless networks. IEEE Trans. Veh. Technol. **57**(2), 1243–1254 (2008)
6. Elayoubi, S.E., Altman, E., Haddad, M., Altman, Z.: A hybrid decision approach for the association problem in heterogeneous networks. In: 2010 Proceedings IEEE INFOCOM, pp. 1–5. IEEE (2010)
7. Aryafar, E., Keshavarz-Haddad, A., Wang, M., Chiang, M.: Rat selection games in Hetnets. In: 2013 Proceedings IEEE INFOCOM, pp. 998–1006. IEEE (2013)
8. Niyato, D., Hossain, E.: Dynamics of network selection in heterogeneous wireless networks: an evolutionary game approach. IEEE Trans. Veh. Technol. **58**(4), 2008–2017 (2009)
9. Damnjanovic, A., Montojo, J., Wei, Y., Ji, T., Luo, T., Vajapeyam, M., Yoo, T., Song, O., Malladi, D.: A survey on 3GPP heterogeneous networks. IEEE Wireless Commun. **18**(3), 10–21 (2011)
10. Bejerano, Y., Han, S.J.: Cell breathing techniques for load balancing in wireless LANs. IEEE Trans. Mobile Comput. **8**(6), 735–749 (2009)
11. Yang, S., Zhang, W., Zhao, X.: Virtual cell-breathing based load balancing in downlink LTE-a self-optimizing networks. In: 2012 International Conference on Wireless Communications Signal Processing (WCSP), pp. 1–6, October 2012
12. Grochla, K., Połys, K.: Dobór parametrów mechanizmu zwielokrotnienia wykorzystania częstotliwości SFR w sieciach LTE. In: Materiały konferencyjne Krajowego Sympozjum Telekomunikacji i Teleinformatyki. KSTiT (2015)
13. Gorawski, M., Grochla, K.: Review of mobility models for performance evaluation of wireless networks. In: Gruca, A., Czachórski, T., Kozielski, S. (eds.) Man-Machine Interactions 3. AISC, vol. 242, pp. 573–584. Springer, Heidelberg (2014)
14. Erceg, V., Greenstein, L.J., Tjandra, S.Y., Parkoff, S.R., Gupta, A., Kulic, B., Julius, A., Bianchi, R., et al.: An empirically based path loss model for wireless channels in suburban environments. IEEE J. Sel. Areas Commun. **17**(7), 1205–1211 (1999)
15. Giambene, G., Yahiya, T.: LTE planning for soft frequency reuse. In: 2013 IFIP Wireless Days (WD), pp. 1–7, November 2013
16. Grochla, K., Połys, K.: Subcarrier allocation for lte soft frequency reuse based on graph colouring. In: Abdelrahman, O.H., Gelenbe, E., Gorbil, G., Lent, R. (eds.) ISCIS 2015. LNEE, pp. 447–454. Springer International Publishing, Switzerland (2016)
17. Lobinger, A., Stefanski, S., Jansen, T., Balan, I.: Load balancing in downlink LTE self-optimizing networks (2010)

18. Fan, J., Yin, Q., Li, G.Y., Peng, B., Zhu, X.: MCS selection for throughput improvement in downlink LTE systems. In: 2011 Proceedings of 20th International Conference on Computer Communications and Networks (ICCCN), pp. 1–5. IEEE (2011)
19. Piro, G., Baldo, N., Miozzo, M.: An LTE module for the NS-3 network simulator. In: Proceedings of the 4th International ICST Conference on Simulation Tools and Techniques, pp. 415–422. ICST (Institute for Computer Sciences, Social-Informatics and Telecommunications Engineering) (2011)
20. Kawser, M.T., Hamid, N.I.B., Hasan, M.N., Alam, M.S., Rahman, M.M.: Downlink SNR to CQI mapping for different multiple antenna techniques in LTE. Int. J. Inf. Electron. Eng. 2(5), 757–760 (2012)

USB 3.1 Gen 1 Hub Time Delay Influence on System Timeout and Bus Clock Synchronization

Wojciech Mielczarek and Michał Sawicki[✉]

Institute of Informatics, Silesian University of Technology,
Akademicka 16, 44-100 Gliwice, Poland
{wojciech.mielczarek,michal.sawicki}@polsl.pl

Abstract. The USB device designer must be aware of the worst case total delay between the host and a peripheral, because this value is necessary for setting the transaction timeout (allowable response packet delay).

In the paper the problem of USB bus timeout is covered for different USB buses: Low/FullSpeed USB 1.x, HighSpeed USB 2.0 and SuperSpeed USB 3.1 Gen 1. The USB 3.1 bus time delay measurement results are presented and supremum of response time is estimated. Additionaly, time delay of isochronous timestamps broadcasting is considered. At last, the demand for adaptive setting of the USB 3.1 bus timeout is expressed.

Keywords: USB 3.1 · Transaction timeout · Total trip delay · Hub time delay

1 Introduction

The SuperSpeed, gigabit USB 3.1 Gen 1 (hereinafter called USB 3.1) brings many advantages against still the most popular 480 MHz USB 2.0 version. This is not just about enormous bit rate that results in bandwidth reduction required by transfers, but above all due to:

- full duplex transmission circuit with separated lines for transmitting and receiving,
- packet routing that transformed USB bus from broadcast to unicast,
- power saving from not used bus segments disabling,
- asynchronous service requests handling.

These modifications required hub redesign from ordinary USB 2.0 repeater forwarding packets and enabling/disabling downstream ports only, into the strongly parametrized device, additionally ready to:

- route packets in downstream port direction,
- managing the link power hierarchy,
- deferring transactions,
- hub error detecting and handling.

© Springer International Publishing Switzerland 2016
P. Gaj et al. (Eds.): CN 2016, CCIS 608, pp. 226–234, 2016.
DOI: 10.1007/978-3-319-39207-3_20

The hub defined in the basic USB 1.x and in USB 2.0 specifications operates at the physical layer only, but the hub 3.0 stack additionally contains the link layer (Fig. 1).

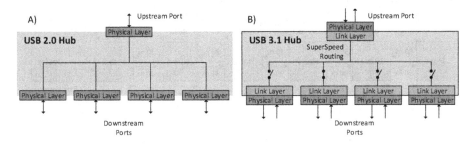

Fig. 1. USB hub connectivity: USB 2.0 packet broadcasting (A), USB 3.1 packet routing (B)

The serious question is, how a new "better" hub influences on USB bus basic parameters, like the transaction timeout and the time (clock) delivery accuracy.

It is a matter of concern, that rapid growth of hub responsibilities in USB bus possibly inhibits performing its basic functions, namely forwarding packets between upstream and downstream ports and bus clock distribution.

2 USB Regular Transaction Timeout

The USB host attempts to simultaneously carry out data transfers via serial bus (to or from USB devices) commissioned by different applications. In that case it is necessary to divide data related to each transfer into multiple units (the process called segmentation) and forwarding successive units using transactions. The USB transaction is the ordered sequence of packets defined just for transporting a data unit. A typical transaction consists of three phases, or packets (Token Packet phase, Data Packet phase and Handshake Packet phase), what really makes it possible to interact data source and data recipient. Parallel performing of transfers is based on multiplexing the transactions related to different transfers.

The transmitter and receiver involved in a transaction must know how long they must wait for response. The transaction timeout is a typical protection against frozen bus due to uncompleted transaction caused by packet error. The timeout for the USB 3.1 is considered in Sect. 3. In this section the timeout for "regular" USB (USB 1.x, USB 2.0) is evaluated in order to compare it further with USB 3.1 timeout.

The USB bus timeout (t_{TTD}) must include some delay related to signal propagation over bus and time needed for preparing a response. The propagation delay depends on the number of cable segments supported downstream from the host. The USB 1.x specification defines the worst case timing when six cable segments are supported via a single downstream path (Fig. 2) and assumes that:

– cable delay is 30 ns maximum,
– hub delay is 40 ns maximum,

so the total delay from the downstream port of the root hub to the downstream port of the first hub is a maximum of 70 ns.

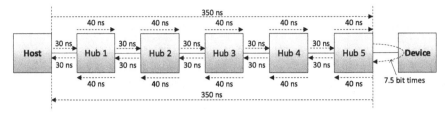

Fig. 2. USB trip delay for Low/FullSpeed

The total round-trip delay between the host port to the upstream end of the device's cable is

$$2 \cdot (5 \cdot 70\,\text{ns}) = 0.7\,\mu\text{s}. \tag{1}$$

Additional components of total trip delay are the signal propagation time over device cable in both directions and the time necessary for preparing a response. It is required that all these components together must take no more than 7.5 bit times[1].

The total trip delay is usually expressed in bit time units. For FullSpeed (bit time = $1/(12\,\text{MHz})$), the total trip delay is equal approximately 16 bit times $(0.7\,\mu\text{s}/(1/(12\,\text{MHz})) + 7.5$ bit times). It is required that the transmitter of the transaction must not timeout before 16 bit times but must timeout by 18 bit times. This is true of both: Low and FullSpeed transactions.

The same value expressed in [ns] is

$$1328\,\text{ns} \leq t_{\text{TTD}} \leq 1494\,\text{ns}. \tag{2}$$

The USB 2.0 interacting data terminal equipment (HighSpeed host and High-Speed peripheral) must not timeout the transaction if the interval between packets is less than 736 bit times, and must timeout the transaction if it is greater than 816 bit times [2]. These limits allow a response to reach a recipient even for the worst case total trip delay. The base for setting HighSpeed bus timeout in the limits mentioned above is the sum of the following components:

– max 12 cables length = 312 ns,
– max 10 delay hubs = 40 ns + 360 bit times,
– max 1 device response time = 192 bit times.

The Worst case total trip delay = 352 ns + 552 bit times. For HighSpeed the bit time is 2.0833 ns $(1/(480\,\text{MHz}))$, so the total trip delay is 721 bit times (1502 ns).

---

[1] This 7.5 bit times delay is defined as the time from the EOP (End of Packet) to idle transition observed on the upstream end of the function cable until the SOP (Start of Packet) transition from the device is returned to upstream end of the target cable [1].

3   USB 3.1 Transaction Timeout

The USB 3.1 specification [3] is very laconic on the host transaction timeout and defines only that:

1. For bulk, interrupt and control transactions it is the time without a response to the last data packet or acknowledge transaction packet that the host sent out before the host assume that the transaction has failed and halt the endpoint.
2. For isochronous input transactions it is the time without a response to the acknowledge transaction packet that the host sent out. The timeout timer must restart counting whenever the host receives data packet that was requested by the acknowledge transaction packet. If a timeout occurs, the host is forbidden to perform any more transactions to the "insecure" endpoint in the current service interval. The host shall not halt the endpoint, but rather restart transactions to the endpoint in the next service interval. No retries of sending damaged data packet are allowed.
3. The timeout min value is 7.6 ms and the timeout max value is 25 ms.

It is, however, difficult not to perceive that the USB 3.1 timeout limits (of about ms) are much more than the USB 2.0 limits (of about μs). Why that is so?

The USB 3.1 hub ports run not only in the physical layer but also in the link layer, which is making propagation through hub rise sharply up to 432 ns (thus about ten times compared with the USB 2.0 hub). Including 30 ns of max cable delay, the set of a hub and cable brings worst case delay of 462 ns. The total round trip delay is proportional to the number of hubs "visited" in both directions, so for 5 hubs between the host and a device, this parameter is equal to 4.6 μs (10 · 462 ns). It is 6.5 times greater than in the USB 2.0 bus, but still much lower than suggested timeout value in the USB 3.1 specification.

It is evident that the timeout of the USB 3.1 bus is not based on the round trip delay only. There is one more important fact that must be taken to account trying to estimate the actual USB 3.1 bus latency. It is power management policy based on putting USB device into a low power consumption mode. The worst case for timeout estimating is the situation when all the hubs as well as endpoint device are in one of the idle states (U1-U3 according to the LTSSM state machine). It needs 2 ms to recover from the idle state [4] and this delay must be included to the total round trip delay. For example if there are 5 hubs between the host and a device, the worst total trip delay is

$$t_{\text{TTD}} = \underbrace{5 \cdot 2000\,\mu s}_{\text{total recovery time}} + \underbrace{10 \cdot 4.6\,\mu s}_{\text{round trip delaty}} + \underbrace{0.4\,\mu s}_{\text{response delay}} \ . \tag{3}$$

It is less from 25 ms (the upper timeout limit) quouted in [4] and it shows that the value in specification is arbitrary to some degree.

4   Time Delay of Isochronous Timestamps Broadcasting

Time (moments of bus interval beginning) on USB 3.1 bus is globally synchronized by host controller which broadcasts isochronous timestamp packet (ITP) over all ports with state U0 at the beginning of each bus interval. This mechanism is based on broadcasting the start of microframe packet (μSOF) introduced in version 2.0. The broadcasting ITP via downstream ports of root hub may be delayed ($0 \leq \Delta\tau \leq 8\,\mu$s). Moreover, the delay is expressed in units equal to time of 8 bits transmission in HighSpeed mode. Peripheral device receives ITP at t'_{ITP} and is able to evaluate the beginning of bus interval (t_{ITP}) using formula:

$$t_{\text{ITP}} = t'_{\text{ITP}} - \Delta\tau - \sum_{i=1}^{h} t^i_{\text{HUB}}. \tag{4}$$

Obviously, peripheral device knows a posteriori the time delay of ITP broadcasting via root hub downstream ports and knows the delays (t^i_{HUB}) added into transmission by h hubs on the route between the host controller and a device.

5   Evaluating USB 3.1 Hub Time Delay

As shown in the preceding sections, total trip delay (t_{TTD}) and the moment of bus interval beginning (t_{ITP}) are depended on hub time delay. Hence, it was decided to evaluate mean of hub time delay. The first step of experiment was to determine the peripheral device response time (t_{DVK}) using system under test shown schematically in Fig. 3.

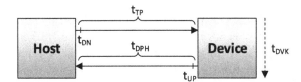

Fig. 3. Scheme of system under test without hub

The evaluation board Cypress EZ-USB FX3 (CYUSB3KIT-001) with the ARM processor and USB 3.1 interface was used as isochronous peripheral device. Real-time operating system ThreadX ran on this evaluation kit [5]. The firmware was responsible for one input isochronous endpoint service. Apart from this, the DMA mechanism was used and ran in Producer-Consumer mode. The DMA simplified passing data from buffer within embedded memory (Producer) to endpoint buffer (Consumer). Emptying the buffer required the CPU action like filling the buffer using random value.

The host controller was a computer equipped with Renesas USB 3.1 controller and Microsoft Windows 7 as operating system. Some application software based on CyUSB library was implemented for executing one large isochronous transfer (read operation) consisting of $N = 150\,000$ data transactions. Furthermore, all links on the route between the host controller and peripheral device were in the state U0 for whole time of experiment, because PING-PING-RESPONSE protocol had been executed before each isochronous transaction.

Transmission time of transaction packet (t_{TP}) and transmission time of data header subpacket (t_{DPH}) are constant and equal 72 ns. Indeed, only first subpacket (data header) of data packet is crucial for further reasoning, because it is forwarded from downstream port to upstream port without waiting for data payload subpacket passing completion. These parameters include not only the bit propagation delay over link but also the transmission time of link commands (credit and acknowledgement mechanisms).

Beagle SuperSpeed 5000 v2 analyzer was connected between the root hub downstream port and the peripheral device upstream port. Accordingly it was possible to record the moments, when first TP and DPH bits of i-th transaction were being sent respectively on root hub downstream port ($t_{DN}(i)$) and device upstream port ($t_{UP}(i)$).

The mean of peripheral device response time may be calculated the using recorded time:

$$\overline{t_{DVK}} = \frac{1}{N} \sum_{i=1}^{N} (t_{UP}(i) - t_{DN}(i) - t_{TP}). \tag{5}$$

In the case of CYUSB3KIT-00 peripheral device $\overline{t_{DVK}}$ equals 271 ns (point "x" in Fig. 4).

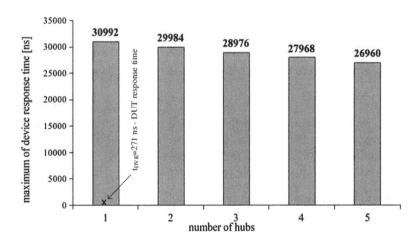

Fig. 4. t_{DVK} against number of hubs between host controller and device

Then, additional hub (Cypress HX3 CY4603) was connected between analyzer and peripheral device (Fig. 5).

Thereby, experiment and calculation were repeated for each of four hub downstream ports. Using the recorded information and calculated response time (5)

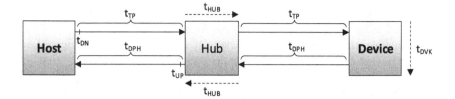

Fig. 5. Scheme of system under test

it was possible to evaluate mean of the hub time delay:

$$\overline{t_{\text{HUB}}} = \frac{1}{2N} \sum_{i=1}^{N} (t_{\text{UP}}(i) - t_{\text{DN}}(i) - 2t_{\text{TP}} - t_{\text{DPH}} - \overline{t_{\text{DVK}}}). \qquad (6)$$

Notice that $\overline{t_{\text{HUB}}}$ is the same for downstream and upstream transmission over hub and this is simplifying assumption.

The experiment was repeated again but for another hub: Logilink UA0170. All results are depicted in Table 1. Last column contains maximum of the hub time delay defined in specification [3].

Table 1. The mean values of hub USB 3.1 time delay

Port number	1	2	3	4	MAX
Cypress HX3 CY4603 hub	277 ns	279 ns	279 ns	278 ns	432 ns
Logilink UA0170 hub	135 ns	134 ns	137 ns	134 ns	432 ns

Finally, whole experiment (with evaluating response time) was repeated for another host controller (Odroid XU3 mobile platform) with Ubuntu 15.04 installed. Similar results were obtained as in previous test.

As shown in Sect. 2, it is possible to calculate the total trip delay if we know:

- hub time delay on route between host controller and peripheral device,
- packets propagation time over each cable,
- device response time.

The supremum ($t_{\text{SUP}} \in [32, 5032 \, \mu\text{s}]$) of total trip delay, called transaction timeout, is defined in specification [3]. Using this value limit, maximum of the response time t_{DVK} (7) can be estimated, and this parameter is important for peripheral device implementator [6].

$$t_{\text{DVK}} \leq t_{\text{SUP}} - 2 \sum_{i=1}^{h} t_{\text{HUB}}^{i} - h(t_{\text{TP}} + t_{\text{DPH}}). \qquad (7)$$

Table 2. USB 2.0 and USB 3.1 bus timeouts

	MIN	MAX
USB 2.0	$1.5\,\mu s$	$1.7\,\mu s$
USB 3.1	$32\,\mu s$	$5032\,\mu s$

In Table 2 bus timeout lower and upper limits for USB 2.0 and USB 3.1 are presented.

Almost 20 times greater USB 3.1 timeout minimal value against USB 2.0 bus timeout is the result of complex packet transport mechanism defined in USB 3.1 specification. The USB 3.1 hubs are not ordinary repeaters like USB 2.0 hubs working in physical layer. They perform much more functions, mainly packets buffering and routing. Furthermore, the USB 3.1 hubs also detect and handle errors (3 attempts to send packet correctly) and perform flow control using credit mechanism.

In Fig. 4 t_{DVK} maximum values calculated for the lower limit of bus timeout ($32\,\mu s$) and different number of hubs in the cascade using formula (7) are shown.

6   The Call for Adaptive USB 3.1 Bus Timeout Setting

The USB 3.1 bus timeout setting is in practice left to the host designers. The limits quoted in the USB 3.1 specification shall be treated rather as a suggestion. It is not sensible to set the bus timeout of about ms, if a USB 3.1 system consists, for example, of the host and the several devices or hubs only (the most common situation). It is not sensible, but now common, to set the bus timeout as constant value.

Our demand is adaptive setting the USB 3.1 bus timeout taking to account the bus structure. The host has all the needed information of the bus topology (branches and leaves), so it can calculate the most appropriate bus timeout value. There is one important condition to implement this idea: the timeout must be recalculated each time a new device or hub is connected or disconnected to or from the bus.

7   Summary

The results of experiments show that hub time delay is not depended on downstream port number and implementation of host controller. Further, all hubs under test add delay into transmission, which is less than maximum defined in specification.

It was confirmed that the USB 3.1 hub time delay is about 7 times greater than the USB 2.0 hub time delay. This is the unavoidable price for routing and buffering packets, detecting and handling errors, flow control performed by each hub between the host controller and peripheral device.

We explained why the USB 3.1 bus timeout is so great and we "discover" that it is hub recovery mechanism primarily responsible for this. Furthermore, we came to conclusion that it is not reasonable to set the USB 3.1 bus timeout as the constant value. Our call is to set it adaptively using information about the bus structure collected by the host controller. This idea is somewhat similar to the IEEE 1394 (FireWire) bus reset mechanism after changing bus topology (plug in or plug out a device) and well known in computer networks, mainly wireless links [7].

References

1. Anderson, D.: Universal Serial Bus System Architecture. MindShare (1997)
2. Universal Serial Bus 2.0 Specification, Rev. 2.0-2000. www.usb.org
3. Universal Serial Bus 3.0 Specification, Rev. 1.0-2011. www.usb.org
4. Universal Serial Bus 3.1 Specification, Rev. 1.0-2013. www.usb.org
5. Hyde, J.: SuperSpeed Device Design By Example (2015)
6. Ramadoss, L., Hung, J.Y.: A study on universal serial bus latency in a real-time control system. In: IEEE Industrial Electronics (2008)
7. Xylomenos, G., Tsilopoulos, C.: Adaptive timeout policies for wireless links. In: 20th International Conference on Advanced Information Networking and Applications (2006)

New Technologies

Quantum Network Protocol for Qudits with Use of Quantum Repeaters and Pauli Z-Type Operational Errors

Marek Sawerwain[1(✉)] and Joanna Wiśniewska[2]

[1] Institute of Control and Computation Engineering, University of Zielona Góra,
Licealna 9, 65-417 Zielona Góra, Poland
M.Sawerwain@issi.uz.zgora.pl
[2] Institute of Information Systems, Faculty of Cybernetics,
Military University of Technology, Kaliskiego 2, 00-908 Warsaw, Poland
jwisniewska@wat.edu.pl

Abstract. In this paper a quantum communication protocol with use of repeaters is presented. The protocol is constructed for qudits i.e. the generalized quantum information units. One-dit teleportation is based on the generalized Pauli-Z (phase-flip) gate's correction. This approach avoids using Pauli-X and Hadamard gates unlike in other known protocols based on quantum repeaters which were constructed for qubits and qudits. It is also important to mention that the repeaters based on teleportation protocol, described in this paper, allow a measurement in the standard base (what simplifies the measurement process) and the use of teleportation causes only Pauli-Z operational errors.

Keywords: Quantum information transfer · Quantum repreater · Qudit teleportation protocol

1 Introduction

One of the solutions currently discussed in the field of quantum communication [1–3] is a quantum repeater [4]. Its construction is not so simple as building classic amplifier because of non-cloning theorem which is one of the most basic and important foundations of quantum mechanics. However, the lack of ability of making perfect copies of quantum information may be compensated with use of teleportation and the phenomenon of entanglement. The protocol of quantum teleportation may be utilized to amplify the signal i.e. a quantum state during the transfer process. This approach may be used, e.g. in a fiber, to improve the quality of transferred information. Applying the teleportation protocol and the entanglement to amplify the quantum information results with building of so-called quantum repeater (QR).

It should be emphasized that the notion of QR is currently in the center of interest of many researchers [5,6]. Apart from the theoretical analysis of this subject there are also physical experiments carried out with use of QRs as

© Springer International Publishing Switzerland 2016
P. Gaj et al. (Eds.): CN 2016, CCIS 608, pp. 237–246, 2016.
DOI: 10.1007/978-3-319-39207-3_21

the elements of quantum networks. These experiments are being accomplished for the transmission of quantum states in an optical fiber [7] and also in the air [8]. Therefore, the progress in the field of quantum computing [9–13] is strongly connected with the progress of quantum communication because of the need to send information.

The notion of QR is currently discussed as a potential solution to the problems of quantum communication and cryptography (especially for quantum key distribution). The three generations of QR were presented do far. The first type utilizes the phenomenon of entanglement and its purification [4,7,14]. The second generation [15] is based on the near-perfectly entangled pairs' generation (however, in this approach still the entanglement purification is used). The third type [16,17] of QR utilizes the protocol of teleportation. The quantum information is transmitted from one point to another, so the communication is organized as a one-way scheme. The solution presented in this paper belongs to the third generation of QR and it uses so-called one-dit teleportation protocol to correct the errors which may appear during the transfer process. The novelty of described solution consists on using only Pauli-Z gate for error correction. More precisely: the result of measurement performed during the teleportation protocol determines the number of Pauli-Z operations which have to realized as an error correction.

The reminder of this paper is organized as follows. In Sect. 2 there is a quantum teleportation protocol presented and the error correction is performed with use of Pauli-Z gate. Section 3 describes the realization of QR with use of previously mentioned teleportation protocol. The interpretation of repeater protocol as a quantum circuit is shown in Sect. 4. The summary and the final conclusions are presented in Sect. 5.

2   Single Dit Teleportation Protocol

The protocols presented in this section are defined with notions of dits and qudits. A concept of dit is a generalization of classic bit. For classic bit there can be distinguished only two states: zero and one. In the case of dit more states are possible. The number of these values is symbolized by the letter d. For example, the classic bit is a dit with $d = 2$.

A unit of quantum information is so-called qubit. A definition of qubit may be presented as:

$$|\psi\rangle = \alpha|0\rangle + \beta|1\rangle, \quad |\alpha|^2 + |\beta|^2 = 1, \quad \alpha, \beta \in \mathbb{C}, \tag{1}$$

where vectors $|0\rangle$, $|1\rangle$ stand for the computational base (in this case the standard base is used):

$$|0\rangle = \begin{bmatrix} 1 \\ 0 \end{bmatrix}, \quad |1\rangle = \begin{bmatrix} 0 \\ 1 \end{bmatrix}. \tag{2}$$

These vectors represent the classic states zero and one.

By analogy, a generalization of qubit is a qudit. In this case more base states are admissible. The state of unknown qudit with d base states is represented as:

$$|\phi\rangle = \alpha_0|0\rangle + \alpha_1|1\rangle + \alpha_2|2\rangle + \ldots + \alpha_{d-1}|d-1\rangle, \quad \sum_{i=0}^{d-1}|\alpha_i|^2 = 1, \quad \alpha_i \in \mathbb{C}. \quad (3)$$

The number of base states for a given qudit, expressed as d, will be also called a qudit's freedom level.

Standard base for qudits requires more vectors e.g. for so-called qutrits (qudits with $d = 3$) standard basis vectors are:

$$|0\rangle = \begin{bmatrix} 1 \\ 0 \\ 0 \end{bmatrix}, \quad |1\rangle = \begin{bmatrix} 0 \\ 1 \\ 0 \end{bmatrix}, \quad |2\rangle = \begin{bmatrix} 0 \\ 0 \\ 1 \end{bmatrix}. \quad (4)$$

The presented repeater protocol is realized as a single-dit teleportation protocol. The described protocol differs from other one-dit protocols [18,19] in this way that only Pauli-Z gate is used as a correction gate. The Pauli-Z (or just Z) gate for qudits may be defined as:

$$Z|j\rangle = \omega^j|j\rangle, \quad (5)$$

where ω stands for the root of unity:

$$\omega_k^d = \cos\left(\frac{2k\pi}{d}\right) + \mathbf{i}\sin\left(\frac{2k\pi}{d}\right) = e^{\frac{2\pi \mathbf{i}k}{d}}, \quad k = 0,1,2,3,\ldots,d-1. \quad (6)$$

More precisely, ω_k^d is a d-th root number k of unity. The symbol $\mathbf{i}$ represents the imaginary unit.

There are also Hadamard gate and CNOT gate used. Definitions of these gates for qudits are:

$$H|j\rangle = \sum_{k=0}^{d-1}\omega^{j\cdot k}|k\rangle, \quad CNOT|ab\rangle = |a, a \overset{d}{\oplus} b\rangle \quad (7)$$

where $a \overset{d}{\oplus} b = (a+b) \mod d$. In the case of CNOT gate the $CNOT^\dagger$ gate may be given as well – this gate is obtained as the Hermitian conjugate of CNOT operation. The $CNOT^\dagger$ gate is useful for a quantum teleportation protocol described in this section.

The exemplary matrix form of CNOT gate for qutrits ($d = 3$) is:

$$CNOT_3 = \begin{pmatrix} 1 & . & . & . & . & . & . & . & . \\ . & 1 & . & . & . & . & . & . & . \\ . & . & 1 & . & . & . & . & . & . \\ . & . & . & 1 & . & . & . & . & . \\ . & . & . & . & 1 & . & . & . & . \\ . & . & . & 1 & . & . & . & . & . \\ . & . & . & . & . & . & 1 & . & . \\ . & . & . & . & . & . & . & 1 & . \\ . & . & . & . & . & . & . & . & 1 \end{pmatrix} \quad (8)$$

where the zeros were replaced by dots to make the notation more legible. In general, the CNOT gate for qudits is constructed in a following way:

$$\text{CNOT}_d = I \oplus X^1 \oplus X^2 \oplus \ldots \oplus X^{d-2} \oplus X^{d-1}, \tag{9}$$

where the symbol $\oplus$ stands for matrix direct sum and X means the negation operation for qudit with freedom level d.

The gates, mentioned above, allow to present the teleportation protocol of an unknown quantum state:

$$|\psi\rangle = \alpha_0|0\rangle + \alpha_2|1\rangle + \alpha_2|2\rangle + \ldots + \alpha_{d-1}|d-1\rangle \quad \text{and} \quad \sum_{i=0}^{d-1}|\alpha_i|^2 = 1. \tag{10}$$

The 3-qudit state $|\psi AB\rangle$, where A belongs to Alice and B to Bob, after performing the Hadamard, CNOT and CNOT† operations on it, due to Fig. 1, may be described as:

$$|00\rangle \left(\omega^{F(0,0)}\alpha_0|0\rangle + \omega^{F(0,1)}\alpha_1|1\rangle + \omega^{F(0,2)}\alpha_2|2\rangle + \ldots + \omega^{F(0,w)}\alpha_w|w\rangle \right)$$

$$+|01\rangle \left(\omega^{F(0,0)}\alpha_0|0\rangle + \omega^{F(0,1)}\alpha_1|1\rangle + \omega^{F(0,2)}\alpha_2|2\rangle + \ldots + \omega^{F(0,w)}\alpha_w|w\rangle \right)$$

$$+\ldots\ldots\ldots\ldots\ldots\ldots\ldots\ldots\ldots\ldots\ldots\ldots$$

$$+|0w\rangle \left(\omega^{F(0,0)}\alpha_0|0\rangle + \omega^{F(0,1)}\alpha_1|1\rangle + \omega^{F(0,2)}\alpha_2|2\rangle + \ldots + \omega^{F(0,w)}\alpha_w|w\rangle \right)$$

$$+\ldots\ldots\ldots\ldots\ldots\ldots\ldots\ldots\ldots\ldots\ldots\ldots$$

$$+|w0\rangle \left(\omega^{F(w,0)}\alpha_0|0\rangle + \omega^{F(w,1)}\alpha_1|1\rangle + \omega^{F(w,2)}\alpha_2|2\rangle + \ldots + \omega^{F(w,w)}\alpha_w|w\rangle \right)$$

$$+|w1\rangle \left(\omega^{F(w,0)}\alpha_0|0\rangle + \omega^{F(w,1)}\alpha_1|1\rangle + \omega^{F(w,2)}\alpha_2|2\rangle + \ldots + \omega^{F(w,w)}\alpha_w|w\rangle \right)$$

$$+|w2\rangle \left(\omega^{F(w,0)}\alpha_0|0\rangle + \omega^{F(w,1)}\alpha_1|1\rangle + \omega^{F(w,2)}\alpha_2|2\rangle + \ldots + \omega^{F(w,w)}\alpha_w|w\rangle \right)$$

$$+\ldots\ldots\ldots\ldots\ldots\ldots\ldots\ldots\ldots\ldots\ldots\ldots$$

$$+|ww\rangle \left(\omega^{F(w,0)}\alpha_0|0\rangle + \omega^{F(w,1)}\alpha_1|1\rangle + \omega^{F(w,2)}\alpha_2|2\rangle + \ldots + \omega^{F(w,w)}\alpha_w|w\rangle \right) \tag{11}$$

where $w = d-1$. The gates CNOT and CNOT† realize the entanglement. Function F is expressed as:

$$F(a,b) = (d - a \cdot b) \bmod d. \tag{12}$$

A correction gate's form depends on the result of measurement performed on the first qudit by Alice (however, both qudits are measured in the standard base what is a consequence of Hadamard gate's use in a quantum circuit – see Fig. 1). If Alice, after the measurement, obtains one of given states

$$|00\rangle, |01\rangle, |02\rangle, \ldots, |0w\rangle,$$

then the Bob's qudit is characterized by the proper values of probability amplitudes and there is no additional operations needed to correct the obtained
state. This also stands with the accordance to the values of function F: $F(0,1)$,
$F(0,2), \ldots$ – they equal zero.

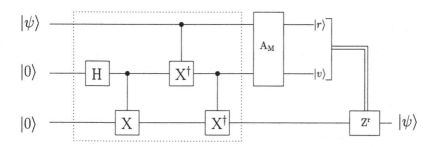

Fig. 1. A quantum circuit realizing the quantum teleportation protocol. The correction
is performed only by the Z gate. The symbols X and $X^\dagger$ stand for CNOT-type gates.
The gate A_M takes part in a process of measurement on the first and the second qudit,
but to perform the correction operation only the value of first qudit's amplitude is
necessary.

If Alice, also after the operation of measurement, obtains one of following
states
$$|10\rangle, |11\rangle, |12\rangle, \ldots, |1w\rangle,$$
then the state of Bob's qudit is
$$\omega^0 \alpha_0 |0\rangle + \omega^w \alpha_1 |1\rangle + \omega^{w-1} \alpha_2 |2\rangle + \ldots + \omega^1 \alpha_w |w\rangle. \tag{13}$$

The values of F function in this case are following:

$$F(1,0) = (d - 1 \cdot 0) \bmod d = d \bmod d = 0, \tag{14}$$
$$F(1,1) = (d - 1 \cdot 1) \bmod d = d - 1 \bmod d = d - 1 = w,$$
$$F(1,2) = (d - 1 \cdot 2) \bmod d = d - 2 \bmod d = d - 2 = w - 1,$$

$$\vdots$$

$$F(1,w) = (d - 1 \cdot w) \bmod d = d - (d-1) \bmod d = 1.$$

When the Z gate is used only once, then the correct values of some amplitudes
may be recovered. In general, the result of measurement determines the number
of operations Z which should be performed. It can be expressed as Z^r if r stands
for the result of measurement performed on the first qudit.

The above description presents the teleportation protocol for an unknown
qudit, where Z gate is needed to correct the final state and only one measurement
decides about the way how the correction is done after the teleportation process.

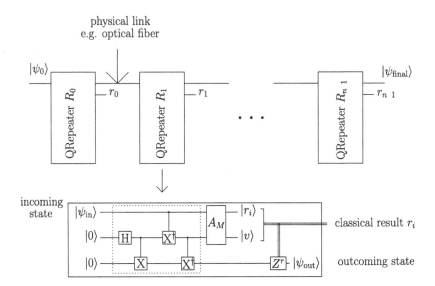

Fig. 2. The general scheme for the one-way communication protocol based on a teleportation protocol and a quantum repeater. The appearing Pauli errors are corrected with a generalized gate Z (QRepeater – Quantum Repeater).

3   Protocol for Quantum Repeater

The presented protocol of quantum communication is based on the idea of teleportation. It belongs to the third generation of protocols utilizing the concept of repeater. It is so-called one-way protocol and the information is transferred from left to right – according to Fig. 2.

The state $|\psi_0\rangle$ is the initial state of the system and it is transferred between repeaters with use of some physical medium. The final state is presented by a vector $|\psi_{\text{final}}\rangle$. The changes of system's state, after passing through following repeaters, may be described as:

$$|\psi_0\rangle \xrightarrow{r_1} |\psi_1\rangle \xrightarrow{r_2} |\psi_2\rangle \xrightarrow{r_3} \ldots \xrightarrow{r_{n-2}} |\psi_{n-1}\rangle \xrightarrow{r_{n-1}} |\psi_n\rangle \xrightarrow{r_n} |\psi_{\text{final}}\rangle. \qquad (15)$$

All states, from $|\psi_1\rangle$ to $|\psi_{\text{final}}\rangle$, need the correction of transferred information because of the protocol's structure. This action may be performed in two ways. The first solution is to correct the state r_i (i.e. the result of measurement performed on Alice's qudit) locally in each repeater by using the gate Z^{r_i}. In this case the local operations guarantee that the state $|\psi_{\text{final}}\rangle$ is equal to the initial state in terms of Fidelity measure:

$$F(|\psi_0\rangle, |\psi_{\text{final}}\rangle) = \langle\psi_0|\psi_{\text{final}}\rangle = 1 \qquad (16)$$

where F represent the Fidelity measure and the states $|\psi_0\rangle$, $|\psi_{\text{final}}\rangle$ are pure.

The second approach needs to collect all the results of local measurement from every repeater:

$$R = \{r_1, r_2, r_3, \ldots, r_{n-1}, r_n\}. \qquad (17)$$

The obtained results should be sent to the last node involved in the communication protocol. In the last node the final correction may be applied according to the following formula:

$$|\psi_{\text{Final}}\rangle = Z^f |\psi_{\text{final}}\rangle \quad \text{where} \quad f = \left(\sum_{i=1}^{n} R_i \right) \mod d \tag{18}$$

where d is a qudit's freedom level and the state on the last repeater is $|\psi_{\text{final}}\rangle$.

4   Quantum Repeaters Network as a Quantum Circuit

Just like in the case of one-dit teleportation protocol, the data transmission with use of QR may be presented as a quantum circuit. The schema of circuit for the teleportation protocol is shown in Fig. 3. The construction of circuit for quantum transmission needs a system consisting of $3 \cdot n$ qudits and n is the number of used repeaters.

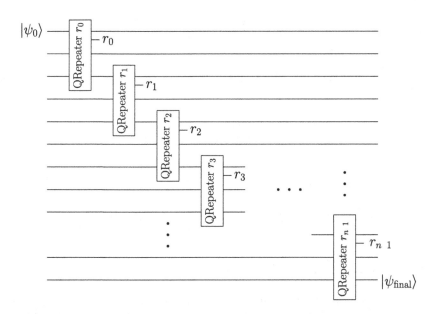

Fig. 3. The general scheme of quantum circuit which realizes the one-way communication protocol. There are $3 \cdot n$ qudits needed to send the information between the initial and the final node. It is assumed that the error correction is performed locally on each repeater. In this one-way protocol the communication is organized from qudit marked as $|\psi_0\rangle$ to qudit $|\psi_{\text{final}}\rangle$.

Although the suggested number of qudits needed to carry a correct simulation is $3 \cdot n$, it should be emphasized that the phenomenon of entanglement occurs inside of repeaters. It means that the entanglement is present in a teleportation process and it occurs between three qudits which consist the repeater itself. There is no entanglement between particular repeaters.

This assumption makes easier the implementation of mentioned quantum phenomena with use of classic computers. It is also important to know that the repeaters are connected physically (e.g. optical fiber) with suitable quantum channel. If the numeric or symbolic simulation is to be carried out on a classic machine, then only the three-qudit repeater needs to be simulated. The features of communication channel, which transmits a quantum state, may be simulated apart from the simulation of QR.

Figure 4 presents the pseudo-code illustrating the simulation's algorithm for a given state. The changes are saved in a classic variable H which collects the quantum states obtained after the calculations in every node. The initial state is assigned to variable $|\psi_0\rangle$ and the final state to $|\psi_{\text{final}}\rangle$.

```
— N : numer of nodes
— i : index of node
— |ψ₀⟩ : initial state
— H : change history of the transmitted state

H ← append(H, |ψ₀⟩, 0)
[|ψₜ⟩, r] ← QRepeater( |ψ₀⟩ )
H ← append(H, |ψₜ⟩, r)
|ψₜ⟩ ← Zʳ|ψₜ⟩
while i < N do
begin
        [|ψₜ⟩, r] ← QRepeater( |ψₜ⟩, r )
        ...
        other operations e.g. noise generation
        entanglement level measuring and etc.
        ...
        H ← append(H, |ψₜ⟩, r)
        |ψₜ⟩ ← Zʳ|ψₜ⟩
        i ← i + 1
end
|ψfinal⟩ ← |ψₜ⟩
```

Fig. 4. The algorithm for classic simulation of quantum transmission protocol based on N nodes with QRs. The symbol $\leftarrow$ stands for the classic assignment operation.

5   Conclusions

The one-dit teleportation protocol with error correction performed by Pauli-Z gate allows the realization of QR for transmitting the unknown qudit state. Two presented strategies for Z-type error correction provide the choice to the user who can run the protocol with the error correction at the end of the whole process or the correction may be performed locally on every QR. The approach with the correction at the end of transmission reduces the use of Z gate application. We use this gate only at the end of the process to complete the communication protocol by correcting the received information. This solution excludes the need of using Pauli-X and Hadamard gates, which are utilized in other known protocols for QR realization. The advantages connected with presented approach are: the influence of possible errors, caused by the imperfect realization of quantum gates, is lower and the construction of QR is easier because only the operation of measurement have to be performed (the error correction is redundant).

Planned further research on presented protocol is to use it as a quantum key distribution protocol. In this case it will be very important to estimate the number of secret bits one can generate.

The another interesting problem for further research is using quantum error correcting codes on transmitted state. The presented protocol concerns the pure states, so it would be also important to analyze the quality of transmitted information taking into account the quality of utilized quantum gates and the influence of environment.

Acknowledgments. We would like to thank for useful discussions with the *Q-INFO* group at the Institute of Control and Computation Engineering (ISSI) of the University of Zielona Góra, Poland. We would like also to thank to anonymous referees for useful comments on the preliminary version of this paper. The numerical results were done using the hardware and software available at the "GPU μ-Lab" located at the Institute of Control and Computation Engineering of the University of Zielona Góra, Poland.

References

1. Stucki, D., Legr, M., Buntschu, F., Clausen, B., Felber, N., Gisin, N., Henzen, L., Junod, P., Litzistorf, G., Monbaron, P., Monat, L., Page, J.-B., Perroud, D., Ribordy, G., Rochas, A., Robyr, S., Tavares, J., Thew, R., Trinkler, P., Ventura, S., Voirol, R., Walenta, N., Zbinden, H.: Long-term performance of the SwissQuantum quantum key distribution network in a field environment. New J. Phys. **13**(12), 123001 (2011)
2. Ritter, S., Nolleke, C., Hahn, C., Reiserer, A., Neuzner, A., Uphoff, M., Mucke, M., Figueroa, E., Bochmann, J., Rempe, G.: An elementary quantum network of single atoms in optical cavities. Nature **484**, 195–200 (2012)
3. Nikolopoulos, G.M., Jex, I.: Quantum State Transfer and Network Engineering. Springer, Heidelberg (2014)
4. Briegel, H.J., Dür, W., Cirac, J.I., Zoller, P.: Quantum repeaters: the role of imperfect local operations in quantum communication. Phys. Rev. Lett. **81**(26), 5932 (1998)
5. van Meter, R., Touch, J., Horsman, C.: Recursive quantum repeater networks. Prog. Inform. **8**, 65–79 (2011)
6. van Meter, R.: Quantum Networking. Wiley, Hoboken (2014)
7. Sangouard, N., Simon, C., de Riedmatten, H., Gisin, N.: Quantum repeaters based on atomic ensembles and linear optics. Rev. Mod. Phys. **83**, 33–80 (2011)

8. Ursin, R., Tiefenbacher, F., Schmitt-Manderbach, T., Weier, H., Scheidl, T., Lindenthal, M., Blauensteiner, B., Jennewein, T., Perdigues, J., Trojek, P., Ömer, B., Fürst, M., Meyenburg, M., Rarity, J., Sodnik, Z., Barbieri, C., Weinfurter, H., Zeilinger, A.: Entanglement-based quantum communication over 144 km. Nat. Phys. **3**, 481–486 (2007)
9. Hirvensalo, M.: Quantum Computing. Springer, Heidelberg (2001)
10. Klamka, J., Węgrzyn, S., Znamirowski, L., Winiarczyk, R., Nowak, S.: Nano and quantum systems of informatics. Bull. Pol. Acad. Sci. Tech. Sci. **52**(1), 1–10 (2004)
11. Klamka, J., Węgrzyn, S., Bugajski, S.: Foundation of quantum computing. Archiwum Informatyki Teoretycznej i Stosowanej **1**(2), 97–142 (2001)
12. Klamka, J., Węgrzyn, S., Bugajski, S.: Foundation of quantum computing. Archiwum Informatyki Teoretycznej i Stosowanej. Part 2 **14**(2), 93–106 (2002)
13. Nielsen, M.A., Chuang, I.L.: Quantum Computation and Quantum Information. Cambridge University Press, New York (2000)
14. Duan, L.M., Lukin, M.D., Cirac, J.I., Zoller, P.: Long-distance quantum communication with atomic ensembles and linear optics. Nature **414**, 413–418 (2011)
15. Jiang, L., Taylor, J.M., Nemoto, K., Munro, W.J., van Meter, R., Lukin, M.D.: Quantum repeater with encoding. Phys. Rev. A **79**, 032325 (2009)
16. Munro, W.J., Stephens, A.M., Devitt, S.J., Harrison, K.A., Nemoto, K.: Quantum communication without the necessity of quantum memories. Nat. Photonics **6**, 777–781 (2012)
17. Muralidharan, S., Kim, J., Lütkenhaus, N., Lukin, M.D., Jiang, L.: Ultrafast and fault-tolerant quantum communication across long distances. Phys. Rev. Lett. **112**, 250501 (2014)
18. Aliferis, P., Leung, D.W.: Computation by measurements: A unifying picture. Phys. Rev. A **70**(6), 062314 (2004)
19. Knill, E.: Quantum computing with realistically noisy devices. Nature **434**, 39–44 (2005)

Multi-level Virtualization and Its Impact on System Performance in Cloud Computing

Paweł Lubomski[1](✉), Andrzej Kalinowski[1], and Henryk Krawczyk[2]

[1] IT Services Centre, Gdańsk University of Technology, Gdańsk, Poland
{lubomski,andrzej.kalinowski}@pg.gda.pl
[2] Faculty of Electronics, Telecommunications and Informatics,
Gdańsk University of Technology, Gdańsk, Poland
hkrawk@eti.pg.gda.pl

Abstract. The results of benchmarking tests of multi-level virtualized environments are presented. There is analysed the performance impact of hardware virtualization, container-type isolation and programming level abstraction. The comparison is made on the basis of a proposed score metric that allows you to compare different aspects of performance. There is general performance (CPU and memory), networking, disk operations and application-like load taken into account. The tested technologies are, inter alia, VMware ESXi, Docker and Oracle JVM.

Keywords: Virtualization · Performance · Benchmark

1 Introduction

For the last few years an intensive growth of popularity in the cloud services has been observed. Nearly all hosting providers now offer such a dynamic cloud service for an affordable cost. The most popular services are Infrastructure-as-a-Service (IaaS) or Platform-as-a-Service (PaaS). There are three main types of cloud: public, private and hybrid [1].

There has also been a significant progress in technology supporting widely understood clouds. Many types of virtualization were developed. They can be divided into 3 main groups [2]. There are hardware virtualizations, containers inside systems, and programming language abstractions such as e.g. Java Virtual Machine (JVM). The last type concerns mainly large scale, service-oriented systems which use .NET or Java Enterprise Edition (JEE) platforms. More detailed classification will be described later on.

The main reason for the popularity of dynamic cloud solutions is comfort and ease of use. Dynamic clouds help share resources among many virtual systems and manage them dynamically depending on the systems' load. This way it is possible to overbook the resources. But the most important advantage is the time of a new virtual server deployment – it is incomparably shorter than buying traditional hardware, placing it in datacenter and operating system installation. Thereby, it is very easy and cheap to add some new virtual machines to a cluster

© Springer International Publishing Switzerland 2016
P. Gaj et al. (Eds.): CN 2016, CCIS 608, pp. 247–259, 2016.
DOI: 10.1007/978-3-319-39207-3_22

for the short time of an increased system load, e.g. academic session, recruitment for the university. When the load decreases, the resources are released and can be used for other purposes.

The deployment process is very fast thanks to, among other reasons, using some prepared templates and snapshots. It may also be a way of releasing new versions of software – the new image is only introduced. Especially Docker images [3,4] are popular for this way of releasing software [5].

There is no rose without a thorn. All these benefits come together with a cost of performance decrease. Of course, the developers of cloud solutions intensively work on the performance improvement, so that this cost is getting lower every day. There are not currently available any reliable reports comparing the costs of these solutions. The last good job was done in 2014 by IBM Research Division [6].

At our university we started to use virtualization intensively, and put it one into another. We use VMware ESXi virtualization on which we run a Debian Linux guest OS. Inside the guest OS we run a Docker image containing the release of our central system. Because our university system is very big, it was designed as a distributed JEE system written in Java language. In this way we have a three-level virtualization used. Thus, we want to check what the real cost of such an approach is.

2   Virtualization Methodologies

As mentioned earlier there are three types (levels) of virtualization: hardware, containers and platform (e.g. JVM). We will briefly describe them below.

2.1   Hardware Virtualization

The hardware virtualization can be divided into two main groups depending on the type of host operating system (hypervisor). The first one is a native hypervisior (also known as "type 1" or "bare-metal") where the guest OS works on virtualized hardware and the host OS is not a standard OS but a highly specialized solution. They are able to take advantage of all hardware capabilities, especially those supporting virtualization. The main representatives of this group are: VMware ESXi [7], XenServer [8], MS Hyper-V [9].

The second type are hosted hypervisors (also known as "type 2"). These solutions cannot work alone – they require a standard operating system and work as a regular application in this system. For better performance they use special kernel modules and/or libraries of host operating system such as KVM (Kernel Virtualization Module) [10,11].

The intention of using this solution is not productive environments – they are rather for testing images of virtual machines (VMs) during the development and testing process. They may also be an easy and cheap solution of a multi-technology environment for testing traditional software on different operating systems, etc. The most famous solutions of this type are: VirtualBox [12], VMware Player [13], QEMU [14].

2.2   Containers

Another approach accompanies the development of containers. This is the oldest type, also known as para-virtualization. The most characteristic aspect of this type of virtualization is that the host OS only separates processes and resources so that it is impossible to run another operating system than the host OS. It started with *chroot*, then there were FreeBSD jails [15] and Solaris zones [16].

Nowadays there are two mainly-used solutions: OpenVZ [17] based on a customized Linux Kernel, and Docker [3] which uses LinuxContainers [18] or libcontainer library (from version 0.9).

Due to the taken architectural assumptions, they should have a lower performance overhead (than hardware virtualization solutions) but they also have some limitations (e.g. impossibility to choose another guest OS) and a worse isolation of processes and resources usage.

2.3   Programming Language Abstraction

In large-scale distributed e-service systems there are mainly used .NET and JEE. Both of them use a specific level of programming language abstraction – .NET Framework and JVM. Especially in the case of JVM the portability is on a high level, however it is done at the cost of a negative impact on performance.

Figure 1 presents the virtualization stack of our university central system. We use either hardware virtualization (hypervisor type 1 – VMware ESXi), containers (Docker) with images containing release versions of software written in Java, so it is run inside a JVM. The use of Docker images ensure that the system is configured and behaves the same in every environment: development, testing and production.

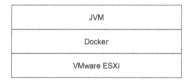

Fig. 1. Virtualization stack of our university system

3   Benchmark Methodology and Tools

The IBM research report [6] focused on the global performance of Docker, baremetal, and pure virtualization environments. They have taken the assumption that so-called *double virtualization* is redundant and has a negative impact on performance. In our paper we want to verify this hypothesis and check if the ease of use (in terms of maintenance, resource reallocation and scalability) of such virtualization infrastructure is worth the cost of the predicted performance drop.

The IBM research team performed comprehensive benchmarking of Docker containers, including both networking mechanisms (i.e. direct host mapping and

Table 1. Benchmarked test configurations

Machine type	OS configuration	Docker configuration
bare-metal	metal.local	metal.aufs
	metal.remote	metal.mnt
		metal.aufs.nat
		metal.aufs.host
		metal.mnt.nat
		metal.mnt.host
virtual machine	vm.local	vm.aufs
	vm.remote	vm.mnt
		vm.aufs.nat
		vm.aufs.host
		vm.mnt.nat
		vm.mnt.host

port forwarding) and two I/O subsystems (direct mounts and *AUFS* – they have omitted: device mapper and btrfs backends). In our paper we want to focus on both the double and triple virtualization problems. In order to achieve that, we need to run our test containers in the same manner as the IBM team did. Therefore, the containers were tested in several different configurations (see Table 1).

3.1 Test Configurations and Suites

For benchmarking purposes we have used three concepts:

test configurations – a set of all available testing environments (i.e. all combinations of virtual/bare-metal and Docker instances),
test suites – a set of *test cases* which measure a similar aspect of performance in a specified *test configuration*,
test case – an individual performance measurement.

Table 1 presents tested configurations. They correspond to the first two lowest levels of virtualization in our university system stack presented in Fig. 1: hardware virtualization (VMware ESXi) and container (Docker). The highest layer of virtualization (JVM) is covered not by the test configuration, but by two test suites: java and rich java application (see Table 2) – this way we could test java applications on every considered type of virtualization. So we benchmarked the 16 following configurations:

(metal|vm).local pure operating system with no Docker container layer, tests are performed locally,

(metal|vm).remote pure operating system with no Docker container layer, tests are performed from a remote location, thus the network layer is also being tested,

(metal|vm).aufs tests are performed on a Docker shared file system – *AUFS*, therefore I/O operations have a serious drawback,

(metal|vm).mnt tests are performed on mounted volume, therefore I/O performance is significantly improved,

(metal|vm).aufs.nat the same as **.aufs** but uses port forwarding,

(metal|vm).aufs.host the same as **.aufs** but uses direct host mapping,

(metal|vm).mnt.nat the same as **.mnt** but uses port forwarding,

(metal|vm).mnt.host the same as **.mnt** but uses direct host mapping.

Table 2. Test configurations and suites matrix

Test configuration	Test suites						
	CPU	multicore	memory	java	disk	network	application
(metal|vm).local	✓	✓	✓	✓	✓		✓
(metal|vm).remote						✓	✓
(metal|vm).aufs	✓	✓	✓	✓	✓		✓
(metal|vm).mnt				✓			✓
(metal|vm).aufs.nat						✓	✓
(metal|vm).aufs.host						✓	✓
(metal|vm).mnt.nat							✓
(metal|vm).mnt.host							✓

The performance of each test configuration was separately measured in the following aspects: CPU, memory, I/O and network. The final test suite covered a typical web application resource usage. This way, we adopted 7 test suites:

1. CPU – overall computation performance tests,
2. multicore – specific tests run to verify multiple CPU cores' performance,
3. memory – overall memory performance tests,
4. java – CPU utilization during java application execution,
5. disk – overall disk performance tests,
6. network – overall network layer performance tests,
7. application – complex performance tests.

Table 2 presents the combinations of test configurations and test suites which were measured during our benchmarking. One can easily notice that only the most representative and not-redundant combinations were taken into account.

3.2   Hardware and Software Configuration

Our tests were launched on Dell R210 hardware with one Intel Xeon E31270 @3.39 GHz (8 cores) processor and 16 GiB of RAM. Network throughput was 1 Gbps. And we used 1 TB Western Digital WD1003FBYX-0 disk.

Our tests were based on a standard GNU/Linux Debian 8.2 distribution which was used as a host and guest OS and also as a base for Docker containers. As for Docker daemon we used a currently available version 1.9. For hardware virtualization tests we used *VMware ESXi 6.1*. For Java applications we used *Oracle Java 1.8.0_45-b14*.

We did not use any limitations like *ulimit, cgroups* or ESXi mechanisms, hence all tests were run on all available resources. On ESXi we used *Hardware version 11* with all default options, for disk storage we used *Local Storage* with thin provisioning and standard 1 MiB data blocks.

To measure our tests we used mainly *Phoronix Test Suite 6.0.1* which was run on *PHP 5.6.14*. Additionally, for measuring network performance we used *netperf 2.6.0*. And finally, for overall testing we used *IBM Daytrader 3* web application with *Apache Jmeter 2.13*.

3.3   General and Disk Performance

As mentioned earlier we divided our tests into 7 test suites. For suites 1–5 we used *Phoronix Test Suite*, with the following standard test suites:

1. pts/cpu (27 different test cases),
2. pts/multicore (19 different test cases),
3. pts/memory (9 different test cases),
4. pts/java (6 different test cases),
5. pts/disk (17 different test cases).

Each of these test cases consisted of three runs. The average score of those runs was selected as the result. The results of each test case were normalized, and finally summed to provide a total score for each test suite.

3.4   Network Performance

For test suite no. 6 we used two test cases:

1. netperf in tcp mode,
2. netperf in udp mode.

Each test case contained 10 identical tests. After running them, the average score was computed, and afterwards the total score was calculated in the same manner as in the previous test suites. In all cases the tested machine was used as a server.

3.5   Application Stack Performance

For test suite no. 7 we used a complex application which tries to utilize all previous test suites to provide a final performance result. It consists of *IBM Daytrader 3* web application which is *a Java EE 6 application built around an online stock trading system. The application allows users to login, view their portfolio, look up stock quotes, buy and sell stock shares, and more* [19]. We deployed it on a *WildFly 9.0.2.Final Application Server*, *H2* standalone database and with EJB3 Mode enabled in the *Daytrader* application (which uses EJBs with JPA to connect to the database).

To measure the total performance of this test suite we used an *Apache JMeter* which was run within the tested system, and also from a remote host, to check network performance. For each run we started with a clean database and a 180 s dry-run. After that, we measured the performance for 180 s, in 6 s intervals.

3.6   Score Calculation

We propose the following score calculation methodology. Each completed test case *produces* an average result. This value is normalized with all average results collected from all test configurations. The acquired values are called test scores. After completion of a particular test suite, those test scores are summed up to present a final performance score in each of the 7 test suites. Most of values collected by test runs are presented as HIB (higher is better) but some (especially in multicore test suite) are presented as LIB (lower is better). In the second case the test score must be subtracted from the final performance score.

Let us assume that our results can be stored in a large 3 dimensional array – t containing all test suites, test cases and test configurations with s, i, c as their indices. Therefore, $t_{s,i,c}$ is a single result of test configuration c on test case i in suite s. $t_{s,i}$ is a vector containing the results of test case i in suite s and $|t_{s,c}|$ is the number of test cases in the test suite s for the particular configuration c. In such a case we can use the following formulae:

$$\text{test\_score}_{s,i,c} = \frac{t_{s,i,c}}{\max(t_{s,i})} \tag{1}$$

$$\text{total\_score}_{s,c} = \sum_{i=1}^{|t_{s,c}|} \text{sgn}_s(i) \cdot \text{test\_score}_{s,i,c} \tag{2}$$

where $\text{sgn}_s(i)$ is "−1" when LIB and "+1" in HIB case.

4   Benchmark Results

4.1   CPU and Memory Performance

Table 3 presents aggregated results for CPU, java and memory test suites. The final results were predictable and are similar to those presented in the IBM report. Docker performance drop in single core configuration is around 3 % for both bare-metal machine and virtualized environment. The multicore performance has a higher drop of 8 %. Java performance tests were based on *scimark*, *bork file encrypter* and *sunflow* rendering system. Therefore, they mainly focused on CPU and memory efficiency. The outcome is similar to the CPU performance. Memory drop is insignificant. ESXi virtualization layer produces around 15 % of CPU performance drop and lowers java application efficiency by 10 %. Memory drop is lower than 2 %.

Table 3. CPU and memory performance – total scores

Test suite	metal.local	metal.aufs	vm.local	vm.aufs
CPU	7.43	7.29	6.36	6.08
multicore	−4.31	−4.64	−4.79	−5.16
java	2.17	2.03	1.96	1.93
memory	9	9	8.83	8.82

When testing Docker CPU/memory efficiency we were really benchmarking the cgrups/namespaces efficiency, thus the Docker overhead is almost nonexistent. The differences between metal-docker test cases were minimal (less than 1 %), but while summing 27 results we ended up with an overhead of 3 %.

4.2   Network Performance

In Table 4 there is presented the overall network performance. The Docker in direct host mapping configuration (**.host**) has no significant performance drop. The port forwarding configuration (**.nat**) lowers performance by around 4 %. The highest efficiency drop (14 %) can be observed in the ESXi environment.

Table 4. Network performance

Test suite	metal.aufs.host	metal.aufs.nat	metal.remote	vm.aufs.host	vm.aufs.nat	vm.remote
netperf tcp	1.00	0.93	1.00	0.86	0.84	0.86
netperf udp	1.00	0.97	1.00	0.88	0.84	0.87
Total score	2.00	1.9	2.00	1.72	1.68	1.73

Table 5. Bare-metal configurations disk performance test scores. In brackets there are in-normalized results denominated in given units.

Test case	Order	Unit	metal.local	metal.mnt	metal.aufs
AIO-Stress	HIB	MB/s	1.00 (2637.12)	0.92 (2430.07)	0.89 (2336.56)
SQLite	LIB	Seconds	1.00 (524.31)	0.88 (463.31)	0.86 (448.47)
FS-Mark	HIB	Files/s	0.34 (15.67)	0.40 (18.43)	0.40 (18.43)
Dbench1	HIB	MB/s	0.13 (39.22)	0.14 (41.72)	0.15 (43.72)
Dbench2	HIB	MB/s	0.13 (66.93)	0.14 (72.27)	0.14 (69.34)
Dbench3	HIB	MB/s	0.28 (67.5)	0.30 (72.43)	0.24 (58.63)
Dbench4	HIB	MB/s	0.36 (9.68)	0.41 (10.76)	0.40 (10.5)
IOzone1	HIB	MB/s	1.00 (8059.31)	0.97 (7780.49)	0.38 (3032.21)
IOzone2	HIB	MB/s	0.60 (69.02)	0.97 (111.48)	0.96 (110.45)
Threaded I/O Tester1	HIB	MB/s	1.00 (13595.01)	0.99 (13453.36)	0.83 (11330.63)
Threaded I/O Tester2	HIB	MB/s	0.42 (0.55)	0.41 (0.53)	0.45 (0.59)
Compile Bench1	HIB	MB/s	0.89 (487.95)	1.00 (548.2)	0.71 (388.47)
Compile Bench2	HIB	MB/s	0.95 (261.22)	1.00 (274.74)	0.56 (154.23)
Compile Bench3	HIB	MB/s	0.80 (1460.93)	0.74 (1343.32)	0.29 (533.12)
Unpacking	LIB	Seconds	0.89 (8.61)	0.89 (8.61)	0.92 (8.95)
PostMark	HIB	TPS	0.99 (5208.0)	1.00 (5282.0)	0.60 (3178.0)
Gzip Compression	LIB	Seconds	0.94 (12.85)	0.94 (12.93)	0.95 (13.05)
Apache Benchmark	HIB	RpS	1.00 (29040.6)	0.89 (25830.51)	0.80 (23352.62)
Total score	HIB		7.06	7.57	5.07

4.3   Disk Performance

The disk performance results are shown in Tables 5 and 6. Oddly, the overall disk performance is significantly greater in the ESXi environment (roughly, by 39 %). This is probably the effect of VMware *VMFS* filesystem features [20], notably the huge block size of 1 MiB. The *ext4* file system used on bare-metal system has a maximum block size of 4 KiB. Therefore, file system buffers will sync more frequently than the *VMFS* counterpart, leading to the observed performance drop.

In some test cases the efficiency of the Docker *AUFS* file system is greater than pure OS. This is the effect of the *COW* (copy on write) mechanism which writes changes to the disk only when it is needed [21,22]. Similar *COW* mechanisms are used while mounting volumes inside Docker (**metal.mnt** and **vm.mnt**).

The bare-metal supremacy performance can be observed in asynchronous AIO-Stress, IOzone1 (8 GiB read test), Threaded I/O tester1 (64 MiB random read by 32 threads), Gzip Compression (2 GiB file) and Apache Benchmark. While analysing the results one can notice that the bare-metal environment shows some serious performance drops on writes (e.g. IOzone2, Threaded I/O tester2). This is the effect of the aforementioned *COW* mechanisms on Docker and *VMFS* buffers.

Table 6. Virtualized configurations disk performance test scores. In brackets there are in-normalized results denominated in given units.

Test case	Order	Unit	vm.local	vm.mnt	vm.aufs
AIO-Stress	HIB	MB/s	0.85 (2239.05)	0.79 (2094.9)	0.75 (1969.52)
SQLite	LIB	Seconds	0.26 (136.46)	0.28 (144.8)	0.28 (144.39)
FS-Mark	HIB	Files/s	1.00 (46.03)	0.97 (44.87)	0.98 (44.97)
Dbench1	HIB	MB/s	1.00 (301.23)	0.98 (295.08)	0.90 (272.4)
Dbench2	HIB	MB/s	1.00 (508.15)	0.93 (472.84)	0.84 (428.83)
Dbench3	HIB	MB/s	1.00 (244.02)	0.80 (194.63)	0.80 (196.13)
Dbench4	HIB	MB/s	1.00 (26.55)	0.96 (25.51)	0.95 (25.24)
IOzone1	HIB	MB/s	0.94 (7559.72)	0.96 (7714.91)	0.69 (5569.87)
IOzone2	HIB	MB/s	1.00 (115.35)	0.99 (114.35)	0.98 (112.72)
Threaded I/O Tester1	HIB	MB/s	0.91 (12390.95)	0.89 (12157.88)	0.81 (11076.75)
Threaded I/O Tester2	HIB	MB/s	0.90 (1.17)	1.00 (1.3)	0.95 (1.24)
Compile Bench1	HIB	MB/s	0.66 (360.75)	0.65 (356.42)	0.61 (336.42)
Compile Bench2	HIB	MB/s	0.73 (199.35)	0.76 (208.75)	0.61 (166.58)
Compile Bench3	HIB	MB/s	1.00 (1825.35)	0.75 (1367.49)	0.76 (1389.03)
Unpacking	LIB	Seconds	0.99 (9.64)	0.98 (9.52)	1.00 (9.69)
PostMark	HIB	TPS	0.89 (4716.0)	0.90 (4746.0)	0.72 (3807.0)
Gzip Compression	LIB	Seconds	1.00 (13.66)	1.00 (13.63)	1.00 (13.69)
Apache Benchmark	HIB	RpS	0.77 (22259.71)	0.67 (19492.16)	0.59 (17089.18)
Total score	HIB		11.4	10.74	9.66

4.4   Application Stack Performance

The results of the most comprehensive test suite are presented in Table 7. We ran this test suite on 16 test configurations, thus benchmarking all useful combinations. The most important finding is the overall low performance of the ESXi network stack (around 25 %, thus lower than in network performance benchmark).

If we compare the local executions of JMeter (thus removing the network stack) we can see that the overall performance of the virtualized environment is almost identical to the bare-metal one (see Fig. 2), with some minor fluctuation of the *AUFS* filesystem (4 %). Docker with mounted volume and direct network mapping is almost identical in performance to its bare-metal counterpart. The only difference can be observed in the virtualized environment where pure OS versions are faster by 6 % (see Fig. 3).

There are no significant differences between Docker network modes (NAT and direct network mapping) in this test suite (see Fig. 4). The performance of *AUFS* and mounted volumes is almost identical to ESXi, some slight differences appear in the bare-metal environment(see Fig. 5).

Table 7. Application stack performance – total scores

	Test configuration	Total score
Local execution	metal.local	0.95
	metal.aufs	0.94
	metal.mnt	0.95
	vm.local	0.95
	vm.aufs	0.91
	vm.mnt	0.95
Remote execution	metal.remote	1
	metal.aufs.host	0.96
	metal.aufs.nat	0.95
	metal.mnt.host	1
	metal.mnt.nat	0.99
	vm.remote	0.79
	vm.aufs.host	0.75
	vm.aufs.nat	0.74
	vm.mnt.host	0.75
	vm.mnt.nat	0.76

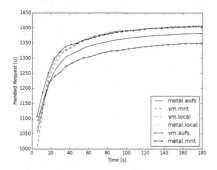

Fig. 2. Application stack performance – Local JMeter executions

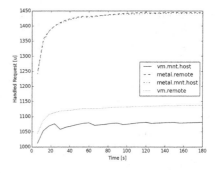

Fig. 3. Application stack performance – Remote JMeter executions

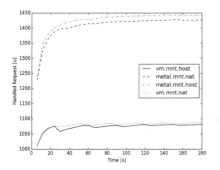

Fig. 4. Docker port forwarding vs direct host mapping comparison

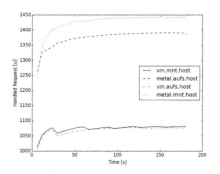

Fig. 5. Docker AUFS vs mounted volumes comparison

5   Conclusions and Future Work

The most significant benchmarking results were obtained in disk and application stack performance measurements. The overall results show that there is no significant decrease in performance while using three-level virtualization configurations. The decrease is from 1 to 9 % with the exception of ESXi network layer overhead. The ESXi network layer is significantly slower when using a standard configuration. The observed network bottleneck in ESXi environment can probably be reduced by changing the default network driver.

Container layer (Docker) has a negligible impact on performance independently of using hardware virtualization or not. This refers to medium loaded systems such as our university system. The Docker default settings are production ready – the overall performance drop is insignificant.

Superior ESXi disk performance may be obtained by using external disk storage. In the future there should be used some larger files (bigger than RAM) for benchmarking disk operations. For additional disk performance measurement we should also focus on a large database.

Finally, we can say that the cost (understood as system performance decrease) of using multi-level virtualization in our university system is acceptable, especially while taking into account the advantages of such configuration (see Introduction). This cost can be balanced by introducing horizontal scaling (clustering) when the system load is higher. It is very quick and easy to do thanks to the usage of virtualization on the first and second levels. There is an overwhelming need to check non-default ESXi network drivers because effective networking is very important while clustering large scale distributed e-service systems, which is the point of cloud configurations.

References

1. Mather, T., Kumaraswamy, S., Latif, S.: Cloud Security and Privacy. An Enterprise Perspective on Risks and Compliance. O'Reilly, Sebastopol (2009)
2. Perez, R., Van Doorn, L., Sailer, R.: Virtualization and hardware-based security. IEEE Secur. Priv. Mag. **6**(5), 24–31 (2008)
3. Merkel, D.: Docker: lightweight Linux containers for consistent development and deployment. Linux J. **2014**(239), 2 (2014)
4. Bernstein, D.: Containers and cloud: from LXC to Docker to Kubernetes. IEEE Cloud Comput. **1**(3), 81–84 (2014)
5. Di Tommaso, P., Palumbo, E., Chatzou, M., Prieto, P., Heuer, M.L., Notredame, C.: The impact of Docker containers on the performance of genomic pipelines. PeerJ PrePr. **3**, e1428 (2015)
6. Felter, W., Ferreira, A., Rajamony, R., Rubio, J.: An updated performance comparison of virtual machines and Linux containers. IBM Research Report (2014)
7. VMware. Performance Best Practices for VMware vSphere® 6.0. http://www.vmware.com/files/pdf/techpaper/VMware-PerfBest-Practices-vSphere6-0.pdf
8. Barham, P., Dragovic, B., Fraser, K., Hand, S., Harris, T., Ho, A., Neugebauer, R., Pratt, I., Warfield, A.: Xen and the Art of Virtualization. http://www.cl.cam.ac.uk/research/srg/netos/papers/2003-xensosp.pdf

9. Microsoft: Why Hyper-V? Competitive Advantages of Microsoft Hyper-V Server 2012 over the VMware vSphere Hypervisor. http://download. microsoft.com/download/5/7/8/578E035F-A1A8-4774-B404-317A7ABCF751/ Competitive-Advantages-of-Hyper-V-Server-2012-over-VMware-vSphere-Hypervisor. pdf

10. Kivity, A., Lublin, U., Liguori, A., Kamay, Y., Laor, D.: kvm: the Linux virtual machine monitor. In: Proceedings of the Linux Symposium, vol. 1, pp. 225–230 (2007)

11. Fenn, M., Murphy, M.A., Martin, J., Goasguen, S.: An evaluation of KVM for use in cloud computing. In: Proceedings of the 2nd International Conference on the Virtual Computing Initiative - ICVCI 2008, vol. V, pp. 1–7 (2008)

12. https://www.virtualbox.org/

13. http://www.vmware.com/products/player/

14. Bartholomew, D.: QEMU: a Multihost, Multitarget Emulator. http://www.ee. ryerson.ca/~courses/coe518/LinuxJournal/elj2006-145-QEMU.pdf

15. Kamp, P., Watson, R.: Jails: confining the omnipotent root. http://phk.freebsd. dk/pubs/sane2000-jail.pdf

16. Oracle: Oracle® Solaris 11.1 Administration: Oracle Solaris Zones, Oracle Solaris 10 Zones, and Resource Management (2013). http://docs.oracle.com/cd/E26502_01/pdf/E29024.pdf

17. https://openvz.org/Features

18. Rosen, R.: Resource management: Linux kernel Namespaces and cgroups (2013). http://www.haifux.org/lectures/299/netLec7.pdf

19. McClure, J.: Measuring performance with the Daytrader 3 benchmark sample (2014). https://developer.ibm.com/wasdev/docs/measuring-performance-daytrader-3-benchmark-sample/

20. https://www.vmware.com/pl/products/vsphere/features/vmfs

21. Kasampalis, S.: Copy On Write Based File Systems Performance Analysis And Implementation. Kongens Lyngby (2010). http://faif.objectis.net/download-copy-on-write-based-file-systems

22. https://docs.docker.com/engine/userguide/storagedriver/aufs-driver/

Ping-Pong Protocol Eavesdropping in Almost Perfect Channels

Piotr Zawadzki[(✉)]

Institute of Electronics, Silesian University of Technology,
Akademicka 16, 44-100 Gliwice, Poland
Piotr.Zawadzki@polsl.pl

Abstract. An undetectable eavesdropping of the entanglement based quantum direct communication in lossy quantum channels has already been demonstrated by Zhang et al. (Phys Lett A 333(12):46–50, 2004). The circuit proposed therein induces losses at a constant 25 % rate. Skipping of some protocol cycles is advised in situations when the induced loss rate is too high. However, such policy leads to a reduction in information gain proportional to the number of skipped cycles.

The entangling transformation, parametrized by the induced loss ratio, is proposed. The new method permits fine-tuning of the loss ratio by a modification of coupling coefficients. The proposed method significantly improves efficiency of the attack operated in the low loss regime. The other properties of the attack remain the same.

Keywords: Ping-pong protocol · Quantum direct communication

1 Introduction

The ping-pong protocol [1] aims to provide confidentiality without encryption. It has been proven that it is asymptotically secure in lossless channels [2–6]. Unfortunately, the existence of the losses is a rule in a quantum world and their exclusion from the analysis is an oversimplification. Zhang et al. [7] presented a circuit than permits successful eavesdropping of 0.311 bits per protocol cycle at the price of loosing one quarter of control photons. The attack targets the only practical implementation [8] of the protocol, so its further analysis and deeper understanding is scientifically justified.

To exemplify the power of the attack, it is frequently argued that the existing quantum channel can be replaced with a better one to mask the presence of the circuit. Although in practice this is usually not the option, the attack still cannot be excluded completely. Legitimate parties usually monitor the average loss ratio of the channel and they assume some safety margin to avoid false alarms. The presence of an adversary is hidden as long as the additional losses stay below that margin, which is usually much lower than 25 % rate. The typical policy of an eavesdropper is to skip some protocol cycles to reduce induced losses. Linear reduction of average information gain with the number of skipped cycles is the price he pays.

© Springer International Publishing Switzerland 2016
P. Gaj et al. (Eds.): CN 2016, CCIS 608, pp. 260–268, 2016.
DOI: 10.1007/978-3-319-39207-3_23

It follows that impact of the attack on practically deployed systems is determined by its properties in the low-loss regime. The new method of an attack that outperforms known solutions in this area is proposed. It permits fine-tuning of the loss ratio by a modification of coupling coefficients of the entangling transformation. The other properties of the attack remain the same.

The paper is constructed as follows. A brief reclaim the ping-pong protocol and the analysis of the Zhang's circuit is provided in Sect. 2. An alternative entangling transformation and key results are introduced in Sect. 3. Concluding remarks are made in the last section.

2   Analysis

The message mode of the ping-pong protocol is composed of three phases: an entanglement distribution, a message encoding and its subsequent decoding. Its further description adheres to the standard cryptographic personification rules: Alice and Bob are the names of the communicating parties, the malevolent eavesdropper is referred as Eve.

Bob starts the communication process through the creation of an EPR pair

$$|\psi_{\text{init}}\rangle = |\psi^+\rangle = (|0_B\rangle|1_A\rangle + |1_B\rangle|0_A\rangle)/\sqrt{2}. \tag{1}$$

The qubits that constitute the pair are further referred to as home and travel/signal qubits, respectively. Bob sends the signal qubit to Alice. Alice applies a phase flip operator $\mathcal{Z}_A = |0_A\rangle\langle 0_A| - |1_A\rangle\langle 1_A|$ to the received qubit to encode a single classic bit

$$|\psi^\nu\rangle = \mathcal{Z}_A|\psi_{\text{init}}\rangle = (\mathcal{Z}_A)^\nu|\psi^+\rangle = ((-1)^\nu|0_B\rangle|1_A\rangle + |1_B\rangle|0_A\rangle)/\sqrt{2}. \tag{2}$$

The signal particle is sent back to Bob, who identifies the transformation that has been applied through the measurement of both qubits (Fig. 1).

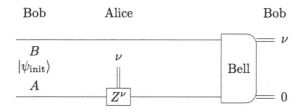

Fig. 1. The schematic diagram of a message mode in the ping-pong protocol

Unfortunately, such a communication scenario is vulnerable to the intercept-resend attack. As a countermeasure, Alice measures the received qubit in randomly selected protocol cycles and asks Bob over an authenticated classic channel to do the same with his qubit (Fig. 2). Her measurement causes the collapse of the shared state. The correlation of the outcomes is preserved only if the qubit measured by Alice is the same one that was sent by Bob. That way Alice and Bob can convince themselves with the confidence approaching certainty that the quantum channel is not spoofed provided that they have executed a sufficient number of control cycles. The above scheme is asymptotically secure in the absence of losses and/or transmission errors i.e., in a perfect quantum channel.

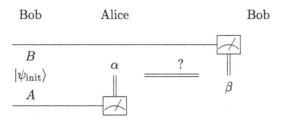

Fig. 2. The schematic diagram of a control mode in the ping-pong protocol

Further, we will consider individual (incoherent) attacks in which Eve attacks each protocol cycle independently. The signal particle travelling back and forth between Alice and Bob can be the subject of any quantum action Q introduced by Eve, as depicted in Fig. 3. Eve's activity can be described as a unitary operation acting on the signal qubit and two additional qubit registers, as follows from Stinespring's dilation theorem.

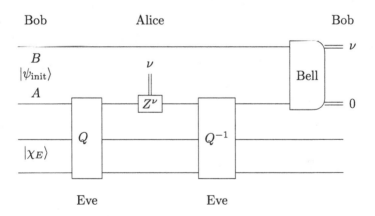

Fig. 3. A schematic diagram of an individual attack

The efficiency of the eavesdropping detection depends on the properties of the control mode. In the seminal version of the protocol, the reliability of quantum communication is estimated by the measurements in the computational basis. The probability of verification failure p_C (the specific form of the operator depends on the assumed initial state) and the probability of a non-conclusive control cycle p_L (it is assumed that Bob's qubit is never lost) can be found as

$$\mathcal{P}_C = |0_B\rangle|0_A\rangle\langle 0_A|\langle 0_B| + |1_B\rangle|1_A\rangle\langle 1_A|\langle 1_B|, \tag{3}$$

$$\mathcal{P}_L = \mathcal{I}_B \otimes |v_A\rangle\langle v_A|, \tag{4}$$

$$p_{C,L} = \text{Tr}_{x,y}\left((\mathcal{P}_{C,L} \otimes \mathcal{I}_E)(\mathcal{I}_B \otimes \mathcal{Q})|\psi_{\text{init}}\rangle|\chi_E\rangle\right). \tag{5}$$

where $|v\rangle$ denotes vacuum state and $|\chi_E\rangle$ is the initial state of the ancilla system.

Zhang's attack [7] addresses the violation of the protocol security in the presence of losses in a quantum channel. The clever circuit permits the detection of phase flip operations at the price of introducing some losses. The expected correlation of outcomes of the conclusive measurements made in the control mode (3) is also preserved so the attack is undetectable.

The Zhang's circuit is composed of two modules [7, Eq. (2)]: the coupling unit $\mathcal{C}_U$ followed by the selective swap $\mathcal{C}_{\text{SWAP}}$ (Fig. 4). The first one entangles the signal qubit from register A with the ancilla registers x and y. The second module swaps the contents of the A and x registers exclusively for signal qubit being in state $|1_A\rangle$. If qubit in register A is equal to $|0_A\rangle$ then no swapping occurs. Both modules are build around Controlled Polarization Beam Splitter (CPBS) originally proposed by Wójcik [9].

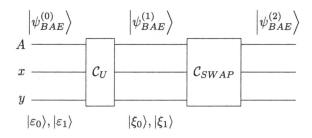

Fig. 4. The block diagram of the Zhang's circuit

Polarization Beam Splitter (PBS) is a two qubit gate which conditionally swaps the port of the input based on its value – the qubit $|0\rangle$ arriving on port x (y) appears on port y (x) of output, but the qubit set to $|1\rangle$ does not change its port. The CPBS acts as normal PBS for control qubit is set to $|0\rangle$. In the complementary situation, i.e., for the control register set to $|1\rangle$, we have the opposite behaviour: $|1\rangle$ ($|0\rangle$) is swapped (hold). The CPBS can be realized as PBS preceded and followed by double CNOT gates (Fig. 5).

The coupling module $\mathcal{C}_U$ is realized as a circuit from Fig. 5. The quantum state of the whole system is given as

$$|\psi_{BAE}^{(0)}\rangle = |\psi_{\text{init}}\rangle|\chi_E\rangle = \frac{1}{\sqrt{2}}\left(|0_B\rangle|1_A\rangle + |1_B\rangle|0_A\rangle\right)|v_x\rangle|0_y\rangle, \tag{6}$$

where it was assumed that the ancilla is initially in the state $|\chi_E\rangle = |v_x\rangle|0_y\rangle$. The $\mathcal{C}_U$ actions are as follows

$$\mathcal{C}_U|0_A\rangle|v_x\rangle|0_y\rangle = \frac{1}{\sqrt{2}}|0_A\rangle\left(|0_x\rangle|v_y\rangle + |v_x\rangle|1_y\rangle\right) = |0_A\rangle|\alpha^+\rangle, \tag{7a}$$

$$\mathcal{C}_U|0_A\rangle|v_x\rangle|1_y\rangle = \frac{1}{\sqrt{2}}|0_A\rangle\left(|0_x\rangle|v_y\rangle - |v_x\rangle|1_y\rangle\right) = |0_A\rangle|\alpha^-\rangle, \qquad (7b)$$

$$\mathcal{C}_U|1_A\rangle|v_x\rangle|0_y\rangle = \frac{1}{\sqrt{2}}|1_A\rangle\left(|v_x\rangle|0_y\rangle + |1_x\rangle|v_y\rangle\right) = |1_A\rangle|\beta^+\rangle, \qquad (7c)$$

$$\mathcal{C}_U|1_A\rangle|v_x\rangle|1_y\rangle = \frac{1}{\sqrt{2}}|1_A\rangle\left(|v_x\rangle|0_y\rangle - |1_x\rangle|v_y\rangle\right) = |1_A\rangle|\beta^-\rangle. \qquad (7d)$$

so the resulting state of the system takes the form (compare with Fig. 4)

$$|\psi_{BAE}^{(1)}\rangle = \frac{1}{\sqrt{2}}\left(|1_B\rangle|0_A\rangle|\alpha^+\rangle + |0_B\rangle|1_A\rangle|\beta^+\rangle\right). \qquad (8)$$

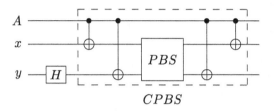

Fig. 5. The coupling circuit $\mathcal{C}_U$

It enters the $\mathcal{C}_{SWAP}$ circuit (Fig. 6) which is also build on the CPBS basis, but this time the y register serves as the control.

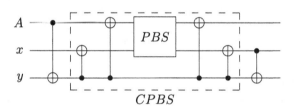

Fig. 6. The selective swap circuit $\mathcal{C}_{SWAP}$

The circuit does nothing for signal qubit set to $|0_A\rangle$. Otherwise, i.e., for signal qubit equal to $|1_A\rangle$, it swaps the contents of A and x registers:

$$\mathcal{C}_{SWAP}|0_A\rangle|\alpha^+\rangle = \frac{1}{\sqrt{2}}\left(|0_A\rangle|0_x\rangle|v_y\rangle + |0_A\rangle|v_x\rangle|1_y\rangle\right) = |0_A\rangle|\alpha^+\rangle, \qquad (9a)$$

$$\mathcal{C}_{SWAP}|1_A\rangle|\beta^+\rangle = \frac{1}{\sqrt{2}}\left(|v_A\rangle|1_x\rangle|0_y\rangle + |1_A\rangle|1_x\rangle|v_y\rangle\right). \qquad (9b)$$

In effect, the state of the system at Alice's end reads (see Fig. 4)

$$|\psi_{BAE}^{(2)}\rangle = \frac{1}{\sqrt{2}}|1_B\rangle|0_A\rangle|\alpha^+\rangle + \frac{1}{\sqrt{2}}|0_B\rangle\left[\frac{1}{\sqrt{2}}\left(|v_A\rangle|1_x\rangle|0_y\rangle + |1_A\rangle|1_x\rangle|v_y\rangle\right)\right]. \qquad (10)$$

Her encoding affects only the term in square brackets

$$\mathscr{Z}_A^{\nu}\left(|v_A\rangle|1_x\rangle|0_y\rangle + |1_A\rangle|1_x\rangle|v_y\rangle\right) = \left(|v_A\rangle|1_x\rangle|0_y\rangle + (-1)^{\nu}|1_A\rangle|1_x\rangle|v_y\rangle\right). \quad (11)$$

The photons that "carry" encoded information and travel back to Bob are affected by C_{SWAP}^{-1}. The component with $|\alpha^+\rangle$ is not changed again but the term (11) sensitive to Alice's encoding operation is transformed to

$$\frac{1}{\sqrt{2}}|1_A\rangle\left(|v_x\rangle|0_y\rangle + (-1)^{\nu}|1_x\rangle|v_y\rangle\right) = |1_A\rangle|\beta^{\pm}\rangle. \quad (12)$$

Thus the forward and backward application of the C_{SWAP} circuit to the state (8) caused that Alice's encoding is effectively applied to the x register. The states visible at this cross-section after applied encoding read (Fig. 4)

$$\nu = 0 \qquad |\xi_0\rangle = \frac{1}{\sqrt{2}}\left(|1_B\rangle|0_A\rangle|\alpha^+\rangle + |0_B\rangle|1_A\rangle|\beta^+\rangle\right), \quad (13a)$$

$$\nu = 1 \qquad |\xi_1\rangle = \frac{1}{\sqrt{2}}\left(|1_B\rangle|0_A\rangle|\alpha^+\rangle + |0_B\rangle|1_A\rangle|\beta^-\rangle\right), \quad (13b)$$

where $|\beta^{\pm}\rangle = (|v_x\rangle|0_y\rangle \pm |1_x\rangle|v_y\rangle)$. These states are transformed by the C_{U}^{-1} to

$$|\varepsilon_0\rangle = C_{\mathrm{U}}^{-1}|\xi_0\rangle = \frac{1}{\sqrt{2}}\left(|1_B\rangle|0_A\rangle|v_x\rangle|0_y\rangle + |0_B\rangle|1_A\rangle|v_x\rangle|0_y\rangle\right), \quad (14a)$$

$$|\varepsilon_1\rangle = C_{\mathrm{U}}^{-1}|\xi_1\rangle = \frac{1}{\sqrt{2}}\left(|1_B\rangle|0_A\rangle|v_x\rangle|0_y\rangle + |0_B\rangle|1_A\rangle|v_x\rangle|1_y\rangle\right), \quad (14b)$$

where expressions (7) have been taken into account. Eve has to discriminate between states

$$\rho_0 = \mathrm{Tr}_{BA}\left(|\varepsilon_0\rangle\langle\varepsilon_0|\right) = |v_x\rangle|0_y\rangle\langle v_x|\langle 0_y|, \quad (15)$$

$$\rho_1 = \mathrm{Tr}_{BA}\left(|\varepsilon_1\rangle\langle\varepsilon_1|\right) = \frac{1}{2}|v_x\rangle|0_y\rangle\langle v_x|\langle 0_y| + \frac{1}{2}|v_x\rangle|1_y\rangle\langle v_x|\langle 1_y|. \quad (16)$$

The information she can draw is limited by the Holevo bound [10]

$$I_{AE} = S\left(\frac{1}{2}\rho_0 + \frac{1}{2}\rho_1\right) - \frac{1}{2}S\left(\rho_0\right) - \frac{1}{2}S\left(\rho_1\right) \quad (17)$$

where $S\left(\cdot\right)$ denotes von Neumann entropy. The straightforward analysis shows that Eve's information gain is equal to $I_{AE} = 0.311$ bits per single message mode cycle. However, the information is intercepted at the price of an induction of a 25 % loss rate in the control mode (10). As long as legitimate parties accept losses above that threshold, the quantum channel can be in theory replaced with a perfect one and the attack can be applied without modification. But the replacement of the quantum channel is impossible in typical real-life scenarios. Moreover, the communicating parties monitor the average losses occurring in the link they use and any abrupt change and/or excess value can trigger an alarm.

However, to avoid false alarms, the estimation procedure cannot be exact and some safety margin have to be assumed. This opens a gap for mounting the attack as long as the induced losses are below that margin. It is advised that Eve should skip some protocol cycles to keep average induced losses below the required threshold. But this decreases her information gain proportionally to the number of omitted cycles. Further, it is shown that such a policy is not optimal. The entangling transformation that provides fine-tuning of the induced loss rate via the control of the coupling coefficients, is proposed. It provides better results than a linear decrease in information gain while reducing losses. The proposed enhancement can be considered as the generalization of the Zhang's attack.

3   Results

The map (7) that defines coupling of the signal qubit with the ancilla has straightforward generalization

$$\mathcal{W}|0_A\rangle|v_x\rangle|0_y\rangle = |0_A\rangle\left(a|0_x\rangle|v_y\rangle + f|v_x\rangle|1_y\rangle\right) = |0_A\rangle|\alpha^+\rangle, \tag{18}$$

$$\mathcal{W}|1_A\rangle|v_x\rangle|0_y\rangle = |1_A\rangle\left(a|1_x\rangle|v_y\rangle + f|v_x\rangle|0_y\rangle\right) = |1_A\rangle|\beta^+\rangle, \tag{19}$$

where $|a|^2 + |f|^2 = 1$. The transformation $\mathcal{W}$ is unitary when

$$\mathcal{W}|0_A\rangle|v_x\rangle|1_y\rangle = |0_A\rangle\left(f|0_x\rangle|v_y\rangle - a|v_x\rangle|1_y\rangle\right) = |0_A\rangle|\alpha^-\rangle, \tag{20}$$

$$\mathcal{W}|1_A\rangle|v_x\rangle|1_y\rangle = |1_A\rangle\left(f|1_x\rangle|v_y\rangle - a|v_x\rangle|0_y\rangle\right) = |1_A\rangle|\beta^-\rangle, \tag{21}$$

$fa^* = f^*a$ and

$$\mathcal{W}|0_A\rangle|0_x\rangle|v_y\rangle = |0_A\rangle|v_x\rangle|0_y\rangle, \qquad \mathcal{W}|0_A\rangle|1_x\rangle|v_y\rangle = |0_A\rangle|1_x\rangle|v_y\rangle, \tag{22}$$

$$\mathcal{W}|1_A\rangle|0_x\rangle|v_y\rangle = |1_A\rangle|0_x\rangle|v_y\rangle, \qquad \mathcal{W}|1_A\rangle|1_x\rangle|v_y\rangle = |1_A\rangle|v_x\rangle|1_y\rangle. \tag{23}$$

The state (10) used for control measurements then takes the form

$$|\psi_{BAE}^{(1)}\rangle = \frac{1}{\sqrt{2}}|1_B\rangle|0_A\rangle|\alpha^+\rangle + \frac{1}{\sqrt{2}}|0_B\rangle\left(f|v_A\rangle|1_x\rangle|0_y\rangle + a|1_A\rangle|1_x\rangle|v_y\rangle\right). \tag{24}$$

The average loss rate observed in control measurements is related to coefficient f as

$$p_L = \frac{1}{2}|f|^2 \tag{25}$$

and Eve is now able to fine-tune induced losses by an appropriate selection of this coupling parameter. However, the above capability does not come without a price. Alice's encoding is still effectively applied to the x register when the system state is observed at the $\mathcal{C}_{\text{SWAP}}$-$\mathcal{C}_{\text{U}}$ cross-section. But this time information encoding does not transform $|\beta^+\rangle$ into $|\beta^-\rangle$ as in (13) but instead into the state

$$|\beta^z\rangle = \left(-a|1_x\rangle|v_y\rangle + f|v_x\rangle|0_y\rangle\right). \tag{26}$$

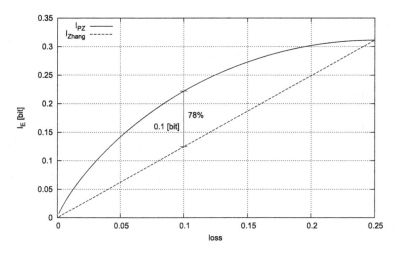

Fig. 7. Eve's information gain as a function of the average loss rate

so the information-encoded states take the form

$$\nu = 0 \qquad |\xi_0\rangle = \frac{1}{\sqrt{2}} \left(|1_B\rangle|0_A\rangle|\alpha^+\rangle + |0_B\rangle|1_A\rangle|\beta^+\rangle \right), \qquad (27a)$$

$$\nu = 1 \qquad |\xi_1\rangle = \frac{1}{\sqrt{2}} \left(|1_B\rangle|0_A\rangle|\alpha^+\rangle + |0_B\rangle|1_A\rangle|\beta^z\rangle \right), \qquad (27b)$$

Consequently, disentangling $\mathcal{W}^{-1} = \mathcal{W}^\dagger$ leads to states

$$|\varepsilon_0\rangle = \mathcal{W}^\dagger|\xi_0\rangle = \frac{1}{\sqrt{2}} \left(|1_B\rangle|0_A\rangle|v_x\rangle|0_y\rangle + |0_B\rangle|1_A\rangle|v_x\rangle|0_y\rangle \right), \qquad (28a)$$

$$|\varepsilon_1\rangle = \mathcal{W}^\dagger|\xi_1\rangle, \qquad (28b)$$

and Eve's information gain is determined by the distinguishability of states

$$\rho_0 = \mathrm{Tr}_{BA} \left(|\varepsilon_0\rangle\langle\varepsilon_0| \right) = |v_x\rangle|0_y\rangle\langle v_x|\langle 0_y|, \qquad (29)$$

$$\rho_1 = \mathrm{Tr}_{BA} \left(\mathcal{W}^\dagger|\varepsilon_1\rangle\langle\varepsilon_1|\mathcal{W} \right). \qquad (30)$$

Figure 7 presents Eve's information gain for the two policies of total induced loss reduction. A key "Zhang" denotes the information gain when the losses are reduced by the plain skipping of protocol cycles. The curve marked as "PZ" illustrates the same quantity computed with the introduced technique and obtained for real-valued coefficient f. The improvement $\Delta I = I_{PZ} - I_{Zhang}$ expressed in bits does not first appear to be impressive as it does not exceed $\Delta I_{\max} = 0.1$ bit. However, the ratio of the additional eavesdropped information to the one received with the traditional technique better exhibits the strength of the contribution. As the graph illustrates, the new way of eavesdropping can be almost 80 % better than methods proposed so far. The other features of the new method remain unchanged compared to Zhang's method. Numerical simulations have shown that maximal information gain is obtained for $a = f = 1/\sqrt{2}$ i.e., for values hard encoded in Zhang's circuit. At the same time, the maximal loss ratio is observed for these coupling constants.

4 Conclusion

The entangling transformation, parametrized by the induced loss ratio, is proposed. In the new method, the eavesdropper's information gain exceeds values offered by other methods. The other key properties of the attack remain the same. The proposed method significantly improves efficiency when attack is operated in low loss regime. It follows, that instead of skipping some protocol cycles, a better policy based on the modification of the entangling transformation parameters should be used to fine tune induced losses.

Acknowledgement. Author acknowledges the support by the Polish Ministry of Science and Higher Education under research project 8686/E-367/S/2015.

References

1. Boström, K., Felbinger, T.: Deterministic secure direct communication using entanglement. Phys. Rev. Lett. **89**(18), 187902 (2002)
2. Boström, K., Felbinger, T.: On the security of the ping-pong protocol. Phys. Lett. A **372**(22), 3953–3956 (2008)
3. Jahanshahi, S., Bahrampour, A., Zandi, M.H.: Security enhanced direct quantum communication with higher bit-rate. Int. J. Quantum Inf. **11**(2), 1350020 (2013)
4. Vasiliu, E.V.: Non-coherent attack on the ping-pong protocol with completely entangled pairs of qutrits. Quantum Inf. Process. **10**, 189–202 (2011)
5. Zawadzki, P.: Security of ping-pong protocol based on pairs of completely entangled qudits. Quantum Inf. Process. **11**(6), 1419–1430 (2012)
6. Zawadzki, P., Puchała, Z., Miszczak, J.: Increasing the security of the ping-pong protocol by using many mutually unbiased bases. Quantum Inf. Process. **12**(1), 569–575 (2013)
7. Zhang, Z., Man, Z., Li, Y.: Improving wójcik's eavesdropping attack on the ping-pong protocol. Phys. Lett. A **333**(1–2), 46–50 (2004)
8. Ostermeyer, M., Walenta, N.: On the implementation of a deterministic secure coding protocol using polarization entangled photons. Opt. Commun. **281**(17), 4540–4544 (2008)
9. Wójcik, A.: Eavesdropping on the ping-pong quantum communication protocol. Phys. Rev. Lett. **90**(15), 157901 (2003)
10. Holevo, A.S.: Bounds for the quantity of information transmitted by a quantum communication channel. Probl. Inform. Transm. **9**(3), 177–183 (1973)

Queueing Theory

A Regeneration-Based Estimation of High Performance Multiserver Systems

Evsey Morozov, Ruslana Nekrasova, Irina Peshkova,
and Alexander Rumyantsev$^{(\boxtimes)}$

Institute of Applied Mathematical Research, Karelian Research Centre
and Petrozavodsk State University, Petrozavodsk, Russia
emorozov@karelia.ru, ruslana.nekrasova@mail.ru, iaminova@mail.petrsu.ru,
ar0@krc.karelia.ru

Abstract. In this paper we develop a novel approach to confidence estimation of the stationary measures in high performance multiserver queueing systems. This approach is based on construction of the two processes which are, respectively, upper and lower (stochastic) bounds for the trajectories of the basic queue size process in the original system. The main feature of these envelopes is that they have classical regenerations. This approach turns out to be useful when the original process is not regenerative or regenerations occur extremely rare to be applied in estimation. We apply this approach to construct confidence intervals for the steady-state queue size in classical multiserver systems and also for the novel high performance cluster (HPC) model, a multiserver system with simultaneous service. Simulation based on a real HPC dataset shows that this approach allows to estimate stationary queue size in a reasonable time with a given accuracy.

Keywords: Regenerative envelopes · Confidence estimation · Multiserver model · High-performance cluster

1 Introduction

Analytical expression of performance measures describing the steady-state multiserver queueing system is available in a few particular cases only. In general, only simulation is the tool available to evaluate the required QoS parameters. However, when a dependence between data exists, the estimation based on simulation faces the problem of the accuracy of the obtained estimates. At the same time a dependence between observed data is a typical situation for the queueing models describing functioning of high performance computer systems. While the standard point estimates, like sample mean, usually provide an accurate estimate of the unknown parameter even for dependent data, the confidence estimation, based on the corresponding form of Central Limit Theorem (CLT), requires in this case a more careful and delicate analysis. In this work, we present a modification of the well-developed and powerful regenerative simulation method,

© Springer International Publishing Switzerland 2016
P. Gaj et al. (Eds.): CN 2016, CCIS 608, pp. 271–282, 2016.
DOI: 10.1007/978-3-319-39207-3_24

see [1–4]. As we show, this approach can be effectively applied to estimate the mean queue size both in the conventional queueing models and in the modern high performance cluster (HPC) in the case, when number of servers is large and classical regenerative approach is not applied in practice. The key feature of this approach is its applicability to non-regenerative queueing processes, for which we construct a *majorant* and *minorant* regenerative systems (called regenerative envelopes). The main idea of the method is to observe the instants when a basic (Markov) process hits a fixed compact set, and then transform the remaining service times in an appropriate way, to obtain classical regeneration. This transformation generates both majorant (upper) and minorant (lower) systems possessing regenerative properties. Then we carry out regenerative-based confidence estimation of the required QoS parameter (the mean queue size, in the present setting) in the new systems. After that we apply a monotonicity property (which, in turn, is based on concept of *coupling*) to construct confidence estimate of the steady-state performance measure in the original system. An essential problem related to this approach is the following: how "rough" are the new regenerative systems comparing with original one? In other words, how tight the obtained confidence interval will be? Hopefully, as we show, the queue size turns out typically to be less sensitive to the transformation of the remaining service times used to construct regenerations of the envelope processes. Thus the key contribution of this work is a new regeneration-based approach allowing correct confidence steady-state estimation of complex models of the multiserver queueing systems, including HPC models.

This paper is organised as follows. In Sect. 2 we give the main notions of the regenerative simulation method and the basic monotonicity properties of the queueing models. In Sect. 3, the new method of regenerative envelopes is described in detail. In Sect. 4, we present numerical examples which cover both the well-known queueing systems ($M/M/m$, $M/G/m$ and $GI/G/m$) and the new HPC model. For the analysis of the latter system, a log-file of the HPC of Karelian Research Centre of the Russian Academy of Sciences is used.

2   Preliminary Results

First we present the basic notions of the regenerative simulation method. Consider a discrete-time regenerative process $\{X_n,\ n \geq 0\}$, with the regeneration instants $\{\beta_k,\ k \geq 1\}$, which describes the dynamics of a stochastic system. It means that the *regeneration cycles* $C_k := \{X_j : \beta_k \leq j < \beta_{k+1}\}$, $k \geq 0$, are independent identically distributed (iid) random elements with the iid cycle lengths $\alpha_k := \beta_{k+1} - \beta_k$ ($\beta_0 := 0$). Consider the iid sequence

$$Y_k = \sum_{i=\beta_k}^{\beta_{k+1}-1} X_i, \quad k \geq 0,$$

and construct the sample mean

$$\overline{r}_n = \frac{\sum_{i=0}^{n-1} X_i}{n}, \quad n \geq 1.$$

(Below we will suppress serial index to denote a generic element of an iid sequence.) Provided $E|Y| < \infty$ and $E\alpha < \infty$, the regenerative process is called *positive recurrent*, and the following limit exists with probability 1 (w.p.1) [1]:

$$r := \lim_{n \to \infty} \overline{r}_n = \frac{\sum_{i=0}^{\alpha-1} X_i}{E\alpha} = \frac{EY}{E\alpha}.$$

Denote the ratio

$$\hat{r}_k := \frac{\sum_{i=0}^{k} Y_i}{\sum_{i=0}^{k} \alpha_i}.$$

As a rule, a stationary QoS measure r is analytically unavailable, and simulation is applied to estimate r. In such cases, if the variance $\sigma^2 := D[Y - r\alpha] \in (0, \infty)$, then the following regenerative variant of the CLT holds ($\Rightarrow$ stands for convergence in distribution):

$$\frac{\sqrt{k}(\hat{r}_k - r)}{\sigma/E\alpha} \Rightarrow \mathbb{N}(0, 1), \quad k \to \infty,$$

implying the following asymptotic $100\,(1 - 2\gamma)\,\%$ confidence interval for the unknown r:

$$\left[\hat{r}_k - \frac{z_{1-\gamma}S(k)}{\hat{\alpha}_k\sqrt{k}}, \; \hat{r}_k + \frac{z_{1-\gamma}S(k)}{\hat{\alpha}_k\sqrt{k}} \right], \tag{1}$$

where $\hat{\alpha}_k$ is the sample mean of $E\alpha$, the estimate $S(k) \to \sigma$, $k \to \infty$, w.p.1, the quantile $z_{1-\gamma}$ is defined as $P(\mathbb{N}(0,1) \le z_{1-\gamma}) = 1 - \gamma$, and $\mathbb{N}(0,1)$ is standard normal variable. It is important to stress that, unlike classic estimation, in this case we deal with the iid groups $\{Y_k\}$ related to regeneration cycles, but not original data $\{X_i\}$, and simulation with a given accuracy requires in general much more simulation time.

Now we consider an infinite buffer FCFS m-server $GI/G/m$ queueing system Σ, with the renewal input with instants t_n, the iid interarrival times $\tau_n = t_{n+1} - t_n$ and the iid service times S_n, $n \ge 1$. Denote ν_n the number of customers at instant t_n^-, Q_n the number of customers waiting in the queue at instant t_n^-, so $Q_n = \max(0, \nu_n - m)$. Finally, denote by S and τ the generic service time and input interval, respectively. Consider another m-server queueing system $\tilde{\Sigma}$ with the same input. The corresponding variables in the system $\tilde{\Sigma}$ we endow with tildes. Assume that the service time distributions in both systems are ordered as $F_S(x) \le F_{\tilde{S}}(x)$, $x \ge 0$, that is $S \ge_{st} \tilde{S}$ (*stochastically*). Moreover, by construction, $\tau =_{st} \tilde{\tau}$. Then the *coupling method* allows to take $\tilde{t}_n = t_n$ and $S_n \ge \tilde{S}_n$ w.p.1. (More on the coupling method see, for instance, in [1].)

The following statement is a key result on the monotonicity of the queueing processes, which we use below, see [1,5].

Theorem 1. *If $\nu_1 = \tilde{\nu}_1 = 0$ and $\tau =_{st} \tilde{\tau}$, $S \ge_{st} \tilde{S}$, then w.p.1,*

$$\tilde{\nu}_n \le \nu_n, \quad \tilde{Q}_n \le Q_n, \quad n \ge 1. \tag{2}$$

Denote (in the system Σ) the input rate $\lambda := 1/\mathsf{E}\tau$, and the service rate $\mu := 1/\mathsf{E}S$. Assume the following conditions hold:

$$\rho := \frac{\lambda}{\mu} < m, \quad \mathsf{P}(\tau > S) > 0. \tag{3}$$

Then it is well known that the instants

$$\beta_{n+1} := \inf_k(k > \beta_n : \nu_k = 0), \quad n \geq 0, \tag{4}$$

are the classic regenerations of the processes (not only the queue size process) describing the system Σ, and form a renewal process (with generic period $\alpha :=_{st} \beta_{n+1} - \beta_n$). Moreover, this process is *positive recurrent*, that is $\mathsf{E}\alpha < \infty$, and aperiodic, since $\mathsf{P}(\alpha = 1) = \mathsf{P}(\tau > S) > 0$. Then, in particular, the weak limits $\nu_n \Rightarrow \nu$, $Q_n \Rightarrow Q$ exist [1]. These limits are important steady-state QoS parameters and are of a great interest for the practitioners dealing with the telecommunications systems. These parameters are analytically available in some particular cases only, and the estimation remains the only way to evaluate the QoS of the system under consideration. Although the values of the queue size process (and other related processes) belonging to the same regeneration cycle are dependent, the enlarged variables Y_n are iid, and it allows to apply above given regenerative variant of the CLT for confidence estimation. (Regenerative simulation method is described in detail in [2–4,6,7].) Notice that the (negative drift) condition $\rho < m$ is necessary for stability, while condition $\mathsf{P}(\tau > S) > 0$ is rather technical and less motivated. Although the latter condition holds if the interarrival time τ has unbounded domain (for instant, if τ is exponential), the probability of an empty state, $\mathsf{P}(\nu_n = 0)$, may be very small, in which case classic regenerations turns out to be too rare to be useful for estimation in practice. As our experiments below show, it is the case when the number of servers m is large enough. Even more difficult problem arises when the opposite condition $\mathsf{P}(\tau > S) = 0$ holds, which precludes the appearance of classic regenerations (4). It motivates a new approach to the estimation, called *regenerative envelopes*, which is presented in the next section.

Remark 1. It is possible to extend Theorem 1 for non-empty initial conditions. The ordering (2) holds true for the weak limits ν, Q as well. Although the monotonicity results given above hold also for the workload process (the *remaining work* in the systems), the coupling used in the method of regenerative envelopes in general does not allow to keep this monotonicity to construct (w.p.1) *upper* and *lower* processes. (We postpone a detailed discussion of this problem to a future paper).

3   The Regenerative Envelopes

Now, given original system Σ, we construct a *majorant* system $\overline{\Sigma}$ and a *minorant* system $\underline{\Sigma}$ with the same input as in the system Σ. (We will supply the corresponding variables in the systems $\overline{\Sigma}$ and $\underline{\Sigma}$ with *overline* and *underline*, respectively). Note that, by assumption, we can take t_n as the arrival instant of customer n in all three systems. First, we construct the system $\overline{\Sigma}$, and denote $\overline{z}_n$

the departure instant of customer n. (In general, in multi-server system, the order of departures is not the same as the order of arrivals.) Define the set $\mathcal{M}_n = \{i : t_i \leq t_n < \overline{z}_i\}$ of the customers, which are being served in the system $\overline{\Sigma}$ at instant t_n. Denote $\overline{S}_i(n)$ the remaining service time of customer i at instant t_n^-. Fix arbitrary integer $\nu_0 \geq 0$, constants $0 \leq a \leq b < \infty$ and define the instant

$$\overline{\beta}_1 = \inf\left\{k : \overline{\nu}_k = \nu_0, \overline{S}_i(k) \in (a, b), i \in \overline{\mathcal{M}}_k\right\}, \tag{5}$$

assuming that the driving sequences in all three systems are identical up to the moment $\overline{\beta}_1$. Then, at the (discrete) instant $k = \overline{\beta}_1$, we replace all the non-zero remaining times $\overline{S}_i(k), i \in \overline{\mathcal{M}}_k$, by the upper bound b. Thus, at the arrival instant $t_{\overline{\beta}_1}$, we increase service times of the customers being served in the system $\overline{\Sigma}$, while the original iid service times $\{S_n\}$ remain unchanged in system Σ. Now we introduce the basic process $\overline{\mathbf{X}}_n := \{\overline{\nu}_n, \overline{S}_i(n), i \in \overline{\mathcal{M}}_n\}, n \geq 1$. It is easy to see that this process is Markov. For $n \geq 1$, we define recursively the instants

$$\overline{\beta}_{n+1} = \inf\left\{k > \overline{\beta}_n : \overline{\nu}_k = \nu_0, \overline{S}_i(k) \in (a, b), \mathcal{M}_{\overline{\beta}_n} \bigcap \mathcal{M}_k = \varnothing, i \in \overline{\mathcal{M}}_k\right\}, \tag{6}$$

and again, at each such instant, replace all remaining service times by the upper bound b. Thus, at each instant $\overline{\beta}_n$, the basic process jumps to the *fixed state* $(\nu_0, \mathbf{b})$, where m-dimensional vector $\mathbf{b}$ contains exactly $\min(\nu_0, m)$ components b and $(m - \nu_0)^+$ zeros. (The order of the components can be arbitrary because of the stochastic equivalence of the servers.) Condition $\mathcal{M}_{\overline{\beta}_n} \bigcap \mathcal{M}_k = \varnothing$ means that all customers *being served* at instant $\overline{\beta}_n$ have left the system before instant $\overline{\beta}_{n+1}$. Thus just after jump, on the event $\{\overline{\beta}_n = k\}$, distribution of $\overline{\mathbf{X}}_k = (\nu_0, \mathbf{b})$ (because of Markovity) is independent of k and the pre-history $\{\overline{\mathbf{X}}_l, l < k\}$. Moreover, the distribution of *post-process*, $\{\overline{\mathbf{X}}_l, l \geq k\}$, is the same for all k. In other words, the process $\{\overline{\mathbf{X}}_n\}$ regenerates at the instants $\{\overline{\beta}_k\}$ and has the iid *regeneration cycles* $\{\overline{\mathbf{X}}_k, \overline{\beta}_n \leq k < \overline{\beta}_{n+1}\}$ and iid cycle lengths $\overline{\beta}_{n+1} - \overline{\beta}_n, n \geq 1$.

Thus the systems $\Sigma, \overline{\Sigma}$ have zero initial state, and moreover (in an evident notation) $\overline{\tau}_n = \tau_n, \overline{S}_n \geq S_n, n \geq 1$. Then it follows from (2) that

$$\overline{\nu}_n \geq \nu_n, \quad \overline{Q}_n \geq Q_n, \quad n \geq 1, \tag{7}$$

and it allows us to use regenerative simulation of $\overline{\Sigma}$ to obtain an upper bound of the mean stationary queue size in original system.

In a similar way, we construct the system $\underline{\Sigma}$ and, at each instant $\{\underline{\beta}_n\}$, replace the remaining service times by the lower bound a. In particular, it implies $\underline{Q}_n \leq Q_n$ and (see (5)–(7))

$$\mathsf{E}\underline{Q} \leq \mathsf{E}Q \leq \mathsf{E}\overline{Q}.$$

Remark 2. Instead of the common upper and lower bounds b, a for all remaining service times one can use different bounds b_i and a_i depending on server i. Also it is expected that a wider interval (a, b) provides more frequent regenerations (in both systems $\overline{\Sigma}, \underline{\Sigma}$) however implies less accurate estimate. Thus, in general there exists a trade-off between given parameters ν_0, a, b for the effective estimation by means of the regenerative envelopes.

4   Simulation Results

4.1 $M/M/m$ System

The multiserver system $M/M/m$ is well-studied and the performance measures are analytically available. Moreover, in this system various types of classical regenerations exist. We stress that for estimation of this model we need not to construct envelope processes, and our purpose is to study the frequency of different types of (classical) regenerations and show how simulation time depends on the type.

It is well known, that under condition $\rho < m$ (see (3)), the process $\{\nu_n\}$ is positive recurrent regenerative and has stationary version ν (stationary queue size Q exists as well). In this case the classic regenerations are generated by the customers meeting an empty system. (We call them 0-regenerations). We recall that the mean 0-regeneration period can be found as $E\alpha = 1/P(\nu = 0)$. Thus, if the stationary probability of an empty system, $P(\nu = 0)$, is very small, then a huge simulation time is required to obtain the number of 0-regenerations to estimate the corresponding parameter with a given accuracy, see (1). By this reason we consider another regenerations which, as we will see, are more effective for estimation.

Recall the expressions for the stationary distribution $\pi_k := P(\nu = k)$ in the system $M/M/m$ [1]:

$$\pi_0 := \left[\sum_{k=0}^{m} \frac{\rho^k}{k!} + \frac{\rho^m}{m!} \frac{m}{m - \rho} \right]^{-1}, \tag{8}$$

$$\pi_k := \begin{cases} \pi_0 \rho^k / k!, & 1 \le k \le m, \\ \pi_0 \rho^k / (m! m^{k-m}), & k > m. \end{cases} \tag{9}$$

It is easy to see from (9) (by induction) that π_k increases in k as long as $k \le \lfloor \rho \rfloor$, implying $\pi_{\lfloor \rho \rfloor} = \max_{k \ge 0} \pi_k$. Moreover, for a fixed ratio ρ/m (close to 1), the probability $\pi_0 \to 0$ exponentially fast, as $m \to \infty$. As a result, the mean 0-regeneration period $E\alpha$ becomes dramatically large. Figure 1 illustrates this effect for $\rho = 0.9m$ depicting the mean 0-regeneration period, $E\alpha = 1/\pi_0$, vs. the number of servers m. Thus, the empty state of the system with large number of servers m appears to be extremely rare.

It is worth to stress that m large corresponds to the modern communication and computer systems, for instance, communication fabrics and server farms.

However, instead of rare 0-regenerations, we can define ρ-*regenerations* as follows:

$$\hat{\beta}_{n+1} := \inf_k (k > \hat{\beta}_n : \nu_k = \lfloor \rho \rfloor), \quad n \ge 0 \quad (\hat{\beta}_0 := 0). \tag{10}$$

which, as follows from discussion above, are the most frequent regenerations in the system. Moreover, it follows from (9) that ρ-regenerations occur in $\rho^{\lfloor \rho \rfloor} / \lfloor \rho \rfloor!$ times faster than 0-regenerations. This gives a practical recommendation to reduce simulation time for steady-state estimation.

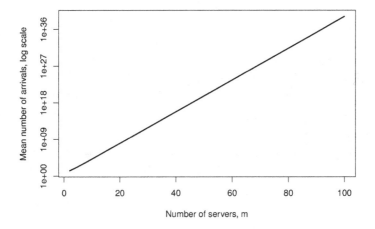

Fig. 1. $M/M/m$ system with fixed ratio $\rho/m = 0.9$: the mean number of arrivals during regeneration cycle vs. the number of servers m (logarithmic scale)

As an example, we consider 10-server system with $E\tau = 1, ES = 9$, implying $\rho = 9$. Figure 2 demonstrates a confidence interval for the mean stationary queue size EQ, based on ρ-regenerations, generated by the events $\{\nu_n = 9\}$. Note that there are no arrivals facing an empty system among 10 000 arrivals, and it does not allow to construct confidence interval. In this regard we mention that various types of regenerations give asymptotically the same confidence interval [8], however, as we see, the required simulation time may be considerably different.

4.2 $M/G/m$ System

In the system $M/G/m$ the process $\{\nu_n\}$ is not Markovian anymore, and we include the remaining service time at each processor to obtain an $(m + 1)$-dimensional Markov process. By the property of the input, the (classical) regenerations still exist but, as we show, are too rare to apply them in the estimation. It motivates construction of the envelops for this system. Moreover, as we show, the frequency of the regenerations in the envelope processes turns out to be enough for accurate estimation in an acceptable time.

We consider a 10-server system with Poisson input, the mean interarrival time $E\tau = 1$, and Pareto service times with the density

$$f_S(x) = \frac{\alpha k^\alpha}{x^{\alpha+1}}, \quad x > k,$$

with given parameters $\alpha > 1$, $k > 0$ and expectation $ES = k\alpha/(\alpha - 1)$. Again we take $\rho = 0.9\,m$ and set $\alpha = 2.5$, implying finite service time variance. Then

$$k = \rho(\alpha - 1)/\alpha = 5.4. \tag{11}$$

In the experiments with 10 000 customers, no one meets an empty system (no 0-regenerations). We apply regenerative envelopes to construct a confidence interval for the stationary mean queue size EQ. To select appropriate low and upper

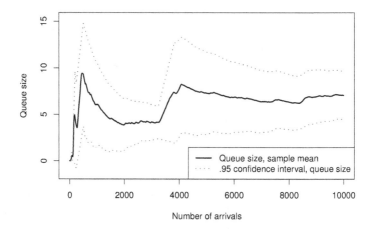

Fig. 2. $M/M/10$ system with $\rho/m = 0.9$: 95 %-confidence interval, based on classic ρ-regenerations, for the stationary mean queue size EQ

bounds a, b, we apply empirical quantiles of the remaining service times. Namely, we take a as 0.2-quantile, so that 80 % of the remaining service time values is not less than a, and b is taken as 0.8-quantile. We also set $\nu_0 = m$. (In $M/M/m$ system we would take $\nu_0 = 9$, see (10).) As Fig. 3 shows, these transformations do not change original system considerably. In other words, the queue size processes in all three systems (Σ, $\overline{\Sigma}$, $\underline{\Sigma}$) turn out to be very close. At the same time, the number of ν_0-regenerations is enough to provide confidence estimation with a high accuracy.

4.3 $GI/G/m$ System

As we saw above, although $M/G/m$ system (theoretically) obeys 0-regenerations (because $P(\tau > S) > 0$), they may not have a practical use because of the rarity. Now we assume that $P(\tau > S) = 0$, that is the system never empties (even theoretically). We consider the same setting as above: 10 servers, 10 000 arrivals, Pareto service time with $\alpha = 2.5$ and k defined as in (11). Now τ is taken uniform in $[0.3, 0.9]$, it implies $P(\tau > S) = 0$ and excludes classical 0-regenerations. We also set the barriers a and b as in the $M/G/m$ system, i.e., as 0.2- and 0.8-quantiles, respectively. Figure 4 demonstrates 95 %-confidence interval for the stationary mean queue size in original system, with the upper bound obtained for $\overline{\Sigma}$, and the lower bound obtained for $\underline{\Sigma}$.

4.4   High Performance Cluster

High performance computing cluster (HPC) is a modern multiserver system, which is used to speedup the computations of computationally-demanding tasks by using the parallel processing capabilities. Such a system is shared among many customers, which submit their tasks into the queue and wait for the results in an asynchronous manner.

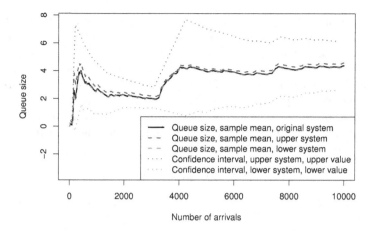

Fig. 3. $M/G/10$ system with $\rho/m = 0.9$: sample means of the queue size in the original, upper and lower systems, and 95 %-confidence interval based on the upper and lower systems

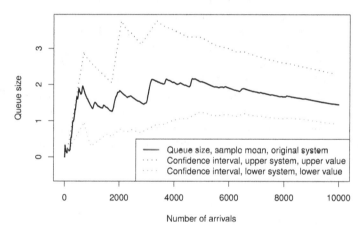

Fig. 4. $GI/G/10$ system with no classic regenerations and with $\rho/m = 0.9$: sample mean of the queue size in original system, and 95 %-confidence interval based on the upper and lower systems

In contrast to classical $GI/G/m$ model, customer i of an HPC may occupy a random number N_i of the servers (processors) simultaneously for a random (but identical) service time. This model is much more difficult to be analyzed. (A detailed description of this model is in [9,10].) Nevertheless, the workload sequence in such a system can be effectively evaluated by a modified Kiefer–Wolfowitz recursion (see [11]). In particular, it allows to determine the departure instants $\{z_i, i \geq 1\}$ by analogue with a classic $GI/G/m$ system. We stress that service discipline in this model turns out to be non-work-conserving, see [9,10,12]. It means that the empty processors may coexist with non-empty queue. This opportunity is reflected in a modified definition of regenerations for the majorant and minorant systems. Namely, in expression (6) we include the requirement $\zeta_n = n_0$, for a constant n_0, where ζ_n is the number of the servers required by the *first customer waiting in the queue* at instant t_n^- ($\zeta_n := 0$ if the queue is empty). Moreover, since the number of customers being served at instant t_n^- depends on the number of processors required by each such a customer, we now replace the number of customers in the system by the number of customers waiting in the queue. In particular, in this case the upper Markov process becomes

$$\overline{\mathbf{X}}_n := \{\overline{Q}_n, \overline{S}_i(n), \zeta_n, i \in \overline{\mathcal{M}}_n\}, \quad n \geq 1,$$

which has the fixed state $(q_0, \mathbf{b}, n_0)$ at each regeneration instant, where q_0 is a predefined fixed value of Q_n.

To illustrate this approach, we set $m = 80$ and take the interarrival times, service times and requested number of servers (cores) from the dataset HPC_KRC available in the CRAN package hpcwld [13] for the R statistical package [14]. This package contains the data from a log-file of the HPC of Karelian Research Centre [15]. We skip the first 2000 customers of the dataset, to exclude a transient period, and consider customers $2000, \ldots, 8000$. These observations cover approximately one year activity of the cluster.

Then we define the regenerations as arrivals into an *empty queue* ($q_0 = 0$), and select the values a, b, respectively, as 0.2- and 0.4-quantile of the observed residual service times in the original system. We note that this choice allows to use in practice an adaptive approach, in which case the quantiles are evaluated online, with the ongoing arrivals into the observed system. Then we construct the corresponding majorant and minorant systems and realize confidence estimation.

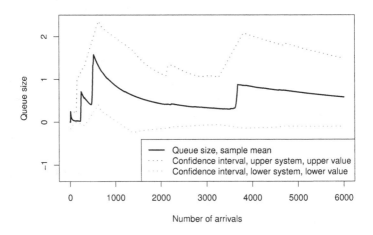

Fig. 5. Simultaneous service system (HPC) based on the dataset HPC_KRC from the HPC of Karelian Research Centre of the Russian Academy of Sciences: sample means of the original system, and 95 %-confidence interval based on the upper and lower systems

The results of the experiments are depicted on Fig. 5. It is worth mentioning that, as a rule, the customer arriving to the HPC never faces an empty system, and it makes impossible confidence estimation based on classical regeneration. We may conclude that the method of regenerative envelops may be used to evaluate a QoS parameter of a HPC with a given accuracy.

5   Conclusion

In this work, we introduce the regenerative envelopes which can be used for QoS estimation of a stationary multidimensional queueing process when classical regenerations do not exist or they are too rare to be useful in simulation. We present simulation results to compare a QoS estimation based on the regenerative envelopes and classic 0-regenerations (when exist). It is shown that this approach can be applied to confidence estimation of a high performance cluster model described by a non-regenerative process.

Acknowledgments. This work is supported by Russian Foundation for Basic research, projects No 15–07–02341, 15–07–02354, 15–07–02360, 15–29–07974, 16–07–00622 and by the Program of strategic development of Petrozavodsk State University.

References

1. Asmussen, S.: Applied Probability and Queues. Wiley, NewYork (1987)
2. Glynn, P.: Some topics in regenerative steady-state simulation. Acta Appl. Math. **34**, 225–236 (1994)
3. Glynn, P., Iglehart, D.: Conditions for the applicability of the regenerative method. Manage. Sci. **39**, 1108–1111 (1993)
4. Sigman, K., Wolff, R.W.: A review of regenerative processes. SIAM Review **35**, 269–288 (1993)
5. Morozov, E.: Coupling and monotonicity of queues, Sci. report No 779. CRM Barcelona, pp. 1–29 (2007). http://www.crm.cat
6. Morozov, E.: An extended regenerative structure and queueing network simulation. Preprint No 1995–08/ISSN 0347–2809. Dept. Math., Chalmers Univ., Gothenburg (1995)
7. Morozov, E., Aminova, I.: Steady-state simulation of some weak regenerative networks. Eur. Trans. Telecommun. ETT. **13**(4), 409–418 (2002)
8. Morozov, E., Sigovtsev, S.: Simulation of queueing processes based on weak regeneration. J. Math. Sci. **89**(5), 1517–1523 (1998)
9. Rumyantsev, A., Morozov, E.: Stability criterion of a multiserver model with simultaneous service. Ann. Oper. Res. 1–11 (2015). doi:10.1007/s10479-015-1917-2
10. Rumyantsev, A., Morozov, E.: Accelerated verification of stability of simultaneous service multiserver systems. In: 7th International Congress on Ultra Modern Telecommunications and Control Systems and Workshops, pp. 239–242. IEEE (2015)
11. Rumyantsev, A.: Simulating Supercomputer Workload with hpcwld package for R. In: 15th International Conference on Parallel and Distributed Computing, Applications and Technologies, pp. 138–143. IEEE (2014)

12. Chakravarthy, S., Karatza, H.: Two-server parallel system with pure space sharing and markovian arrivals. Comput. Oper. Res. **40**(1), 510–519 (2013)
13. CRAN - Package hpcwld. http://cran.r-project.org/web/packages/hpcwld/
14. R Foundation for Statistical Computing, Vienna, Austria. ISBN 3-900051-07-0. http://www.R-project.org/
15. Centre for collective use of Karelian Research Centre of Russian Academy of Sciences. http://cluster.krc.karelia.ru

Queuing Model of the Access System in the Packet Network

Sławomir Hanczewski[1]([⊠]), Maciej Stasiak[1], Joanna Weissenberg[2], and Piotr Zwierzykowski[1]

[1] Faculty of Electronics and Telecommunications,
Poznan University of Technology, Poznań, Poland
`slawomir.hanczewski@put.poznan.pl`
[2] Faculty of Mechanics and Applied Computer Science,
Kazimierz Wielki University, Bydgoszcz, Poland

Abstract. This article proposes a new multi-dimensional Erlang's Ideal Grading (EIG) model with queues that can service a number of call classes with differentiated access to resources. The model was used to determine delays and packet loss probabilities in the access system composed of a node in the operator's network and a number of users. The analytical results obtained in the study were then compared with the results of a simulation, which confirmed the essential and required accuracy of the proposed model. The model developed in the study can be used to analyse, design and optimize present-day operator's networks.

Keywords: Erlang's Ideal Grading · Queue

1 Introduction

Local area network and Ethernet technology has shown a dynamic development in recent years and is expanding at a fast pace [1]. Present-day operator's networks are more and more often based on the Ethernet technology [1]. These networks make use of the hierarchical visualization of network resources. This means that virtual networks are created both at local network levels of providers and companies that are serviced by a given operator and at the level of the backbone network.

Such an approach ensures that traffic introduced to a service provider's network by individual providers or firms can be separated. As a result, a construction of a logical structure at the core network level for pre-defined types of services offered to end-users is feasible [2]. The working principle of this structure is based on appropriate relevant standards that allow multistage and hierarchical systems of tagging Ethernet frames to be introduced [3]. Hierarchical tagging has been widely used, for example, in the development of the architecture of the Ethernet service provider's network known as Carrier or Metro Ethernet [2].

The scope, quality and responsibilities to support the cooperation between network operators and users are underlain in SLA (*Service Level Agreement*) [4].

© Springer International Publishing Switzerland 2016
P. Gaj et al. (Eds.): CN 2016, CCIS 608, pp. 283–293, 2016.
DOI: 10.1007/978-3-319-39207-3_25

These agreements also include technical parameters that define quality of service for network services (QoS – *Quality of Service*). A good example of these parameters can be the maximum delay or the acceptable level of packet loss. To guarantee appropriate values for the QoS parameters it is necessary to take these parameters into consideration at the network designing stage and to monitor their values when the operation of the network is in progress.

The present article proposes a simplified analytical model of a system composed of a number of users connected to the node of a service provider's network, further in the text referred to as the access system. In the study, a model of Erlang's Ideal Grading (EIG) is used to analyse the access system. Ideal grading in its canonical form was proposed by Erlang [5,6]. The author of [7] proposes a model of EIG with infinite queues in each of the load groups. Gradings with different structures, with finite and infinite queues and different types of call streams, are addressed in [8,9], among others. [10] proposes a model of EIG without queues in which the system can service a number of call classes. The assumption in models [11,12] is that call classes can have differentiated, non-integer availability. In [13–17] the model is used to analyse a VoD system, UMTS and LTE systems, and mobile overflow system. It should be noted at this point that both the model of EIG with losses [6,12,18] and EIG with queues [19] are good at approximating other structures of gradings. This article proposes an EIG queueing model that can service a number of call classes with differentiated access to resources. This model, applied to an analysis of access systems, will allow us to perform approximate evaluation of the value of delays and packet losses in such systems.

The article is structured as follows. Following the Introduction, Sect. 2 describes the multi-dimensional EIG model with a queues. In Sect. 3, the analysed access model is parametrized in line with the EIG model with queues proposed earlier. In Sect. 4, the results of the analytical calculations are compared with the results of a digital simulation for selected structures of the access network. Section 5 sums up the article.

2   Erlang's Ideal Grading Model with Queues

The starting point for the analysis of the access system is the Erlang's Ideal Grading model (EIG) that is offered a traffic stream composed of a number of call classes [12]. The group has the capacity of V allocation units (AU), i.e. units that have been adopted in the discretization process of the capacity of the system [20]. Traffic of class i is offered to g_i load groups. A call of class i appearing at the incoming of a single load group has access to d_i AUs from among the number of all V AUs. The d_i parameter is called the availability of class i. The number of load groups g_i for calls of class i in EIG is equal to:

$$g_i = \binom{V}{d_i}. \tag{1}$$

(a)

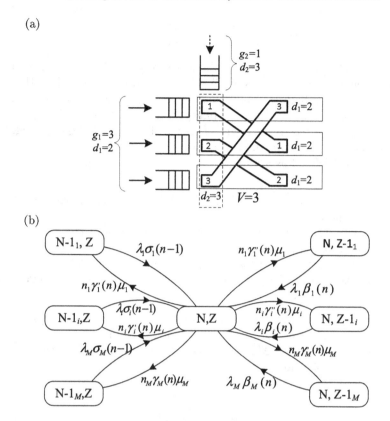

(b)

Fig. 1. Erlang's Ideal Grading with queues: (a) schematic diagram, (b) fragment of a Markov process

Figure 1a shows a simple model of EIG with the capacity $V = 3$ AU's. The group services two classes of calls with the availability $d_1 = 2$, $d_2 = 3$. Hence, the number of load groups for relevant call classes is equal $g_1 = 3$ and $g_2 = 1$.

The number of load groups in EIG is then equal to the number of possible ways in which d_i AUs, from their general number V, can be selected, while the two load groups differ from each other in at least one AU. With uniform traffic offered to all load groups, occupancy distributions in each of the groups are the same. This property enables the conditional blocking probability $\beta_i(n)$ for calls of class i in EIG servicing n AUs, to be determined in a very simple way. The $\beta_i(n)$ parameter is equal to the probability of the appearance of a new call of class i in these load groups in which all AUs are busy:

$$\beta_i(n) = \begin{cases} \binom{n}{d_i} \Big/ \binom{V}{d_i} & \text{for } d_i \leq n \leq V, \\ 0 & \text{for } 0 \leq n < d_i. \end{cases} \tag{2}$$

A complementary event to blocking phenomenon is the event of service acceptance for a call of class i in the occupancy state n. The probability of such an event is defined as the conditional transition probability $\sigma_i(n)$, and is equal to:

$$\sigma_i(n) = \begin{cases} 1 - \beta_i(n) & \text{for } d_i \leq n \leq V, \\ 1 & \text{for } 0 \leq n < d_i. \end{cases} \tag{3}$$

Observe that the parameters $\sigma_i(n)$ and $\beta_i(n)$ for calls of class i depend on the total number n of serviced calls of all classes and do not depend on the distribution of these calls between individual classes. A call of class i that cannot be admitted for service due to the occupancy of all AUs available in a given load group will be redirected to the queue of class i of a given load group (Fig. 1(b)). The basic assumptions applying to the Markov processes occurring in EIG with a queues can be defined as follows:

- system services M call classes;
- call stream of class i $(0 \leq i \leq M)$ has Poisson character with the average intensity equal to λ_i;
- service stream of class i $(0 \leq i \leq M)$ has exponential character with the average intensity equal to μ_i;
- upon the termination of the service process of a call of class i, a call of the same class is retrieved from the queue of any randomly chosen load group that has access to released resources.

The last assumption expresses the fact that the process taking part in EIG with queues is independent of the service discipline for queues of individual classes. Microstate (N, Z) of the Markov process that occurs in the system can be described by the following parameters:

$$(N, Z) = (n_1, \ldots, n_i, \ldots, n_M, z_1, \ldots, z_i, \ldots, z_M), \tag{4}$$

where n_i and z_i are the numbers of serviced calls and those calls that have been placed in queues of class i. In line with the adopted notation, the microstates that are adjacent to microstate (N, Z), in which service calls of class i are decreased one, or alternatively there is one call of class i less in the queue of class i, can be written as follows:

$$(N - 1_i, Z) = (n_1, \ldots, n_i - 1, \ldots, n_M, z_1, \ldots, z_i, \ldots, z_M), \tag{5}$$

$$(N, Z - 1_i) = (n_1, \ldots, n_i, \ldots, n_M, z_1, \ldots, z_i - 1, \ldots, z_M), \tag{6}$$

where 1_i denotes one call of class i. The probability of microstate (N, Z) will be denoted by the symbol $p(N, Z)$. The macrostate of Markov process (n, z) is defined by the total number of serviced calls and the total number of calls waiting to be serviced, regardless of the distribution of these calls between individual classes. Therefore, the probability of macrostate $[P_{n,z}]_{V,U}$, where U is the total capacity of the queues in the system, can be determined on the basis of the probabilities of microstates in the following way:

$$[P_{n,z}]_{V,U} = \sum_{\Omega(n,z)} p(N, Z), \tag{7}$$

where: $\Omega(n, z)$ is the set of all microstates (N, Z) that fulfil equations:

$$n = \sum_{i=1}^{M} n_i, \quad z = \sum_{i=1}^{M} z_i. \tag{8}$$

Consider the portion of a Markov process at the microstate level shown in Fig. 1(b). In microstate $(N - 1_i, Z)$, the intensity of transition to microstate (N, Z) is determined by the conditional transition probability $\sigma_i(n - 1)$, and is equal $\lambda_i \sigma_i(n-1)$, whereas the intensity of transition from microstate $(N, Z - 1_i)$ to microstate (N, Z) results from the conditional blocking probability $\beta_i(n)$, and equals $\lambda_i \beta_i(n)$. The $\gamma_i'(n)$ parameter is the conditional probability of an event in which, after the termination of service of a call of class i, none of the queues of class i have access to released resources. The conditional probability $\gamma_i''(n)$ is, in turn, the probability of an event that after a termination of service of a call of class i at least one queue of class i has access to released resources, which, in consequence, leads to the admission of a next call of class i taken from the queue for service. Since events determined by the probabilities $\gamma_i'(n)$ and $\gamma_i''(n)$ are complementary events, then:

$$\gamma_i'(n) + \gamma_i''(n) = 1. \tag{9}$$

The values of the parameters $\gamma_i'(n)$ and $\gamma_i''(n)$ will be omitted in further considerations. The proposed model is based on the assumption of the reversibility of the Markov process in EIG with queues. This means that local balances between the microstates presented in Fig. 1(b) are fulfilled. We can thus write:

$$\gamma_i''(n) n_i \mu_i \, p(N, Z) = \lambda_i \beta_i(n) \, p(N, Z - 1_i), \tag{10}$$

$$\gamma_i'(n) n_i \mu_i \, p(N, Z) = \lambda_i \sigma_i(n - 1) \, p(N - 1_i, Z). \tag{11}$$

By dividing both sides of Eqs. (10) and (11) by μ_i and then by adding them, with (9) taken into consideration, we get:

$$n_i p(N, Z) = A_i \beta_i(n) \, p(N, Z - 1_i) + A_i \sigma_i(n - 1) \, p(N - 1_i, Z), \tag{12}$$

where A_i is the intensity of traffic of class i:

$$A_i = \lambda_i / \mu_i. \tag{13}$$

Since the general assumption is that the process under consideration is a reversible process, then we add side by side M equations of the type (12). As a result, we get:

$$p(N, Z) \sum_{i=1}^{M} n_i = \sum_{i=1}^{M} A_i \beta_i(n) \, p(N, Z - 1_i)$$
$$+ \sum_{i=1}^{M} A_i \sigma_i(n - 1) \, p(N - 1_i, Z). \tag{14}$$

Equation (14) expresses the sum of all balances of microstate (N, Z) with younger microstates $(N - 1_i, Z)$ and $(N, Z - 1_i)$. This equation provides a basis for the determination of the occupancy distribution at the macrostate level. This can be done by summing up both sides of Eq. (14) over all microstates that belong to the set $\Omega(n, z)$ and fulfil the conditions (8):

$$n \sum_{\Omega(n,z)} p(N, Z) = \sum_{i=1}^{M} A_i \beta_i(n) \sum_{\Omega(n,z)} p(N, Z - 1_i)$$
$$+ \sum_{i=1}^{M} A_i \sigma_i(n - 1) \sum_{\Omega(n,z)} p(N - 1_i, Z). \tag{15}$$

The sum over $\Omega(n, z)$, on the left side of Eq. (15) defines, according to definition (7), the probability of microstate (n, z). The sums over $\Omega(n, z)$, on the right side of Eq. (15) define probabilities of macrostates $(n, z - 1)$ and $(n - 1, z)$. Except for the case with $z = 0$, Eq. (15) can be re-written in the following way:

$$n\,[P_{n,0}]_{V,U} = [P_{n-1,0}]_{V,U} \sum_{i=1}^{M} A_i \sigma_i(n-1) \text{ for } 1 \leq n \leq V, z = 0, \tag{16}$$

$$n\,[P_{n,z}]_{V,U} = [P_{n,z-1}]_{V,U} \sum_{i=1}^{M} A_i \beta_i(n)$$

$$+ [P_{n-1,z}]_{V,U} \sum_{i=1}^{M} A_i \sigma_i(n-1) \text{ for } 1 \leq n \leq V, 1 \leq z \leq U. \tag{17}$$

Equations (16) and (17), along with the normalisation condition:

$$\sum_{n=0}^{V} \sum_{z=0}^{U} [P_{n,z}]_{V,U} = 1, \tag{18}$$

create a system of equations that allows a recurrent determination of the occupancy distribution at the macrostate level in EIG with queues to be recursively determined. Observe that for one class of calls ($M = 1$), the distributions (16) and (17) are reduced to the result obtained in [7,19]. The average length of the queue $q_i(n, z)$ for calls of class i in macrostate (n, z) can be expressed by the following formula:

$$q_i(n, z) = \sum_{\Omega(n,z)} z_i p(N, Z). \tag{19}$$

The parameter $q_i(n, z)$ can be therefore defined on the basis of (12), i.e. on the occupancy distribution at the microstate level. A determination of the distribution $p(N, Z)$ at the microstate level is, however, far more complex than the determination of the occupancy $[P_{n,z}]_{V,U}$ at the macrostate level. Using the occupancy $[P_{n,z}]_{V,U}$ as a basis for our considerations, the average length of a queue of class i in macrostate (n, z) can be expressed approximately on the basis of the following reasoning: a transition from any randomly chosen macrostate (n, k) to $(n, k+1)$ causes an increase in the queue length of a queue of class i by one, with the probability $\pi_i(n, k)$ that is equal to the participation of a stream of class i in the total stream that increases the total queue size by one waiting call:

$$\pi_i(n, k) = A_i \beta_i(n) / \sum_{j=1}^{M} A_j \beta_j(n). \tag{20}$$

Therefore, the total average length of queues of class i in macrostate (n, z) can be determined by the following formula:

$$q_i(n, z) = \sum_{k=0}^{z-1} 1 \cdot \pi_i(n, k) = z \frac{A_i \beta_i(n)}{\sum_{j=1}^{M} A_j \beta_j(n)}. \tag{21}$$

Now, we are in position to determine the total average length of queues of class i in the system:

$$Q_i = \sum_{n=d_i}^{V} \sum_{z=0}^{U} q_i(n, z) [P_{n,z}]_{V,U}. \tag{22}$$

Since there are g_i queues of class i in the system (the number of queues is equal to the number of load groups of a given call class), then the average length of one queue q_i of class i and the average waiting time t_i in one queue of class i (determined on the basis of Little's formula [21]) are respectively as follows:

$$q_i = Q_i/g_i, \quad t_i = Q_i/\lambda_i g_i. \tag{23}$$

3   Resource Access Model

Access to resources of the system with the capacity of V AUs is available for G users (access links), whereas user j has access to a link with capacities V_j AUs, where $1 \leq j \leq G$ (Fig. 2). Each link can be offered a mixture of M packet streams. If a packet of class i cannot be serviced due to the occupancy of the access link or the occupancy of resources, then it is redirected to a queue of class i of a given user. Availability in the system for a given class of packets is a measure of the accessibility to resources and can be evaluated on the basis of the following reasoning [13]: in the access link j, the resources are co-shared by M packet streams. It can thus be assumed that the number of occupied resources in the link by individual classes will be directly proportional to the traffic participation of these classes in the total traffic offered to the access link. Therefore, availability $d_{i,j}$ for packets of class i via the access link j can be expressed as follows:

$$d_{i,j} = V_j A_i / \sum_{k=0}^{M} A_k. \tag{24}$$

The total average availability for packets of class i is equal to:

$$d_i = \frac{1}{G} \sum_{j=1}^{G_i} d_{i,j}, \tag{25}$$

where G_i is the number of access links that service packets of class i ($1 \leq G_i \leq G$). The presented scheme for the operation of an access system can be modelled on the basis of EIG with queues. This means that a fictitious value of the number of load groups g_i, defined by Formula (1) on the basis of the availability (25), will be used for modelling. The real value of the access links G_i will be then used to determine the average queue length and the average waiting time in a single queue by substituting G_i in place of g_i in Formula (23).

4   Numerical Evaluation

In order to verify the proper functioning of the analytical model of an access system proposed in the paper, the results obtained on its basis were compared with the results of simulation experiments. To do so, an access system simulator was developed. The operator's network node capacity was assumed to be $C = 10$ Gbps. 12 clients ($G = 12$) were connected with the node through access links. The assumed speed of access links was 1 Gbps. The clients were offered $M = 3$ service classes. The simulator's input parameters were as follows:

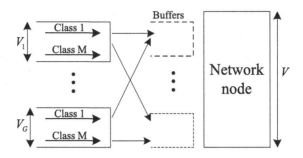

Fig. 2. A diagram of an access system

- λ_i – intensity of packets of class i,
- μ_i – intensity of service stream of class i,
- l_i – average length of packets of class i.

It was assumed that subscribers generated data with the speed of 1 Gbps.

In order to determine the bit length of each packet belonging to a stream of class i, a pseudo-random number following an exponential distribution with parameter μ_i was generated, corresponding to packet duration τ_i. Next, the packet bit length was calculated:

$$l_{\tau_i} = \tau_i c_i, \tag{26}$$

where c_i is the average bitrate of the packet stream:

$$c_i = \lambda_i l_i. \tag{27}$$

The access system capacity was expressed by the number of resource allocation units according to the following formula [20]:

$$V = C/c_{\text{AU}}, \tag{28}$$

where c_{AU} is the system resource allocation unit. For the purpose of the experiment, it was assumed that an allocation unit was equal to $c_{\text{AU}} = 10$ kbps. The value was determined on the basis of the maximum packet possible to be sent via the Ethernet (1526 bytes). The adopted total buffers capacity for all services was 1000 AUs.

The experiment also assumed that the traffic offered in each access network fulfilled the following condition: $A_1 : A_2 : A_3 = 3 : 2 : 1$. The obtained results are presented in graphs in the function of traffic offered per one AU of the system:

$$a = \sum_{i=1}^{M} A_i/V. \tag{29}$$

The results obtained on the basis of the analytical model are shown by sold line, whereas those obtained in the simulation by appropriate symbols.

(a)

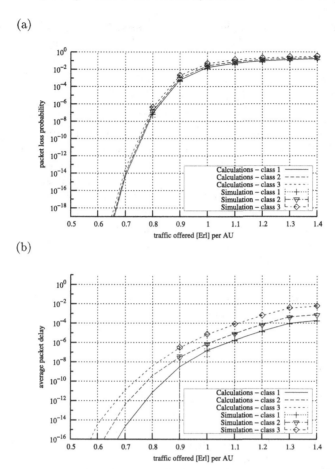

(b)

Fig. 3. Numerical results: (a) average packet delay of individual traffic classes, (b) packet loss probability of individual traffic classes

The study made it possible to evaluate the average value of packet delays and the levels of packet losses for individual traffic classes. Figure 3(a) shows the values for the packet loss probability, whereas Fig. 3(b) presents the values for the average waiting time for packets in one queue. The study was conducted for changing traffic intensities.

To determine one point in the graph, 5 series of simulation runs, each involving 10 000 000 packets, was performed. This made it possible for confidence intervals to be determined at the level of 95 %. These intervals are small enough that, in the graphs, they do not exceed the values of the symbol that defines the result of a simulation.

By analysing the presented results it can be seen that when values of traffic offered to the system do not exceed 0.7 Erl. per AU, packet losses are low in the networks so that they can be ignored, and their transmission delays are also negligible. Packet losses can be further decreased by increasing the buffers for individual traffic classes. This will be followed, however, by an increase in delays of packets.

5   Conclusion

This article proposes a new analytical model of Ideal Erlang's Grading with queues for traffic classes that are differentiated by their availability to the resources of the group. The proposed model is used to analyse a designated section of the operator's network. The study confirms that satisfactory accuracy can be obtained by the application of the analytical model presented in the article. This means that the developed model can be used in evaluating QoS parameters considered as early as the designing stage of present-day operator's networks. The model can be also used for solving practical problems in design and optimization of modern networks.

Acknowledgements. The presented work has been funded by the Polish Ministry of Science and Higher Education within the status activity task *"Structure, analysis and design of modern switching system and communication networks"* in 2016.

References

1. Gero, B., Farkas, J., Kini, S., Saltsidis, P., Takacs, A.: Upgrading the metro ethernet network. IEEE Commun. Mag. **51**(5), 193–199 (2013)
2. Checko, A., Ellegaard, L., Berger, M.: Capacity planning for carrier ethernet LTE backhaul networks. In: IEEE Wireless Communications and Networking Conference, pp. 2741–2745 (2012)
3. Diab, W.W., Frazier, H.M.: Ethernet in the First Mile: Access for Everyone. Wiley, New York (2006)
4. Wikisource: Telecommunications act of 1996 (2009). http://en.wikisource.org
5. Brockmeyer, E.: A survey of A.K. Erlang's mathematical works. Danish Acad. Tech. Sci. **2**, 101–126 (1948)
6. Lotze, A.: History and development of grading theory. In: 5th International Teletraffic Congress, pp. 148–161 (1967)
7. Thierer, M.: Delay system with limited accessibility. In: 5th International Teletraffic Congress, pp. 203–213 (1967)
8. Gambe, E.: A study on the efficiency of graded multiple delay systems through artificial traffic trials. In: 3rd International Teletraffic Congress, doc. 16 (1961)
9. Kühn, P.: Combined delay and loss systems with several input queues, full and limited accessibility. In: 6th International Teletraffic Congress, pp. 323/1–323/7 (1970)
10. Stasiak, M.: An approximate model of a switching network carrying mixture of different multichannel traffic streams. IEEE Trans. Commun. **41**(6), 836–840 (1993)
11. Stasiak, M., Hanczewski, S.: Approximation for multi-service systems with reservation by systems with limited-availability. In: Thomas, N., Juiz, C. (eds.) EPEW 2008. LNCS, vol. 5261, pp. 257–267. Springer, Heidelberg (2008)

12. Głąbowski, M., Hanczewski, S., Stasiak, M., Weissenberg, J.: Modeling Erlang's ideal grading with multirate BPP traffic. Math. Probl. Eng. **2012**, 35 (2012). Article ID 456910
13. Hanczewski, S., Stasiak, M.: Performance modelling of Video-on-Demand systems. In: 17th Asia-Pacific Conference on Communications, pp. 784–788 (2011)
14. Stasiak, M., Głąbowski, M., Hanczewski, S.: The application of the Erlang's ideal grading for modelling of UMTS cells. In: 8th International Symposium on Communication Systems, Networks Digital Signal Processing, pp. 1–6 (2012)
15. Hanczewski, S., Stasiak, M., Zwierzykowski, P.: A new model of the soft handover mechanism in the UMTS network. In: 9th International Symposium on Communication Systems, Networks Digital Signal Processing, pp. 84–87 (2014)
16. Hanczewski, S., Stasiak, M., Zwierzykowski, P.: Modelling of the access part of a multi-service mobile network with service priorities. EURASIP J. Wirel. Commun. Netw. **2015**(1), 1–14 (2015)
17. Głąbowski, M., Hanczewski, S., Stasiak, M.: Modelling of cellular networks with traffic overflow. Math. Probl. Eng. **2015**, 15 (2015). Article ID 286490
18. Šneps, M.: Sistemy raspredeleniâ informacii. ser. Metody rasčёta. Radio i Swâz', Moskva (1979)
19. Thierer, M.: Delay-tables for limited and full availability according to the interconnection delay formula. Institute for Switching and Data Technics, Technical University Stuttgart, Germany, Technical report (1968)
20. Roberts, J. (ed.): Performance Evaluation and Design of Multiservice Networks, Final Report COST 224. Commission of the European Communities, Brussels (1992)
21. Little, J.: A proof for the queueing formula: L = λw. Oper. Res. **9**(3), 383–387 (1961)

A Study of IP Router Queues with the Use of Markov Models

Tadeusz Czachórski[1]([envelope]), Adam Domański[2], Joanna Domańska[1], and Artur Rataj[1]

[1] Institute of Theoretical and Applied Informatics,
Polish Academy of Sciences, Baltycka 5, 44-100 Gliwice, Poland
{tadek,joanna,arataj}@iitis.gliwice.pl
[2] Institute of Informatics, Silesian Technical University,
Akademicka 16, 44-100 Gliwice, Poland
adamd@polsl.pl

Abstract. We investigate the use of Markov chains in modeling the queues inside IP routers. The model takes into account the measured size of packets, i.e. collected histogram is represented by a linear combination of exponentially distributed phases. We discuss also the impact of the distribution of IP packets size on the loss probability resulting from the limited size of a router memory buffer. The model considers a self similar traffic generated by on-off sources. A special interest is paid to the duration of a queue transient state following the changes of traffic intensity as a function of traffic Hurst parameter and of the utilization of the link. Our goal is to see how far, taking into account the known constraints of Markov models (state explosion) we are able to refine the queueing model.

Keywords: Markov queueing models · Self-similarity · IP packets length distribution

1 Introduction

It is well known that the distribution of the size of packets and traffic self-similarity have both an impact on the transmission quality of service (QoS) determined by transmission time, jitter, and loss probability and that they influence also the dynamics of changes of number of packets waiting in routers to be forwarded. This latter problem is important for implementation of active queue management (AQM) traffic control algorithms performed at IP routers. These issues are usually investigated with the use of discrete-event simulations which, especially in case of self similar traffic, demand very long runs and are time consuming. Simulation applied to transient states studies needs the repetition of multiple (millions) runs to obtain a smooth time-dependent distribution of a queue length. Therefore, analytical models are welcome.

We present here an approach based on continuous time Markov chains which, since the advent of queueing theory a hundred years ago, are a very popular tool

© Springer International Publishing Switzerland 2016
P. Gaj et al. (Eds.): CN 2016, CCIS 608, pp. 294–305, 2016.
DOI: 10.1007/978-3-319-39207-3_26

in queueing models despite their constraints – only very simple models have a straightforward analytical solution, the other models should be solved numerically and the size of a model (the number of states equal to the number of equations to be solved) is naturally limited. However, this limit is constantly shifting due to the increase of computer power and size of memory and also development of better software and numerical methods able to solve very large systems of equations. Apart from the formulated above goals we test the usefulness of known tools which may be helpful in such models, HyperStar [1] to fit a system of exponentially distributed phases to a real distribution, and Prism [2] to study transient states at a complex Markov model. This is a quantitative approach: several fine analytical results for similar Markov models were already obtained e.g. [3] but at the end they also need complex numerical computations to furnish quantitative results, hence we prefer to start numerical part of the model as early as possible, i.e. on the level of Chapman-Kolmogorov equations defining state probabilities. This is a pure engineering approach, a kind of Markov chain applied to IP routers cookbook.

2   Distribution of the Size of IP Packets

The size of a packet determines the time of sending this packet by a router, hence in a realistic queueing model the service time distribution should correspond to the distribution of packets size. However, in existing purely analytical queueing Markovian models, there are difficulties to correlate the distribution of the size of packets and the distribution of service time, see e.g. [4]. An approach to this problem based on diffusion approximation was presented in [5]. The distribution of the size of IP packets depends on the version of IP protocol and the type of traffic; numerous measurements and statistics are stocked, among others in repositories of CAIDA [6]. In the examples that follow we used CAIDA measurements of the size of IP v4 packets from (IPv4 equinix-chicago 2008) file with typical data.

   In a Markov model we should represent any real distribution with the use of a system of exponentially distributed phases (PH). The PH fitting is extensively investigated problem, see e.g. [7] for details. The fitting is done numerically, e.g. with the use of Expectation Maximization Algorithms. Numerous tools exist [8–16]. We used HyperStar [1] which is reported to be much more successful at fitting peaks (as in our case) than other tools.

   To determine service time distribution in our model, the real histogram where the size is expressed in bytes (up to 1500 bytes) was rescaled to have the average value of 1 time unit. Therefore, traffic intensity λ is at the same time the utilisation factor ϱ of the server. Figure 1 presents the real histogram and the distribution (probability density and probability distribution functions) of its approximation: three parallel Erlang distributions having 21, 1387 and 2 phases. The corresponding probabilities of an Erlang choice are 0.4078, 0.2288, 0.3634, and the parameters of exponential distributions of phases are 616.24, 954.42, 3.26. We call it below tcp distribution and use it in Kendal's notation for queueing modells, e.g. M/tcp/1/N.

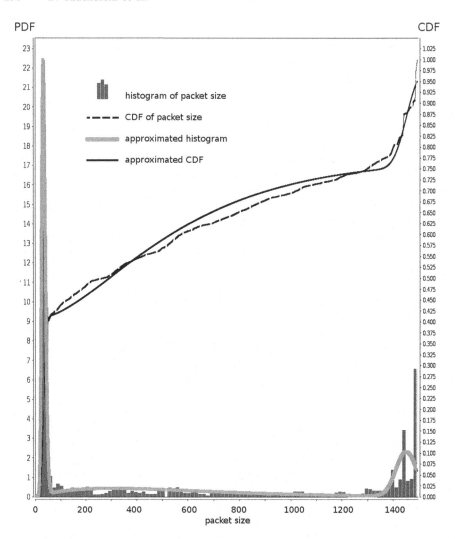

Fig. 1. Approximation of IP packet length histogram by a combination of exponentially distributed phases: its probability density function (pdf) – left vertical axis, and cumulative distribution function (CDF) – right vertical axis

3   Self-Similar Traffic

During the last two decades, self-similarity became an important research domain [17,18]. Extensive measurements demonstrated the self-similarity of network traffic on several levels of communication protocols. Various studies made also evident that ignoring these phenomena in the analysis of computer networks leads to an underestimation of important performance measures as queue lengths at buffers and packet loss probability [19,20]. Therefore, it is necessary to take self-similarity into account in realistic models of traffic [21].

The suitability of Markovian models to describe IP network traffic that exhibits self-similarity were discussed e.g. in [21,22]. The conclusion is that we do not need perfect self similar processes in all time scales but the one that exhibits self similarity in a finite number of time scales, and Markov processes may be used for this purpose, [23,24]. In particular, the input traffic may be modeled by Markov Modulated Poisson Process (MMPP) – a Poisson process having parameter λ defined by a state of a separate Markov process called modulator. If the modulator is at state i, the parameter λ is λ_i. The simplest modulator has only 2 states and if in one of these states there is no traffic, $\lambda = 0$, we speak about ON-OFF process.

We use here a sum of ON-OFF processes, as proposed in [25], see details in [26]. This model consists of d ON-OFF processes, each of them may be parametrized by two square matrices:

$$\mathbf{D_0^i} = \begin{bmatrix} -(c_{1i} + \lambda_{1i}) & c_{1i} \\ c_{2i} & -(c_{2i} + \lambda_{2i}) \end{bmatrix}, \quad \mathbf{D_1^i} = \begin{bmatrix} \lambda_{1i} & 0 \\ 0 & \lambda_{2i} \end{bmatrix}, \quad i = 1, \ldots d.$$

The element c_{1i} is the transition rate from state ON to OFF of the i-th source, and c_{2i} is the rate out of state OFF to ON, λ_{1i} is the traffic rate when the i-th source is ON. In the state OFF a source is inactive. The sum of $\mathbf{D_0}^i$ and $\mathbf{D_1}^i$ is an irreducible infinitesimal generator $\mathbf{Q}^i$ with the stationary probability vector π_i

$$\pi_i = \left(\frac{c_{2i}}{c_{1i} + c_{2i}}, \frac{c_{1i}}{c_{1i} + c_{2i}} \right).$$

The superposition of these two-state Markov chains is a new Markov process with 2^d states and its parameter matrices, $\mathbf{D_0}$ and $\mathbf{D_1}$, can be computed using the Kronecker sum of those of the d two-state processes [27]:

$$(\mathbf{D_0}, \mathbf{D_1}) = \left(\oplus_{i=1}^{d} \mathbf{D_0}^i, \oplus_{i=1}^{d} \mathbf{D_1}^i \right).$$

Article [25] demonstrates that a superposition of four described above two-state MMPP models is sufficient to replicate second-order self-similar behaviour over several time scales. Table 1 presents the parameters of superpositions of four two-state MMPP's used in our experiments. These parameters were obtained following algorithm given in [25] for chosen traffic intensities and degrees of self-similarity expressed by Hurst parameter H. In notation we use S-S symbol to denote self-similar input, e.g. S-S/M/1/64.

4   Buffer Occupation and Loss Probability

In majority of queueing models the limitation of a system capacity is given by the maximum number of customers that may be allowed inside the system, waiting in the queue or being served. This approach ignores the distribution of the size of customers. In fact, the joint distribution of the size i packets gives us the information if there is still place for the next one. In Markov models this issue may be modelled by batch arrivals, namely by Batch Markovian Arrival Process

Table 1. The parameters of superpositions of four two-state MMPP's; $\lambda_{2i} = 0$

Input parameters		λ_{1i}	c_{1i}	c_{2i}
$\lambda = 0.5\ H = 0.6$	$MMPP_1$	43.7	7.923×10^{-1}	7.707×10^{-3}
	$MMPP_2$	6.926	7.923×10^{-3}	7.707×10^{-5}
	$MMPP_3$	1.096	7.923×10^{-5}	7.707×10^{-7}
	$MMPP_4$	0.18	7.923×10^{-7}	7.707×10^{-9}
$\lambda = 0.5\ H = 0.8$	$MMPP_1$	12.145	7.798×10^{-1}	2.016×10^{-2}
	$MMPP_2$	4.945	7.798×10^{-3}	2.016×10^{-4}
	$MMPP_3$	1.798	7.798×10^{-5}	2.016×10^{-6}
	$MMPP_4$	0.945	7.798×10^{-7}	2.016×10^{-8}
$\lambda = 0.8\ H = 0.6$	$MMPP_1$	43.952	7.877×10^{-1}	1.226×10^{-2}
	$MMPP_2$	6.966	7.877×10^{-3}	1.226×10^{-4}
	$MMPP_3$	1.102	7.877×10^{-5}	1.226×10^{-6}
	$MMPP_4$	0.181	7.877×10^{-7}	1.226×10^{-8}
$\lambda = 0.8\ H = 0.8$	$MMPP_1$	12.329	7.682×10^{-1}	3.179×10^{-2}
	$MMPP_2$	5.02	7.682×10^{-3}	3.179×10^{-4}
	$MMPP_3$	1.826	7.682×10^{-5}	3.179×10^{-6}
	$MMPP_4$	0.96	7.682×10^{-7}	3.179×10^{-8}

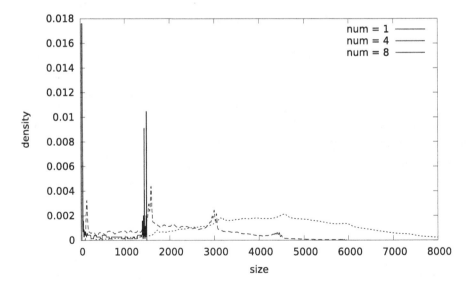

Fig. 2. The distribution of a sum of $n = 4, 8$ packets having tcp distribution

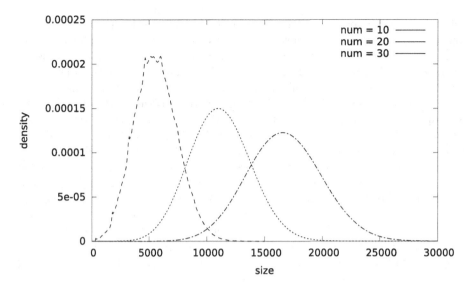

Fig. 3. The distribution of a sum of $n = 10, 15, 25, 40$ packets having tcp distributions and corresponding Erlang distributions

(BMAP) – customers arrive in groups and the size of group is determined by a probability distribution, see e.g. [3] for details. Here we assume that the size of a packet is given by previously determined distribution and determine the distribution of the size of $i = 2, \ldots, K$ packets together, see Figs. 2 and 3. This distribution is converging with the increase of K to the normal distribution the mean and variance of which are determined by the mean and variance of packets distribution. This way we may define probability that the size of i packets exceeds a given volume V of the buffer where packets are stored and we take this value as $p_{\text{loss}}(i)$ – probability that a packet is refused when there are already i packets in the buffer. The rate of the input flow is thus $\lambda(i) = \lambda(1 - p_{\text{loss}}(i))$. It is an approximation: in fact, if the buffer is nearly full, there is a zone in its capacity where smaller packets are still allowed to enter.

Because the packets do not exceed 1500 bytes, the considered distribution, although irregular, has no heavy tail and its squared coefficient of variation (ratio of variation to squared mean value) is closed to one as in exponential distribution. Therefore in our case the distribution of the sum of n packets is closed to the Erlang order n distribution, see Fig. 3. Hence, in this case, we may also use Erlang distribution to estimate loss probability.

5 Numerical Solutions, Transient States, Network Dynamics

Most frequently, queueing models are limited to the analysis of steady states. It means that flows of customers considered in models are constant and obtained solutions do not depend on time. Numerous Markovian solvers, e.g. [28,29] follow this approach. It is in contrast with the flows observed in real networks where the perpetual changes of traffic intensities are due to the nature of users, sending variable quantities of data, cf. multimedia traffic, and also due to the performance of traffic control algorithms which are trying to avoid congestion in networks, e.g. the algorithm of congestion window used in TCP protocol which is adapting the rate of the sent traffic to the observed losses or transmission delays. To study transient states we essentially dispose, apart from simulation, three analytical methods: diffusion approximation, fluid-flow approximation, and Markov models. First and second method see traffic on the level of time-variable flows, Markov chains models consider transmissions in detail, on the packet level. In diffusion approximation a diffusion equation (second order partial differential equation) solved with the use of semi-analytical, semi-numerical approach is used to give the probability distribution of a queue length [30], in fluid-flow approximation we use in this purpose first order linear ordinary differential equations; this simplification allows us to consider very large network topologies, having hundreds thousands on nodes and flows [31].

Theoretically, for any continuous time Markov chain the Chapman-Kolmogorov equations with transition matrix $\mathbf{Q}$

$$\frac{d\boldsymbol{\pi}(t)}{dt} = \boldsymbol{\pi}(t)\mathbf{Q}, \tag{1}$$

have the analytical transient solution $\boldsymbol{\pi}(t) = \boldsymbol{\pi}(0)e^{\mathbf{Q}t}$, where $\boldsymbol{\pi}(t)$ is the probability vector and $\boldsymbol{\pi}(0)$ is the initial condition. However, it is not easy to compute the expression $e^{\mathbf{Q}t}$ where $\mathbf{Q}$ is a large matrix, see e.g. [32,33]. It may be done by its expansion to Taylor's series

$$e^{\mathbf{Q}t} = \sum_{k=0}^{\infty} \frac{(\mathbf{Q}t)^k}{k!},$$

but the task is numerically unstable, especially for large $\mathbf{Q}$. *Nineteen dubious ways to compute matrix exponential* are discussed in [32,33] this work indicates that some of the methods are preferable to others, but that none are completely satisfactory. Otherwise, it may be wasteful to compute the solution of a mathematical model correct to full machine precision when the model itself contains errors of the order, say 10 %.

In practice, to solve Eq. 1 we should apply either the uniformization (Jensen's) method which studies discrete Markov chain (it is much easier) embedded inside the considered continuous time one or use readily available efficient equations solvers as Runge-Kutta or the Adams formulae and backward differentiation formulae (BDF). In a general opinion, the most efficient approach is to use projection methods where the original problems is projected to a space (e.g. Krylov subspace) where it has considerably smaller dimension, solve it there and then retransform this solution to the original state space, [34]. We may find its implementation in tools [2,35]. Here we used the latter: a well known probabilistic model checker Prism [2]. We supplemented it with a kind of preprocessor to ease the formulation of more complex queueing models.

6   Numerical Examples

The examples below illustrate the results furnished by the model and give an idea on the impact of the distribution of packet size – if we compare a service station with real distribution of service times with an analogous with exponentially distributed service time – as well as on the impact of self-similarity of the performance of stations. We see how both self-similarity and non-exponential distribution increase the queue length and the duration of transient period. In the examples the input flow starts at $t = 0$ to arrive to the empty station. This impact increases with the utilization ϱ of the stations. Note the influence of self-similarity on the distribution of the queue size: it is not decreasing monotonically with the number of queued customers.

Numerical examples:

Example 1. The impact of packet size distribution on the queue size and the length of transient period: comparison of M/M/1/N and M/TCP/1/N stations, two levels of congestion $\varrho = 0.5$, Fig. 4 and $\varrho = 0.8$, Fig. 5.

Example 2. The impact of self similar input flow on the queue size and the length of transient period: comparison of S-S/M/1/64 and M/M/1/64 stations, two levels of congestion $\varrho = 0.5$, Fig. 6 and $\varrho = 0.8$, Fig. 7.

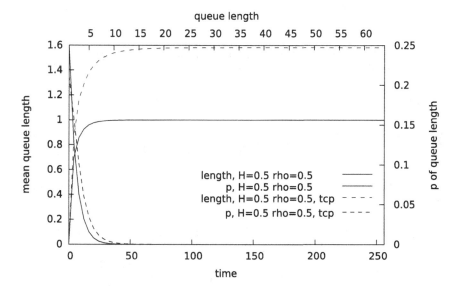

Fig. 4. Mean queue as a function of time and steady state distribution of the queue length: comparison of M/M/1/64 and M/tcp/1/64 stations, $\varrho = 0.5$

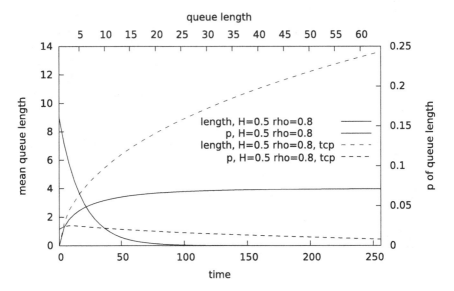

Fig. 5. Mean queue as a function of time and steady state distribution of the queue length: comparison of M/M/1/64 and M/tcp/1/64 stations, $\varrho = 0.8$

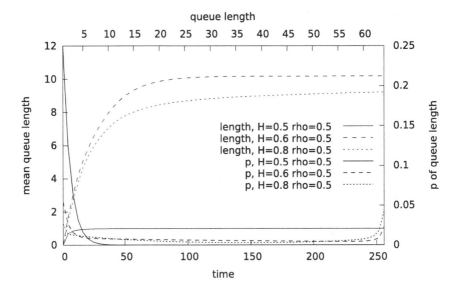

Fig. 6. Steady state distribution of the queue length and mean queue as a function of time: comparison of M/M/1/64 and S-S/M/1/64 ($H = 0.6, 0.8$) stations, $\varrho = 0.5$

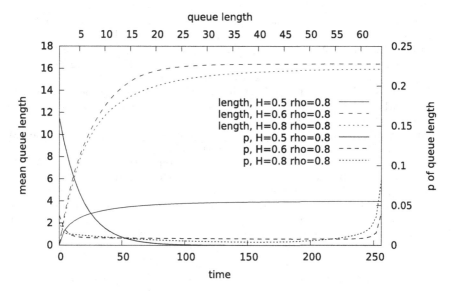

Fig. 7. Steady state distribution of the queue length and mean queue as a function of time: comparison of M/M/1/64 and S-S/M/1/64 ($H = 0.6, 0.8$) stations, $\varrho = 0.8$

7 Conclusions

The presented numerical examples confirm that the proposed approach that unifies in a Markovian model (i) a real IP packet distribution which is a basis to define both the losses due to a finite buffer volume and the service time distribution (ii) self similar traffic, is feasible and may be used also to study transient behavior of router queues. Quantitative results may be obtained with the use of well known public software tools. We plan to test the approach when applied to more complex models.

References

1. Reinecke, P., Krauß, T., Wolter, K.: HyperStar: phase-type fitting made easy. In: 9th International Conference on the Quantitative Evaluation of Systems (QEST 2012), pp. 201–202 (September 2012)
2. Kwiatkowska, M., Norman, G., Parker, D.: PRISM 4.0: verification of probabilistic real-time systems. In: Gopalakrishnan, G., Qadeer, S. (eds.) CAV 2011. LNCS, vol. 6806, pp. 585–591. Springer, Heidelberg (2011). http://www.prismmodelchecker.org/
3. Chydziński, A.: Nowe modele kolejkowe dla węzłów sieci pakietowych. Pracownia Komputerowa Jacka Skalmierskiego, Gliwice (2013)
4. Tikhonenko, O., Kawecka, M.: Total volume distribution for multiserver queueing systems with random capacity demands. In: Kwiecień, A., Gaj, P., Stera, P. (eds.) CN 2013. CCIS, vol. 370, pp. 394–405. Springer, Heidelberg (2013)

5. Czachórski, T., Nycz, T., Pekergin, F.: Queue with limited volume, a diffusion approximation approach. In: Gelenbe, E., Lent, R., Sakellari, G., Sacan, A., Toroslu, H., Yazici, A. (eds.) Computer and Information Sciences. Lecture Notes in Electrical Engineering, vol. 62, pp. 71–74. Springer, Netherlands (2010)
6. https://data.caida.org/datasets/passive-2008/equinix-chicago/20080319/
7. Buchholz, P., Kriege, J., Felko, I.: Input Modeling with Phase-Type Distributions and Markov Models: Theory and Applications. SpringerBriefs in Mathematics. Springer, Heidelberg (2014)
8. Asmussen, S., Nerman, O., Olsson, M.: Fitting phase-type distribution via the EM algorithm. Scand. J. Stat. **23**, 419–441 (1996)
9. Horváth, A., Telek, M.: PhFit: a general phase-type fitting tool. In: Field, T., Harrison, P.G., Bradley, J., Harder, U. (eds.) TOOLS 2002. LNCS, vol. 2324, p. 82. Springer, Heidelberg (2002)
10. Riska, A., Diev, V., Smirni, E.: Efficient fitting of longtailed data sets into phase-type distributions. In: SIGMETRICS Performance Evaluation Review, vol. 30, pp. 6–8, December 2002. http://doi.acm.org/10.1145/605521.605525
11. Pérez, J.F., Riaño, G.: jPhase: an object-oriented tool for modeling phase-type distributions. In: Proceeding From the 2006 Workshop on Tools for Solving Structured Markov Chains, ser. (SMCtools 2006), New York, NY, USA. ACM (2006)
12. Thümmler, A., Buchholz, P., Telek, M.: A novel approach for phase-type fitting with the EM algorithm. IEEE Trans. Dependable Secur. Comput. **3**(3), 245–258 (2006)
13. Casale, G., Zhang, E.Z., Smirni, E.: KPC-toolbox: Simple yet effective trace fitting using markovian arrival processes. In: Proceedings of the 2008 Fifth International Conference on Quantitative Evaluation of Systems, pp. 83–92. Computer Society, IEEE, Washington, DC (2008)
14. Wang, J., Liu, J., She, C.: Segment-based adaptive hyper-erlang model forlong-tailed network traffic approximation. J. Supercomput. **45**, 296–312 (2008)
15. Sadre, R., Haverkort, B.: Fitting heavy-tailed HTTP traces with the new stratified EM-algorithm. In: 4th International Telecommunication Networking Workshop on QoS in Multiservice IP Networks (IT-NEWS), pp. 254–261. IEEE Computer Society Press, Los Alamitos, February 2008
16. Bause, F., Buchholz, P., Kriege, J.: ProFiDo - the processes fitting toolkit dortmund. In: Proceedings of the 7th International Conference on Quantitative Evaluation of Systems (QEST 2010), pp. 87–96. IEEE Computer Society (2010)
17. Loiseau, P., Gonçalves, P., Dewaele, G., Borgnat, P., Abry, P., Primet, P.V.-B.: Investigating self-similarity and heavy-tailed distributions on a large-scale experimental facility. IEEE/ACM Trans. Netw. **18**(4), 1261–1274 (2010)
18. Bhattacharjee, A., Nandi, S.: Statistical analysis of network traffic inter-arrival. In: 12th International Conference on Advanced Communication Technology, USA, pp. 1052–1057 (2010)
19. Kim, Y.G., Min, P.S.: On the prediction of average queueing delay with self-similar traffic. In: Proceedings of the IEEE Globecom 2003, vol. 5, pp. 2987–2991 (2003)
20. Gorrasi, A., Restino, R.: Experimental comparison of some scheduling disciplines fed by self-similar traffic. In: Proceedings of the IEEE International Conference on Communication, vol. 1, pp. 163–167 (2003)
21. Muscariello, L., Mellia, M., Meo, M., Marsan, M.A., Cigni, R.L.: Markov models of internet traffic and a new hierarchical MMPP model. Comput. Commun. **28**, 1835–1851 (2005)
22. Clegg, R.G.: Markov-modulated on/off processes for long-range dependent internet traffic. Computing Research Repository, CoRR (2006)

23. Grossglauser, M., Bolot, J.C.: On the relevance of long-range dependence in network traffic. IEEE/ACM Trans. Netw. **7**(5), 629–640 (1999)
24. Nogueira, A., Valadas, R.: Analyzing the relevant time scales in a network of queues. In: SPIE Proceedings, vol. 4523 (2001)
25. Andersen, A.T., Nielsen, B.F.: A markovian approach for modeling packet traffic with long-range dependence. IEEE J. Sel. Areas in Commun. **16**(5), 719–732 (1998)
26. Domańska, J., Domański, A., Czachórski, T.: Modeling packet traffic with the use of superpositions of two-state MMPPs. In: Kwiecień, A., Gaj, P., Stera, P. (eds.) CN 2014. CCIS, vol. 431, pp. 24–36. Springer, Heidelberg (2014)
27. Fischer, W., Meier-Hellstern, K.: The markov-modulated poisson process (MMPP) cookbook. Perform. Eval. **18**(2), 149–171 (1993)
28. PEPS. www-id.imag.fr/Logiciels/peps/userguide.html
29. Potier, D.: New User's Introduction to QNAP2. Rapport Technique no. 40, INRIA, Rocquencourt (1984)
30. Czachórski, T.: A method to solve diffusion equation with instantaneous return processes acting as boundary conditions. Bull. Pol. Acad. Sci. Tech. Sci. **41**(4), 417–451 (1993)
31. Nycz, M., Nycz, T., Czachórski, T.: Modelling dynamics of TCP flows in very large network topologies. In: Abdelrahman, O.H., Gelenbe, E., Gorbil, G., Lent, R. (eds.) Information Sciences and Systems 2015. Lecture Notes in Electrical Engineering, vol. 363, pp. 251–259. Springer, Switzerland (2016)
32. Moler, C., Van Loan, C.: Nineteen dubious ways to compute the exponential of a matrix. SIAM Rev. **20**, 801–836 (1978)
33. Moler, C., Van Loan, C.: Nineteen dubious ways to compute the exponential of a matrix twenty-five years later. SIAM Rev. **45**(1), 30–49 (2003)
34. Stewart, W.: Introduction to the Numerical Solution of Markov Chains. Princeton University Press, Chichester (1994)
35. Pecka, P., Deorowicz, S., Nowak, M.: Efficient representation of transition matrix in the markov process modeling of computer networks. In: Czachórski, T., Kozielski, S., Stańczyk, U. (eds.) Man-Machine Interactions 2. AISC, vol. 103, pp. 457–464. Springer, Heidelberg (2011)

Analysis of a Queueing Model with Contingent Additional Server

Chesoong Kim[1(✉)] and Alexander Dudin[2]

[1] Sangji University, Wonju, Kangwon 220-702, Korea
dowoo@sangji.ac.kr
[2] Belarusian State University, 4, Nezavisimosti Avenue, 220030 Minsk, Belarus
dudin@bsu.by

Abstract. We consider a two-server queueing system with a finite buffer. Customers arrive to the system according to the Markovian arrival process. Normally, only one server is active. The service time of a customer has a phase-type distribution. An additional server is activated only if the queue length exceeds some fixed preassigned threshold. The service time by the additional server also has a phase-type distribution with the same state space. While the underlying Markov chains of service at two servers have non-coinciding states, service in two servers is provided independently. But if it occurs that the underlying Markov chain for one server, say, server 1, needs transition to the state, at which the underlying Markov chain for server 2 is currently staying, service in the server 1 is postponed until the Markov chain for server 2 transits to another state. Dynamics of the system is described by the multi-dimensional Markov chain. The generator of this Markov chain is written down. Expressions for computation of performance measures are derived. Problem of numerical determination of the optimal threshold is solved.

Keywords: Markovian arrival process · Phase-type service time distribution · Dependent service processes · Congestion avoidance

1 Introduction

Queueing theory is suitable for solving mathematical problems arising in capacity planning, performance evaluation and optimization in computer networks. In this paper, we consider a two-server queueing system with a finite buffer. Due to the random nature of arrival process, especially when this process exhibits positive correlation, congestion situations may occur. Queue length becomes large and some customers are lost due to the buffer overflow. Aiming to reduce average value and to smooth the variance of a queue length as well as to decrease customer's loss probability, it is reasonable to switch on the additional server whenever congestion occurs. The literature about the queues with controlled number of servers is not poor, for references see, e.g., [1–4]. It is assumed in these papers that service processes in all servers are mutually independent and do not interfere. However, in some real world systems service processes in active

© Springer International Publishing Switzerland 2016
P. Gaj et al. (Eds.): CN 2016, CCIS 608, pp. 306–315, 2016.
DOI: 10.1007/978-3-319-39207-3_27

servers may be dependent. In particular, if service is provided to applications or customers by the servers with help of some common hardware or software, service processes become dependable. E.g., if service of a customer on computer consists of implementation of a sequence of I/O operations and operations with CPU (or a sequence of operations with tables and indices of a relational database) and service to at least two customers is provided at the same time, collisions may occur. Such collisions are explained by the conflicts arising when the same device or logical unit (CPU, table or index) is required by some application when it is already busy (or locked) by another application. To the best of our knowledge, such kind of queueing models with dependent service is not addressed in the queueing literature. In this paper, we analyse a single server queue with contingent additional server that activates on demand when congestion occurs.

2   Mathematical Model

A queueing system with a finite buffer of capacity N is considered. The arrival flow is described by the Markovian arrival process (MAP). Arrivals may occur at instants of transitions of continuous time Markov chain ν_t, $t \geq 0$, which is called as underlying process of arrivals. This process has a finite state space $\{0, \ldots, W\}$. The intensities of transitions of the process ν_t that lead (do not lead) to a customer arrival are defined as the entries of the matrix D_1 (D_0) of size $\bar{W} \times \bar{W}$ where $\bar{W} = W + 1$. The stationary distribution vector $\boldsymbol{\theta}$ of the process ν_t satisfies the system of equations $\boldsymbol{\theta}(D_0 + D_1) = \mathbf{0}$, $\boldsymbol{\theta}\mathbf{e} = 1$. Here and throughout this paper, $\mathbf{0}$ is a zero row vector, $\mathbf{e}$ is the column vector consisting of 1's. In case if the dimension of a vector is not clear from the context, it is indicated as a lower index. The average intensity λ (fundamental rate) of the MAP is defined by $\lambda = \boldsymbol{\theta}D_1\mathbf{e}$. Generally speaking, inter-arrival times in the MAP may be correlated. For more information about the MAP, its properties, definition of coefficients of variation and correlation of inter-arrival times as well as about the partial cases and generalizations see, e.g., [5].

The service process is of phase (PH) type. This means that the duration of service is governed by the underlying process n_t which is a continuous time Markov chain with state space $\{1, \ldots, M\}$. The initial state of the process n_t at the epoch of starting the vacation is determined by the probabilistic row-vector $\boldsymbol{\beta} = (\beta_1, \ldots, \beta_M)$. The intensities of transitions of the process n_t within the state space that do not lead to the service completion, are defined by the square irreducible matrix (subgenerator) S of size M. The intensities of transitions, which lead to the service completion, are given by the column-vector $\mathbf{S}_0 = -S\mathbf{e}$. The pair $(\boldsymbol{\beta}, S)$ of a vector and a matrix is called as irreducible representation of PH distribution. For more information about PH distribution see [6].

Customers service discipline is described as follows. Some threshold, say, K, $K = \overline{1, N+1}$, is fixed. If the number of customers in the system is less or equal to K, only one server is working. Otherwise, the second server is switched on. The service time distribution at this server also has PH distribution with an irreducible representation $(\boldsymbol{\beta}, S)$. If, at an arbitrary moment, underlying processes of service at two servers reside at different phases from the set $\{1, \ldots, M\}$, durations of these phases are independent of each other as well as the choice of the next phase of service. But, if service of customers by the l-th server, $l = 1, 2$, is at some phase while the underlying process of service at the l'-th server, $l' = 1, 2$, $l' \neq l$, finishes residing at some phase of service and the next phase of service by this server is the one, at which service is provided by the l-th server, the transition of the phase of service by the l'-th server is not performed (the server is considered as blocked) until service at this phase by the l-th server will be finished. We suggest that the servers enumerated in order of their occupation. During the time when service of customer by some server is blocked, this customer can be impatient and leave the system without continuation of service after the random time having the exponential distribution with parameter (intensity) γ. If, at some customer service completion moment, the number of customers drops to the value K, the server, which just finished service, is switched off.

The goal of our analysis presented below is the determination of the optimal value K^* of the threshold K which ensures the smallest value of the loss probability of an arbitrary customer in the system.

3   Process of System States

It is easy to see that the behavior of the system under study is described in terms of the following regular irreducible continuous-time Markov chain

$$\xi_t = \{i_t, r_t, \nu_t, n_t, m_t\}, \ t \geq 0,$$

where, during the epoch t, $t \geq 0$,

- i_t is the number of customers in the system, $i_t = \overline{0, N+2}$;
- r_t is an indicator that indicates whether some server is blocked or not: $r_t = 0$ corresponds to the case when the server is not blocked and $r_t = 1$ otherwise;
- ν_t is the state of the underlying process of the MAP, $\nu_t = \overline{0, W}$;
- n_t is the state of PH service process in the first server, $n_t = \overline{1, M}$;
- m_t is the state of PH service process in the second server, $m_t = \overline{1, M}$.

The Markov chain ξ_t, $t \geq 0$, has the following state space:

$$\left(\{0, 0, \nu\} \right) \bigcup \left(\{i, 0, \nu, n\}, \ i = \overline{1, K}, \ n = \overline{1, M} \right) \bigcup$$

$$\left(\{i, 0, \nu, n, m\}, \ i = \overline{K+1, N}, \ n = \overline{1, M}, \ m = \overline{1, M}, \ m \neq n \right)$$

$$\bigcup \left(\{i, 1, \nu, n\}, \ i = \overline{K+1, N}, \ n = \overline{1, M} \right), \quad \nu = \overline{0, W}.$$

For further use throughout this paper, we introduce the following notation:

- I is the identity matrix, and O is a zero matrix of appropriate dimension;
- $\otimes$ and $\oplus$ indicate the symbols of Kronecker product and sum of matrices, respectively;
- $\delta_{a=b} = \begin{cases} 1, & \text{if } a = b, \\ 0, & \text{otherwise}; \end{cases}$
- I_{l_1,l_2}, $l_1, l_2 = \overline{1,M}$, $l_1 \neq l_2$, is the square matrix of size $M - 1$ with all zero entries except the entries $(I_{l_1,l_2})_{k,k}$, $k = \overline{0, M - 2}$, $k \neq l_2 - 2$, in the case $l_1 < l_2$ and $(I_{l_1,l_2})_{k,k}$, $k = \overline{0, M - 2}$, $k \neq l_2 - 1$, in the case $l_1 > l_2$ which are equal to 1;
- $\tilde{S}_l$, $l = \overline{1,M}$, is the square matrix of size $M - 1$ that is obtained from the matrix S by removing the l-th column and the l-th row;
- $\mathbf{e}_{l_1,l_2}$, $l_1, l_2 = \overline{1,M}$, $l_1 \neq l_2$, is the column vector of size $M - 1$ with all zero entries except the entries $(\mathbf{e}_{l_1,l_2})_{l_2-1}$ in the case $l_1 > l_2$ and $(\mathbf{e}_{l_1,l_2})_{l_2-2}$ in the case $l_1 < l_2$ which are equal to 1;
- $\mathbf{c}_{l_1,l_2}$, $l_1, l_2 = \overline{1,M}$, $l_1 \neq l_2$, is the row vector of size $M - 1$ with all zero entries except the entry $(\mathbf{c}_{l_1,l_2})_{l_1-2}$ in the case $l_1 > l_2$ and $(\mathbf{c}_{l_1,l_2})_{l_1-1}$ in the case $l_1 < l_2$ which are equal to 1;
- $\boldsymbol{\beta}_l$, $l = \overline{1,M}$, is the row vector that is obtained from the vector $\boldsymbol{\beta}$ by deleting the l-th component;
- $\mathbf{a}_l$, $l = \overline{1,M}$, is the column vector of size $M - 1$ that is obtained from the l-th column of the matrix S by removing the l-th entry;
- I_l^+, $l = \overline{1,M}$, is the matrix of size $(M - 1) \times M$ which obtained from the identity matrix of size $M - 1$ by adding the zero column in position l;
- $\mathbf{S}_0^l$, $l = \overline{1,M}$ is a column vector of size $M - 1$ which is obtained from the vector $\mathbf{S}_0$ by removing the $l - 1$-th component;
- $\tilde{\mathbf{a}}_l$, $l = \overline{1,M}$, is a row vector of size M with all zero components except the component $(\tilde{\mathbf{a}}_l)_{l-1}$ which is equal to 1;
- B_l, $l = \overline{1,M}$, is the matrix of size $(M - 1) \times (M - 1)M$ which is obtained from the matrix $\mathrm{diag}\{\boldsymbol{\beta}_1, \ldots, \boldsymbol{\beta}_M\}$ by deleting the l-th row;
- C_l, $l = \overline{1,M}$, is the matrix of size $(M - 1) \times M$ which is obtained from the matrix $\mathrm{diag}\{\boldsymbol{\beta}_1, \ldots, \boldsymbol{\beta}_M\}$ by deleting the l-th row.

Let us enumerate the states of the Markov chain ξ_t, $t \geq 0$, in the direct lexicographic order of the components k, ν, ζ, η and refer to the set of the states of the Markov chain having values (i, r) of the first two components of the chain as a macro-state (i, r).

Let $\mathbf{Q}$ be the generator of the Markov chain ξ_t, $t \geq 0$, consisting of the blocks $\mathbf{Q}_{i,j}$, which, in turn, consist of the matrices $(\mathbf{Q}_{i,j})_{r,r'}$ of the transition rates of this chain from the macro-state (i, r) to the macro-state (j, r'), $r, r' = 0, 1$. The diagonal entries of the matrices $\mathbf{Q}_{i,i}$ are negative, and the modulus of the diagonal entry of the blocks $(\mathbf{Q}_{i,i})_{r,r}$ defines the total intensity of leaving the corresponding state of the Markov chain ξ_t, $t \geq 0$.

Analysing all transitions of the Markov chain ξ_t, $t \geq 0$, during an interval of an infinitesimal length and rewriting the intensities of these transitions in the block matrix form we obtain the following result.

Lemma 1. *The infinitesimal generator* $\mathbf{Q} = (\mathbf{Q}_{i,j})_{i,j\geq 0}$ *of the Markov chain* ξ_t, $t \geq 0$, *has a block-tridiagonal structure:*

$$\mathbf{Q} = \begin{pmatrix} \mathbf{Q}_{0,0} & \mathbf{Q}_{0,1} & O & \cdots & O & O \\ \mathbf{Q}_{1,0} & \mathbf{Q}_{1,1} & \mathbf{Q}_{1,2} & \cdots & O & O \\ O & \mathbf{Q}_{2,1} & \mathbf{Q}_{2,2} & \cdots & O & O \\ \vdots & \vdots & \vdots & \ddots & \vdots & \vdots \\ O & O & O & \cdots & \mathbf{Q}_{N+1,N+1} & \mathbf{Q}_{N+1,N+2} \\ O & O & O & \cdots & \mathbf{Q}_{N+2,N+1} & \mathbf{Q}_{N+2,N+2} \end{pmatrix}.$$

The non-zero blocks $\mathbf{Q}_{i,j}$, $i, j \geq 0$, *have the following form:*

$$\mathbf{Q}_{0,0} = D_0, \quad \mathbf{Q}_{i,i} = D_0 \oplus S, \quad i = \overline{1, K},$$

$$\mathbf{Q}_{i,i} = \begin{pmatrix} \mathbf{Q}^{(0,0)} & \mathbf{Q}^{(0,1)} \\ \mathbf{Q}^{(1,0)} & \mathbf{Q}^{(1,1)} \end{pmatrix}, \quad i = \overline{K+1, N+1},$$

$$\mathbf{Q}_{N+2,N+2} = \begin{pmatrix} \mathbf{Q}^{(0,0)} + D_1 \otimes I_{M(M-1)} & \mathbf{Q}^{(0,1)} \\ \mathbf{Q}^{(1,0)} & \mathbf{Q}^{(1,1)} + D_1 \otimes I_M \end{pmatrix},$$

$$\mathbf{Q}^{(0,0)} = D_0 \otimes I_{M(M-1)} + I_{\bar{W}} \otimes (\mathcal{S} + \operatorname{diag}\{\tilde{S}_1, \ldots, \tilde{S}_M\}),$$

$$\mathcal{S} = \begin{pmatrix} S_{1,1}I_{M-1} & S_{1,2}I_{1,2} & \cdots & S_{1,M}I_{1,M} \\ S_{2,1}I_{2,1} & S_{2,2}I_{M-1} & \cdots & S_{2,M}I_{2,M} \\ \vdots & \vdots & \ddots & \vdots \\ S_{M,1}I_{M,1} & S_{M,2}I_{M,2} & \cdots & S_{M,M}I_{M-1} \end{pmatrix},$$

$$\mathbf{Q}^{(0,1)} = I_{\bar{W}} \otimes \begin{pmatrix} \mathbf{a}_1 & S_{1,2}\mathbf{e}_{1,2} & \cdots & S_{1,M}\mathbf{e}_{1,M} \\ S_{2,1}\mathbf{e}_{2,1} & \mathbf{a}_2 & \cdots & S_{2,M}\mathbf{e}_{2,M} \\ \vdots & \vdots & \ddots & \vdots \\ S_{M,1}\mathbf{e}_{M,1} & S_{M,2}\mathbf{e}_{M,2} & \cdots & \mathbf{a}_M \end{pmatrix},$$

$$\mathbf{Q}^{(1,0)} = I_{\bar{W}} \otimes \begin{pmatrix} \mathbf{0} & S_{1,2}\mathbf{c}_{1,2} & \cdots & S_{1,M}\mathbf{c}_{1,M} \\ S_{2,1}\mathbf{c}_{1,M} & \mathbf{0} & \cdots & S_{2,M}\mathbf{c}_{1,M} \\ \vdots & \vdots & \ddots & \vdots \\ S_{M,1}\mathbf{c}_{1,M} & S_{M,2}\mathbf{c}_{1,M} & \cdots & \mathbf{0} \end{pmatrix},$$

$$\mathbf{Q}^{(1,1)} = D_0 \oplus \operatorname{diag}\{S_{1,1}, \ldots, S_{M,M}\} - \gamma I_{\bar{W}M},$$

$$\mathbf{Q}_{0,1} = D_1 \otimes \boldsymbol{\beta}, \quad \mathbf{Q}_{i,i+1} = D_1 \otimes I_M, \quad i = \overline{1, K-1},$$

$$\mathbf{Q}_{K,K+1} = \left(D_1 \otimes \operatorname{diag}\{\boldsymbol{\beta}_1, \ldots, \boldsymbol{\beta}_M\} \mid D_1 \otimes \operatorname{diag}\{\boldsymbol{\beta}_1, \ldots, \boldsymbol{\beta}_M\} \right),$$

$$\mathbf{Q}_{i,i+1} = \begin{pmatrix} D_1 \otimes I_{M(M-1)} & O \\ O & D_1 \otimes I_M \end{pmatrix}, \quad i = \overline{K+1, N+1},$$

$$\mathbf{Q}_{1,0} = I_{\bar{W}} \otimes \mathbf{S}_0, \quad \mathbf{Q}_{i,i-1} = I_{\bar{W}} \otimes \mathbf{S}_0\boldsymbol{\beta}, \quad i = \overline{2, K},$$

$$Q_{K+1,K} = \begin{pmatrix} I_{\bar{W}} \otimes \left(\begin{pmatrix} (\mathbf{S_0})_1 I_1^+ \\ \vdots \\ (\mathbf{S_0})_M I_M^+ \end{pmatrix} + \mathrm{diag}\{\mathbf{S}_0^l, l = \overline{1,M}\} \right) \\ I_{\bar{W}} \otimes \left(\begin{pmatrix} (\mathbf{S_0})_1 \tilde{a}_1 \\ \vdots \\ (\mathbf{S_0})_M \tilde{a}_M \end{pmatrix} + \gamma I_M \right) \end{pmatrix},$$

$$Q_{i,i-1} = \begin{pmatrix} Q_-^{(0,0)} & Q_-^{(0,1)} \\ Q_-^{(1,0)} & Q_-^{(1,1)} \end{pmatrix}, \quad i = \overline{K+2, N+2},$$

$$Q_-^{(0,0)} = I_{\bar{W}} \otimes \left(\begin{pmatrix} (\mathbf{S_0})_1 B_1 \\ \vdots \\ (\mathbf{S_0})_M B_M \end{pmatrix} + \mathrm{diag}\{\mathbf{S}_0^l \beta_l, l = \overline{1,M}\} \right),$$

$$Q_-^{(0,1)} = I_{\bar{W}} \otimes \left(\begin{pmatrix} (\mathbf{S_0})_1 C_1 \\ \vdots \\ (\mathbf{S_0})_M C_M \end{pmatrix} + \mathrm{diag}\{\mathbf{S}_0^l \beta_l, l = \overline{1,M}\} \right),$$

$$Q_-^{(1,0)} = I_{\bar{W}} \otimes \mathrm{diag}\{((\mathbf{S_0})_l + \gamma)\beta_l, l = \overline{1,M}\},$$

$$Q_-^{(1,1)} = I_{\bar{W}} \otimes \mathrm{diag}\{((\mathbf{S_0})_l + \gamma)\beta_l, l = \overline{1,M}\}.$$

Because the Markov chain ξ_t is regular, irreducible and has a finite state space, the following limits (stationary probabilities) always exist:

$$\pi(i, r, \nu, n, m) = \lim_{t \to \infty} P\{i_t = i, \ r_t = r, \ \nu_t = \nu, \ n_t = n, \ m_t = m\},$$

$$i = \overline{0, N+2}, \quad r = \overline{0,1}, \quad \nu = \overline{0, W}, \quad n = \overline{1, M}, \quad m = \overline{1, M}.$$

Then let us form the row vectors of the stationary probabilities π_i as follows:

$$\pi_0 = (\pi(0,0,0), \pi(0,0,1), \ldots, \pi(0,0,W)),$$

$$\pi_i = (\pi(i,0,0), \pi(i,0,1), \ldots, \pi(i,0,W)), \quad i = \overline{1, K},$$

where

$$\pi(i,0,\nu) = (\pi(i,0,\nu,1), \pi(i,0,\nu,2), \ldots, \pi(i,0,\nu,M)) \ , \quad \nu = \overline{0, W} \ , \quad i = \overline{1, K}.$$

$$\pi_i = (\pi(i,0), \pi(i,1)), \quad i = \overline{K+1, N+2},$$

where

$$\pi(i,r) = (\pi(i,r,0), \pi(i,r,1), \ldots, \pi(i,r,W)), \quad i = \overline{K+1, N+2} \ ,$$

$$\pi(i,0,\nu) = (\pi(i,0,\nu,1), \pi(i,0,\nu,2), \ldots, \pi(i,0,\nu,M)), \quad \nu = \overline{0, W},$$

$$\pi(i,0,\nu,n) = (\pi(i,0,\nu,n,1), \pi(i,0,\nu,n,2), \ldots, \pi(i,0,\nu,n,M)), \quad n = \overline{1, M},$$

$$\boldsymbol{\pi}(i,1,\nu) = (\pi(i,1,\nu,1), \pi(i,1,\nu,2), \ldots, \pi(i,1,\nu,M)), \quad \nu = \overline{0,W}.$$

It is well known that the probability vectors $\boldsymbol{\pi}_i$, $i = \overline{0, N+2}$, satisfy the following system of linear algebraic equations (equilibrium or Kolmogorov's equations):

$$(\boldsymbol{\pi}_0, \boldsymbol{\pi}_1, \ldots, \boldsymbol{\pi}_{N+2})Q = 0, \quad (\boldsymbol{\pi}_0, \boldsymbol{\pi}_1, \ldots, \boldsymbol{\pi}_{N+2})\mathbf{e} = 1. \tag{1}$$

To solve system (1), we use the numerically stable algorithm for computation of the probability vectors $\boldsymbol{\pi}_i$, $i = \overline{0, N+2}$, developed in [7].

4   Performance Measures

The average number N_{buffer} of customers in the buffer is computed by

$$N_{\text{buffer}} = \sum_{i=2}^{K} (i-1)\boldsymbol{\pi}_i\mathbf{e} + \sum_{i=K+1}^{N+2} (i-2)\boldsymbol{\pi}_i\mathbf{e}.$$

The average number N_{busy} of busy servers is computed by

$$N_{\text{busy}} = \sum_{i=1}^{K} \boldsymbol{\pi}_i\mathbf{e} + 2\sum_{i=K+1}^{N+2} \boldsymbol{\pi}_i\mathbf{e}.$$

The probability P_{blocked} that, at an arbitrary moment, a server is blocked is computed by

$$P_{\text{blocked}} = \sum_{i=K+1}^{\infty} \boldsymbol{\pi}(i,1)\mathbf{e}.$$

The probability $P_{\text{ent-loss}}$ that an arbitrary customer will be lost upon arrival is computed by

$$P_{\text{ent-loss}} = \frac{1}{\lambda}\left(\boldsymbol{\pi}(N+2,0)(D_1 \otimes I_{M(M-1)})\mathbf{e} + \boldsymbol{\pi}(N+2,1)(D_1 \otimes I_M)\mathbf{e}\right).$$

The probability $P_{\text{imp-loss}}$ that an arbitrary customer will be lost due to impatience on blocked server is computed by

$$P_{\text{imp-loss}} = \frac{\gamma P_{\text{blocked}}}{\lambda}.$$

The average intensity λ_{out} of flow of customers who receive service is computed by

$$\lambda_{\text{out}} = \sum_{i=1}^{K} \boldsymbol{\pi}_i(\mathbf{e}_{\bar{W}} \otimes \mathbf{S}_0) + \sum_{i=K+1}^{N+2}\sum_{\nu=0}^{W}\sum_{n=1}^{M}\left(\sum_{m=1}^{n-1} \boldsymbol{\pi}(i,0,\nu,n,m)((\mathbf{S}_0)_n + (\mathbf{S}_0)_m)\right.$$
$$\left. + \sum_{m=n}^{M-1} \boldsymbol{\pi}(i,0,\nu,n,m)((\mathbf{S}_0)_n + (\mathbf{S}_0)_{m+1})\right) + \sum_{i=K+1}^{N+2} \boldsymbol{\pi}(i,1)(\mathbf{e}_{\bar{W}} \otimes \mathbf{S}_0).$$

The loss probability P_{loss} of an arbitrary customer is computed by

$$P_{\text{loss}} = 1 - \frac{\lambda_{\text{out}}}{\lambda} = P_{\text{imp-loss}} + P_{\text{ent-loss}}.$$

5   Numerical Examples

We assume that the buffer capacity is $N = 20$. To illustrate importance of taking into account correlation in the arrival process, let us introduce three $MAPs$ defined by the matrices D_0 and D_1 which have the same average arrival rate $\lambda = 0.6$, but different coefficients of correlation and variation.

The first process coded as MAP_0 is defined by the matrices $D_0 = -0.6$ and $D_1 = 0.6$. It has the coefficient of correlation $c_{cor} = 0$ and the coefficient of variation $c_{var} = 1$.

The second process coded as $MAP_{0.2}$ has the coefficient of correlation $c_{cor} = 0.2$ and the coefficient of variation $c_{var} = 12.3$, and is defined by the matrices

$$D_0 = \begin{pmatrix} -0.81098 & 0 \\ 0 & -0.02632 \end{pmatrix}, \quad D_1 = \begin{pmatrix} 0.80559 & 0.00539 \\ 0.01466 & 0.01166 \end{pmatrix}.$$

The third process coded as $MAP_{0.4}$ has the coefficient of correlation $c_{cor} = 0.4$ and the coefficient of variation $c_{var} = 12.3$, and is defined by the matrices

$$D_0 = \begin{pmatrix} -2.03861 & 0 \\ 0.00061 & -0.06613 \end{pmatrix}, \quad D_1 = \begin{pmatrix} 2.01738 & 0.02123 \\ 0.00728 & 0.05824 \end{pmatrix}.$$

The impatience rate is $\gamma = 0.6$.
The service time distribution is defined by the matrix

$$S = \begin{pmatrix} -2 & 0.2 & 0.3 & 0.3 & 0.4 \\ 0.5 & -3 & 0.2 & 0.1 & 1.2 \\ 1.1 & 2 & -4 & 0.7 & 0.2 \\ 0.5 & 0.6 & 0.7 & -3 & 1 \\ 1.1 & 1 & 0.4 & 0.5 & -4 \end{pmatrix}$$

and the vector $\beta = (0.2, 0.3, 0.2, 0.2, 0.1)$.
The average service rate is computed as $b_1 = \beta(-S)^{-1}\mathbf{e} = 1.4137$.
Let us vary the threshold K in the interval $[1, 21]$.
The dependence of the average number N_{buffer} of customers in the buffer on the threshold K for different $MAPs$ is illustrated on Fig. 1. The dependence of the probability $P_{ent\text{-}loss}$ that an arbitrary customer will be lost upon arrival on the threshold K for different $MAPs$ is illustrated on Fig. 2.

The dependence of the probability $P_{imp\text{-}loss}$ that an arbitrary customer will be lost due impatience on blocked server on the threshold K for different $MAPs$ is presented on Fig. 3. Figure 4 shows The dependence of the probability P_{loss} that an arbitrary customer will be lost on the threshold K for different $MAPs$.

As it may be seen from Fig. 4, for all considered arrival flows, there exists the optimal value K^* of the threshold that provides the minimal value of the loss probability. This minimal value is 0.0019286 and the optimal threshold is $K^* = 19$ for MAP_0; the minimal value of the loss probability is 0.0296258 and the optimal threshold is $K^* = 16$ for $MAP_{0.2}$; and the minimal value of the loss probability is 0.2845692 under the optimal threshold $K^* = 2$ for $MAP_{0.4}$.

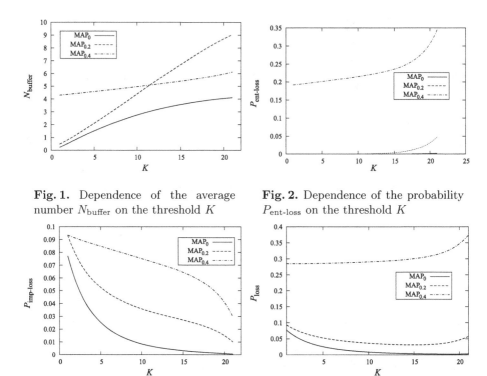

Fig. 1. Dependence of the average number N_{buffer} on the threshold K

Fig. 2. Dependence of the probability $P_{\text{ent-loss}}$ on the threshold K

Fig. 3. Dependence of the probability $P_{\text{imp-loss}}$ on threshold K

Fig. 4. Dependence of the probability P_{loss} on the threshold K

Intuitive explanation of existence of the optimal number of threshold K is as follows. When K is large, the additional server is activated only when the number of customers in the queue is large. So, loss of an arbitrary customer upon arrival is quite likely, the probability $P_{\text{ent-loss}}$ may be quite large. When K is small, two servers often provide service in parallel, so the probability $P_{\text{imp-loss}}$ to have a conflict and to lose the customer due to the long waiting for releasing the required phase of service is pretty high. Because the probability P_{loss} that an arbitrary customer will be lost is equal to the sum of $P_{\text{ent-loss}}$ and $P_{\text{imp-loss}}$, some trade-off in the choice of the value of K exists. This implies the existence of the optimal value K^* of the threshold.

6   Conclusion

We considered the queueing model with a finite buffer, Markovian arrival process and PH type distribution of the service time. It is assumed that when the queue length exceeds some fixed threshold an additional server is switched on. However, the use of the second server implies reduction of the rate of service in the first server and may cause the loss of the customer in service. Under the fixed value of the threshold, we analyzed the stationary distribution of the multi-dimensional

Markov chain that describe behavior of the system. In numerical experiments, we discovered existence of optimal value of the threshold which provides the minimal value of the loss probability of an arbitrary customer. Results may be extended to the systems with several servers and to more simple cases of PH distribution, e.g., generalized Erlangian distribution where numerical analysis may give more insight into the system behavior.

Acknowledgments. This research was supported by Basic Science Research Program through the National Research Foundation of Korea (NRF) funded by the Ministry of Education (Grant No. 2014K2A1B8048465) and by Belarusian Republican Foundation of Fundamental Research (Grant No. F15KOR-001).

References

1. Artalejo, J., Orlovsky, D.S., Dudin, A.N.: Multiserver retrial model with variable number of active servers. Comput. Indus. Eng. **28**, 273–288 (2005)
2. Li, H., Yang, T.: Queues with a variable number of servers. Eur. J. Operational Res. **124**, 615–628 (2000)
3. Chou, C.F., Golubchik, L., Lui, J.C.S.: Multiclass multiserver threshold-based systems: a study of noninstantaneous server activation. IEEE Trans. Parallel Distrib. Syst. **18**, 96–110 (2007)
4. Rykov, V., Efrosinin, D.: Optimal control of queueing systems with heterogeneous servers. Queueing Syst. **46**, 389–407 (2004)
5. Chakravarthy, S.R.: The batch Markovian arrival process: a review and future work. In: Krishnamoorthy, A., Raju, N., Ramaswami, V. (eds.) Advances in Probability Theory and Stochastic Processes, pp. 21–29. Notable Publications Inc., New Jersey (2001)
6. Neuts, M.F.: Matrix-Geometric Solutions in Stochastic Models. The Johns Hopkins University Press, Baltimore (1981)
7. Dudin, S., Dudina, O.: Help desk center operating model as two-phase queueing system. Prob. Inf. Transm. **49**(1), 66–82 (2013)

A Dual Tandem Queue with Multi-server Stations and Losses

Valentina Klimenok[1]($\boxtimes$) and Vladimir Vishnevsky[2]

[1] Department of Applied Mathematics and Computer Science,
Belarusian State University, 220030 Minsk, Belarus
{klimenok,dudin}@bsu.by
[2] Institute of Control Sciences of Russian Academy of Sciences and Closed
Corporation "Information and Networking Technologies", Moscow, Russia
vishn@inbox.ru

Abstract. In this paper, a tandem queue consisting of two multi-server stations without buffers is investigated. Customers of two different types arrive to the first station in accordance with $MMAP$ (Marked Markovian Arrival Process). The first type customers are satisfied with service at the first station only while the second type customers should be served successively at both stations. The system is studied in steady state. The stationary distribution of the system is calculated. A number of useful performance measures is derived. Decomposition and optimization problems are discussed.

Keywords: Tandem queueing system · Heterogeneous customers · Stationary distribution · Optimization · Decomposition

1 Introduction

Queueing networks are widely used in capacity planning and performance evaluation of computer and communication systems, service centers, and manufacturing systems among several others. Tandem queues can be used for modeling real-life two-node networks as well as for the validation of general decomposition algorithms in networks (see, e.g., [1,2]). Thus, tandem queueing systems have found much interest in the literature. Most of early papers on tandem queues are devoted to exponential queueing models. Over the last two decades efforts of many investigators in tandem queues were directed to weakening the distribution assumptions on the service times and arrival pattern. In particular, the arrival process should be able to capture correlation and difference in kind of customers because real traffic in modern communication networks exhibits [2] these features.

In the model under study, the correlation and the variability of inter-arrival times in input flow of the heterogeneous calls are taken into account via the consideration of a *Marked Markovian Arrival Process* ($MMAP$). A tandem queue with $MMAP$ and two multi-server stations without buffers is considered.

© Springer International Publishing Switzerland 2016
P. Gaj et al. (Eds.): CN 2016, CCIS 608, pp. 316–325, 2016.
DOI: 10.1007/978-3-319-39207-3_28

A queuing model under consideration can be used, for example, for design of optimal topology, technology, hardware and software of an office local area network with restricted access to an external network (e.g., Internet). It is assumed that a part of users (officers) have access to the external network while the rest of users can work only with local resources. To start access to the external network authorized officers should first go through the local recourses to reach the router to the external network. Sometimes, the use of external network is quite expensive. So, important problem is to match architecture of the local network to the bandwidth of access to the external network. The absence of control by the access can cause congestion in the network. The excessive bandwidth and too strict restriction of access of users of the local network to the external network leads to poor utilization of the bandwidth while it can be rather expensive.

A queuing model which is studied in this paper may be applied for optimization of the structure of an office computer network with partial access of officers to the external network. Also, this model is used to investigate the problem of decomposition of the tandem under consideration.

2   Model Description

A tandem queueing system consisting of two stations is considered. The first station is represented by the N-server queue without a buffer. The input flow to this station consists of two types of customers which arrive in a correlated flow described as a $MMAP$ (Marked Markovian Arrival Process). The first type customers aim to be served at the first station only while the second type customers should be served at both stations.

Arrivals of customers in the $MMAP$ is directed by the underlying process ν_t, $t \geq 0$, which is an irreducible continuous time Markov chain with state space $\{0, 1, \ldots, W\}$. The intensity of transitions, which are accompanied by an arrival of the type k customers, are combined into the matrices $D_k, k = 1, 2$, of size $(W + 1) \times (W + 1)$. The intensity of transitions, which are not accompanied by an arrival, are combined into the matrix D_0. The matrix $D(1) = D_0 + D_1 + D_2$ is a generator of the process $\nu_t, t \geq 0$. The fundamental arrival rate λ is defined by $\lambda = \boldsymbol{\theta}(D_1 + D_2)\mathbf{e}$ where $\boldsymbol{\theta}$ is a row vector of stationary probabilities of the Markov chain $\nu_t, t \geq 0$. The vector $\boldsymbol{\theta}$ is the unique solution to the system $\boldsymbol{\theta}D(1) = \mathbf{0}$, $\boldsymbol{\theta}\mathbf{e} = 1$. Here and in the sequel $\mathbf{e}$ is a column-vector of 1's and $\mathbf{0}$ is a row-vector of zeroes. The arrival rate λ_r of type r customers is defined by $\lambda_r = \boldsymbol{\theta} D_r \mathbf{e}, r = 1, 2$. The coefficient of variation, $c_{\text{var}}^{(r)}$, of inter-arrival times for type r customers is given by $(c_{\text{var}}^{(r)})^2 = 2\lambda_r \boldsymbol{\theta}(-D_0 - D_{\bar{r}})^{-1}\mathbf{e} - 1$, $r, r = 1, 2, \bar{r} \neq r$. The coefficient of correlation, $c_{\text{corr}}^{(r)}$, of two successive intervals between type r customers arrival is computed by $c_{\text{corr}}^{(r)} = [\boldsymbol{\theta}\lambda_r(D_0 + D_{\bar{r}})^{-1}D_r(D_0 + D_{\bar{r}})^{-1}\mathbf{e} - 1]/(c_{\text{var}}^{(r)})^2$, $\bar{r} \neq r$. The coefficient of variation, c_{var}, of intervals between successive arrivals is defined by $c_{\text{var}}^2 = 2\lambda\boldsymbol{\theta}(-D_0)^{-1}\mathbf{e} - 1$. The coefficient of correlation, c_{corr}, of successive intervals between arrivals is given by $c_{\text{corr}} = (\lambda\boldsymbol{\theta}(-D_0)^{-1}(D(1) - D_0)(-D_0)^{-1}\mathbf{e} - 1)/c_{\text{var}}^2$. For more information about a $MMAP$ see, e.g., [3].

An assumption that an arrival process of customers of two types is defined by a $MMAP$ instead of the more simple model with two independent stationary Poisson arrival processes allows to take into account variance of inter-arrival times, correlation of successive inter-arrival times of customers of the same type and cross-correlation between two types.

If an arriving customer of any type sees all servers of the first station busy, it leaves the system. Otherwise the customer occupies any idle server and its service starts immediately. Service time is assumed to be exponentially distributed with parameter μ_r depending on the type r of the customer, $r = 1, 2$. After the service completion, a customer of type 2 leaves the system forever while a customer of type 1 proceeds to the second station. The second station is represented by R-server queue without a buffer. If a customer proceeding from the first station finds an idle server, his/her service is started immediately. In the opposite case this customer is lost. Service time at the second station is exponentially distributed with parameter γ.

Our aim is to calculate the steady state distribution and the main performance measures of the system. The obtained results are used to solve numerically the problem of optimal choice of the number of servers at the stations and to discuss the decomposition problem.

3   Process of the System States

The process of the system states is described in terms of the irreducible multi-dimensional continuous-time Markov chain $\xi_t = \{i_t, r_t, n_t, \nu_t\}, t \geq 0$, where i_t is the number of busy servers at the first station; r_t is the number of busy servers at the second station, n_t is the number of servers of the first station occupied by customers of type 1, ν_t is the state of the $MMAP$ underlying process at time t, $i_t \in \{0, \ldots, N\}$; $r_t \in \{0, \ldots, R\}$; $n_t \in \{0, \ldots, i_t\}$; $\nu_t \in \{0, \ldots, W\}$.

Let us enumerate the states of the Markov chain $\xi_t, t \geq 0$, in the lexicographic order. The matrices $Q_{i,l}$, $i, l \in \{0, \ldots, N\}$, contain the transition rates from the states having the value i of the component i_t to the states having the value l of this component.

Theorem 1. *Infinitesimal generator A of the Markov chain ξ_t, $t \geq 0$, has the following block structure:*

$$\begin{pmatrix} Q_{0,0} & Q_{0,1} & 0 & \cdots & 0 & 0 \\ Q_{1,0} & Q_{1,1} & Q_{1,2} & \cdots & 0 & 0 \\ \vdots & \vdots & \vdots & \ddots & \vdots & \vdots \\ 0 & 0 & 0 & \cdots & Q_{N,N-1} & Q_{N,N} \end{pmatrix}$$

where

$$Q_{0,0} = \begin{pmatrix} D_0 & 0 & 0 & \cdots & 0 & 0 \\ \gamma I & D_0 - \gamma I & 0 & \cdots & 0 & 0 \\ \vdots & \vdots & \vdots & \ddots & \vdots & \vdots \\ 0 & 0 & 0 & \cdots & R\gamma I & D_0 - R\gamma I \end{pmatrix},$$

and the matrices $Q_{i,j}, i + j \neq 0$, of size

$$[(R+1)(i+1)(W+1)] \times [(R+1)(j+1)(W+1)]$$

are represented in the block form

$$Q_{i,j} = (Q_{i,j}(r,r'))_{r,r' \in \{0,...,R\}}$$

where the non-zero blocks $Q_{i,j}(r,r')$ are defined as follows:

$$Q_{i,i-1}(r,r) = A_i = \mu_2 \begin{pmatrix} i & 0 & \dots & 0 \\ 0 & i-1 & \dots & 0 \\ \vdots & \vdots & \ddots & \vdots \\ 0 & 0 & \dots & 1 \\ 0 & 0 & \dots & 0 \end{pmatrix} \otimes I_{W+1}, \ i \in \{1,\dots,N\}, \ r \in \{0,\dots,R-1\},$$

$$Q_{i,i-1}(R,R) = \begin{pmatrix} i\mu_2 & 0 & \dots & 0 & 0 \\ \mu_1 & (i-1)\mu_2 & \dots & 0 & 0 \\ \vdots & \vdots & \ddots & \vdots & \vdots \\ 0 & 0 & \dots & (i-1)\mu_1 & \mu_2 \\ 0 & 0 & \dots & 0 & i\mu_1 \end{pmatrix} \otimes I_{W+1}, \ i \in \{1,\dots,N\},$$

$$Q_{i,i-1}(r,r+1) = F_i = \mu_1 \begin{pmatrix} 0 & 0 & 0 & \dots & 0 \\ 1 & 0 & 0 & \dots & 0 \\ 0 & 2 & 0 & \dots & 0 \\ \vdots & \vdots & \vdots & \ddots & \vdots \\ 0 & 0 & 0 & \dots & i \end{pmatrix} \otimes I_{W+1}, \ i \in \{1,\dots,N\}, \ r \in \{0,\dots,R-1\},$$

$$Q_{i,i}(r,r-1) = r\gamma I_{(i+1)(W+1)}, \ i \in \{1,\dots,N\}, \ r \in \{1,\dots,R\},$$

$$Q_{i,i}(r,r) = B_i = I_{i+1} \otimes (D_0 + \delta_{i,N} D_1) - diag\{n\mu_1 + (i-n)\mu_2, \ n = 0,\dots,i\} \otimes I_{W+1}$$
$$- Q_{i,i}(r,r-1), \ i \in \{1,\dots,N\}, \ r \in \{0,\dots,R\},$$

$$Q_{i,i+1}(r,r) = C_i = \begin{pmatrix} D_2 & D_1 & 0 & \dots & 0 & 0 \\ 0 & D_2 & D_1 & \dots & 0 & 0 \\ \vdots & \vdots & \vdots & \ddots & \vdots & \vdots \\ 0 & 0 & 0 & \dots & D_2 & D_1 \end{pmatrix}, \ i \in \{0,\dots,N-1\}, \ r \in \{1,\dots,R\}.$$

Here $\otimes$ is a symbol of Kronecker's product of matrices; $\delta_{i,N}$ is equal to 1 if $i = N$ and is equal to 0 otherwise, $diag\{a_i, i \in \{1,\dots,N\}\}$ stands for the diagonal matrix defined by the diagonal entries listed in the brackets.

4   Stationary Distribution and Performance Measures

Because the four-dimensional Markov chain $\xi_t = \{i_t, r_t, n_t, \nu_t\}, \ t \geq 0$, is an irreducible and regular and has the finite state space, the following limits (stationary probabilities) exist for any set of the system parameters:

$$p(i, r, n, \nu) = \lim_{t \to \infty} P\{i_t = i, r_t = r, n_t = n, \nu_t = t\},$$

$$i_t \in \{0, \ldots, N\}; \ r_t \in \{0, \ldots, R\}; \ n_t \in \{0, \ldots, i_t\}; \ \nu_t \in \{0, \ldots, W\}.$$

Let us enumerate the probabilities corresponding the value i of the first component in lexicographic order and form from these probabilities the row vectors $\mathbf{p}_i$, $i \in \{0, \ldots, N\}$. These vectors satisfy Chapman-Kolmogorov's equations (equilibrium equations)

$$\sum_{i=0}^{N} \mathbf{p}_i Q_{i,j} = \mathbf{0}, \quad j \in \{0, \ldots, N\}, \quad \sum_{i=0}^{N} \mathbf{p}_i \mathbf{e} = 1. \tag{1}$$

The rank of System (1) is equal to $(N + 1)(N + 2)(R + 1)(W + 1)/2$ and can be very large. E.g., in case $N = 10$, $R = 2$, $W = 1$ the rank is equal to 396, in case $N = 20$, $R = 2$, $W = 1$ it is equal to 2310, in case $N = 30$, $R = 2$, $W = 1$ it is equal to 6944 etc. Thus, the direct solution of system (1) can be time and resource consuming.

In the case when at least one of values N, R, W is large, the numerically stable algorithm based on probabilistic meaning of the matrix Q is used. This algorithm is presented in [4].

Having the stationary distribution $\mathbf{p}_i$, $i \geq 0$, calculated, a number of stationary performance measures of the system can be derived. Below formulas for computing some of them are presented.

- Mean number of busy servers at the first station $N_{\text{busy}}^{(1)} = \sum\limits_{i=1}^{N} i \mathbf{p}_i \mathbf{e}.$

- Mean number of type 1 customers at the first station

$$N_{\text{busy},1}^{(1)} = \sum_{i=1}^{N} \mathbf{p}_i (\mathbf{e}_{R+1} \otimes I_{i+1} \otimes \mathbf{e}_{W+1}) \text{diag}\{0, 1, \ldots, i\} \mathbf{e}.$$

- Mean number of type 2 customers at the first station

$$N_{\text{busy},2}^{(1)} = N_{\text{busy}}^{(1)} - N_{\text{busy},1}^{(1)}.$$

- Probability that a customer of type k will be lost at the first station

$$P_{\text{loss},k}^{(1)} = \lambda_k^{-1} \mathbf{p}_N (\mathbf{e}_{(R+1)(N+1)} \otimes I_{W+1}) D_k \mathbf{e}, \quad k = 1, 2.$$

- Probability that a customer of type k will be successfully served at the first station $P_{\text{succ},k}^{(1)} = 1 - P_{\text{loss},k}^{(1)}$, $k = 1, 2$.
- The rate of input flow at the second station (coincides with the rate of output flow of type 1 customers from the first station) $\lambda_{\text{inp}}^{(2)} = \lambda_1 (1 - P_{\text{loss},1}^{(1)})$.
- The rate of output flow from the second station

$$\lambda_{\text{out}}^{(2)} = \gamma \sum_{i=0}^{N} \mathbf{p}_i (I_{R+1} \otimes \mathbf{e}_{(i+1)(W+1)}) \text{diag}\{0, 1, \ldots, R\} \mathbf{e}.$$

- The vector $\mathbf{q} = (q_0, q_1, \ldots, q_R)$ of the stationary distribution of the number of busy servers at the second station

$$\mathbf{q} = \sum_{i=0}^{N} \mathbf{p}_i (I_{R+1} \otimes \mathbf{e}_{(i+1)(W+1)}).$$

- Mean number of busy servers at the second station $N_{\text{busy}}^{(2)} = \frac{\lambda_{\text{out}}^{(2)}}{\gamma}$.
- Probability that a customer of type 1 will be lost at the second station $P_{\text{loss}}^{(2)} = 1 - \frac{\lambda_{\text{out}}^{(2)}}{\lambda_{\text{inp}}^{(2)}}$.
- Probability that a customer of type 1 will be successfully served at both stations $P_{\text{succ}} = P_{\text{succ},1}^{(1)} (1 - P_{\text{loss}}^{(2)})$.

5   Numerical Experiments

5.1   Optimization Problem

In this experiment, the problem of optimal choice of the number N of servers at the first station and the number R of servers at the second station is solved. To this end, the following cost criterion (an average gain per unit time under the steady-state operation of the system) of the system operation is considered:

$$I = c_1 \lambda_1 P_{\text{succ},1} + c_2 \lambda_2 P_{\text{succ},2} - aR$$

where a is a cost of utilization of a server at the second station per unit time (maintenance cost), c_r is a gain for successful service of each customer of type r, $r = 1, 2$.

Our aim is to find numerically the optimal number R of servers at the second station that provides the maximal value to the cost criterion for different number N of servers at the first station.

The input flow to the tandem is described by the $MMAP$ defined by the following matrices:

$$D_0 = \begin{pmatrix} -13.49076 & 0.0000109082 \\ 0.0000109082 & -0.43891 \end{pmatrix}, \quad D_1 = \begin{pmatrix} 2.2335616667 & 0.0148965153 \\ 0.040809 & 0.032340848633 \end{pmatrix},$$

$$D_2 = \begin{pmatrix} 11.1678083333 & 0.0744825765 \\ 0.204045 & 0.161704243167 \end{pmatrix}.$$

Intensities of arrival of type-r customers are equal to $\lambda_1 = 1.66671$ and $\lambda_2 = 8.33357$, respectively. Coefficient of correlation of two successive inter-arrival times is equal to 0.2. Coefficients of correlation of two successive intervals between arrivals of type-1 and type-2 customers are equal to 0.0425364 and 0.178071, respectively.

The service rates of customers at the first station are defined by $\mu_1 = 0.5$, $\mu_2 = 1$. The service rate of customers at the second station is $\gamma = 0.1$.

The values of the cost coefficients are assumed to be as follows: $c_1 = 15$, $c_2 = 1$, $a = 1$.

The value of criterion I as a function of the numbers R and N of servers at the second station and the first station is presented in Table 1. The optimal values of I as a function of R for different values of N are printed in bold font.

Table 1. Dependence of the cost criterion I on R and N

	$R = 1$	$R = 3$	$R = 5$	$R = 8$	$R = 10$	$R = 13$	$R = 17$	$R = 20$
$N = 5$	2.877	**3.133**	3.001	1.804	0.332	−2.447	−6.423	−9.423
$N = 10$	5.319	5.871	6.274	**6.495**	6.294	5.308	2.634	−0.080
$N = 15$	7.212	7.851	8.393	8.979	**9.178**	9.091	8.028	6.3898
$N = 20$	8.306	8.974	9.558	10.249	10.559	**10.736**	10.268	9.243
$N = 35$	8.699	9.373	9.967	10.681	11.016	**11.251**	10.922	10.066

For better understanding the behavior of cost criterion we present the values of the criterion as function of R under different but fixed values of N in Fig. 1 and the values of the criterion as function of N under different but fixed values of R in Fig. 2.

Table 2 contains the relative profit provided by the optimal value R^* of a parameter R comparing to other values of a parameter. This relative profit is calculated by formula $\frac{I(R^*)-I(R)}{I(R^*)} 100\,\%$.

Table 2. Dependence of the the relative profit on R and N

	$R = 1$	$R = 3$	$R = 5$	$R = 8$	$R = 10$	$R = 13$	$R = 17$	$R = 20$
$N = 5$	8.167	**0.00**	4.193	42.409	89.386	178.10	305.04	400.79
$N = 10$	18.103	9.604	3.399	**0.00**	3.089	18.268	59.447	101.23
$N = 15$	21.412	14.453	8.551	2.165	**0.00**	0.943	12.524	30.385
$N = 20$	22.631	16.415	10.976	4.539	1.651	**0.00**	4.368	13.903
$N = 35$	22.681	16.686	11.407	5.059	2.086	**0.00**	2.919	10.530

It is seen from Fig. 2 and Table 1 that, for each R, the value $I(N)$ of the gain increases in the beginning and then becomes constant. It is explained by the fact that, for large value of N, all entering customers get service at the first station so that the gain from the successful service at this station and the rate of input flow to the second station become constant. Since the number R of servers at the second station is constant, the gain from the successful service at the second station becomes constant too. Thus, the value of total gain does not vary for large N.

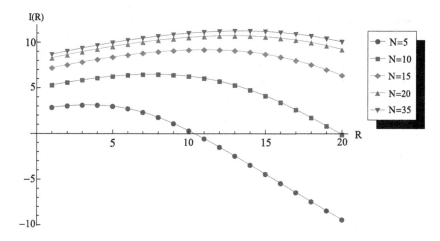

Fig. 1. The cost criterion as a function of the number of servers at the second station under different number of servers at the first station

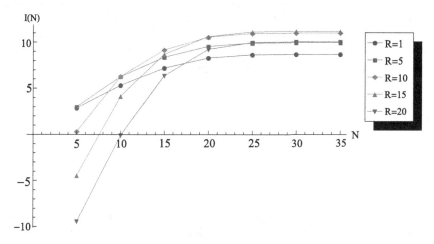

Fig. 2. The cost criterion as a function of the number of servers at the first station under different number of servers at the second station

5.2   Decomposition Issues

In this section, the problem of decomposition of the original tandem system into two multi-server systems is investigated numerically. To solve properly this problem, the output flow from the first station (input flow to the second station) should be described exactly. We find out that this flow is a MAP of order $(W + 1)(N + 1)(N + 2)/2$. To avoid the dimensionality problem, we make an attempt to approximate the output flow from the first station as a stationary Poisson one. To evaluate the error of the approximation, performance measures of the second station calculated for the stationary Poisson input and for exact MAP input flow are compared.

Theorem 2. *Output flow from the first station is a MAP which is described by the following square matrices* $\tilde{D}_0, \tilde{D}_1$ *of size* $(W+1)(N+1)(N+2)/2$:

$$
\tilde{D}_0=
\begin{pmatrix}
B_0 & C_0 & 0 & 0 & \cdots & 0 & 0 & 0 \\
A_1 & B_1 & C_1 & 0 & \cdots & 0 & 0 & 0 \\
0 & A_2 & B_2 & C_2 & \cdots & 0 & 0 & 0 \\
\vdots & \vdots & \vdots & \vdots & \ddots & \vdots & \vdots & \vdots \\
0 & 0 & 0 & 0 & \cdots & A_{N-1} & B_{N-1} & C_{N-1} \\
0 & 0 & 0 & 0 & \cdots & 0 & A_N & B_N
\end{pmatrix},
\quad
\tilde{D}_1=
\begin{pmatrix}
0 & 0 & 0 & \cdots & 0 & 0 & 0 \\
F_1 & 0 & 0 & \cdots & 0 & 0 & 0 \\
0 & F_2 & 0 & \cdots & 0 & 0 & 0 \\
\vdots & \vdots & \vdots & \ddots & \vdots & \vdots & \vdots \\
0 & 0 & 0 & \cdots & F_{N-1} & 0 & 0 \\
0 & 0 & 0 & \cdots & 0 & F_N & 0
\end{pmatrix},
$$

where the matrices A_i, B_i, C_i, F_i *are defined in Theorem 1.*

Using Theorem 2, the second station of the tandem can be modeled as a separate $MAP/M/R/R$ queue. Of course, the first station is modeled as $MAP/M/N/N$ queue. Thus, we state the fact that *the original tandem queue can be decomposed into two separate multi-server queues with MAP inputs.* Naturally, this fact simplifies the investigation of the original system. At the same time, in case of large values of N the dimensionality problem which takes place in the original tandem queue is still open because the state space of underlying process of the MAP of arrivals to the second station can be very large. In this situation, we can try to approximate the input MAP to the second station by the stationary Poisson one with the rate $\lambda_{\text{inp}}^{(2)}$ (see Sect. 4, Performance measures). Under such an approximation the stationary distribution $p_r, r = \overline{0, R}$, of the number of busy servers at the second station is calculated by Erlang formulas

$$
p_r = \frac{\frac{a^r}{r!}}{\sum\limits_{l=0}^{R} \frac{a^l}{l!}}, \quad r = \overline{0, R}, \quad a = \frac{\lambda_{\text{inp}}^{(2)}}{\gamma}.
$$

To evaluate the error of the approximation, we compare this distribution with the corresponding distribution calculated for exact MAP input to the second station defined in Theorem 2.

Consider the numerical experiment with the following input data: the number of servers at the stations $N = 5$, $R = 4$. The service rates an the stations $\mu = 0.5$, $\gamma = 0.1$. The $MMAP$ at the first station is defined by the matrices

$$
D_0=\begin{pmatrix} -1.349076 & 0.00000109082 \\ 0.00000109082 & -0.043891 \end{pmatrix}, \quad D_1=\begin{pmatrix} 1.340137 & 0.00893790918 \\ 0.0244854 & 0.01940450918 \end{pmatrix}, \quad D_2=0.
$$

Under this data $\lambda = 1.00003$, $c_{\text{corr}} = 0.2$, $\lambda_{\text{inp}}^{(2)} = 0.918877$.

In Table 3 we present the approximate stationary distribution $p_r, r = \overline{0, R}$, of the number of busy servers at the second station calculated for the stationary Poisson input and the corresponding exact stationary distribution $q_r, r = \overline{0, R}$, calculated for the MAP input.

It is seen from Table 3 that an approximation of the MAP input flow to the second station by the stationary Poisson flow can be very bad. For example,

Table 3. Stationary distribution of the number of busy servers at the second station

	q_0	q_1	q_2	q_3	q_4	$N^{(2)}_{\text{busy}}$
Poisson	0.00208876	0.0191931	0.0881804	0.27009	0.620448	3.4876259
MAP	0.133934	0.0691116	0.0735184	0.189183	0.534253	2.9207094

in the above experiment the relative error in the calculation of the steady state probabilities may reach 98 % and the relative error in the calculation of the mean number of busy servers at the second station is about 20 %.

6 Conclusion

In the paper, the tandem queueing system with $MMAP$ and two stations defined by the multi-server queues without buffers is studied. A part of arriving customers are served only at the first station while the others have to be served at both stations. Optimization problem is considered and solved numerically. Cost criterion includes the gain of the system obtained from the service of customers and the maintenance cost of the servers at the second station. Decomposition problem is discussed. It was shown that the original tandem can be reduced to two multi-server queues with $MMAP$ and MAP input. The approximation problem is considered. It was demonstrated numerically that a simple approximation of the input flow to the second station by the stationary Poisson one can be very poor. The obtained results can be applied for optimization of the structure of an office computer network with partial access of officers to the external network. Results can be extended to the case of more general so called PH – phase type distribution of the service times, see, e.g., [5] and the case when not only the number of servers at both stations should be optimized, but additionally the share of customers, which are permitted to go to the second station after the service at the first station, is a subject of optimization.

References

1. Ferng, H.W., Chang, J.F.: Connection-wise end-to-end performance analysis of queueing networks with $MMPP$ inputs. Perform. Eval. **43**, 39–62 (2001)
2. Heindl, A.: Decomposition of general tandem networks with $MMPP$ input. Perform. Eval. **44**, 5–23 (2001)
3. He, Q.M.: Queues with marked customers. Adv. Appl. Probab. **28**, 567–587 (1996)
4. Klimenok, V., Kim, C.S., Orlovsky, D., Dudin, A.: Lack of invariant property of Erlang loss model in case of the MAP input. Queueing Syst. **49**, 187–213 (2005)
5. Neuts, M.: Matrix-Geometric Solutions in Stochastic Models - An Algorithmic Approach. Johns Hopkins University Press, Baltimore (1981)

Innovative Applications

Adjustable Sampling Rate – An Efficient Way to Reduce the Impact of Network-Induced Uncertainty in Networked Control Systems?

Michał Morawski$^{(\boxtimes)}$ and Przemysław Ignaciuk

Institute of Information Technology, Lodz University of Technology,
215 Wólczańska St., 90-924 Łódź, Poland
{michal.morawski,przemyslaw.ignaciuk}@p.lodz.pl

Abstract. The practical realization of Network Control Systems (NCSs) enforces handling network-induced effects: information transfer and processing delay, delay variability, packet loss and reordering, etc. The impact of these phenomena can be limited by using dedicated, expensive real-time networks and control systems, or by implementation of algorithmic methods. This paper presents an adjustable sampling rate (ASR) algorithm to alleviate negative influence of network-induced uncertainty in NCSs, and compares it with previously developed predictor based (PB) one in a common experimental framework. Since the experiments are conducted using standard modules and communication technologies, the reported results are applicable to a relatively broad class of remote control applications and NCSs.

Keywords: Network control systems · Discrete-time systems · Time-delay systems · Experimental evaluation

1 Introduction

While local plant control may nowadays be considered well investigated, the problems of remote regulation and steering, or the control of distributed and networked control systems (NCSs) continue to attract significant attention of researchers and practitioners [1,2]. Unlike local plant control, influencing the state of objects placed in separate locations requires signal exchange over a communication network, which is inherently uncertain. The information passed among the communicating parties may be lost, discarded as a result of errors, temporarily misplaced, and arrives with delay. Therefore, although gaining in popularity owing to the ease of installation, fault tolerance, economic benefits, or pure necessity (e.g. in hazardous environments), remote control settings need extra measures to ensure consistency of the control process in the presence of network-induced perturbations [3–5].

A remote control system consists of at least two subsystems: a local computer that interacts with the plant through actuators, gathers sensor measurements, and provides a communication interface for the networked information exchange

© Springer International Publishing Switzerland 2016
P. Gaj et al. (Eds.): CN 2016, CCIS 608, pp. 329–343, 2016.
DOI: 10.1007/978-3-319-39207-3_29

and a remote computer implementing the control logic and interacting with the plant via a data transmission network. Obviously, in order to elevate the quality of networked information exchange, e.g. in the effort to fulfill real-time requirements, robust communication media and high performance electronics can be used. However, the cost of such approach in terms of money and resources is prohibitive in many prospective applications. One would like to achieve similar control system performance using inexpensive devices and general-purpose networks rather than recur to large monolithic constructs with dedicated connectivity solutions. The primary drawback of low-end solutions, in turn, is the increased level of system uncertainty related to longer information transfer delays, larger delay variability (jitter), packet reordering, and higher loss rate.

While in the circumstances of aggravated uncertainty the system stability can be preserved by implementing robust, sophisticated control laws [6–9], or throttling the responsiveness of linear controllers [10], the control quality cannot be improved unless the severity of perturbations themselves is mitigated. The current literature [3] advocates solutions based on either assessment of delay (e.g. [11]) or by incorporating intelligence in the local control system to select a desired control value at a given time instant. The paper focuses on the latter case and presents an adjustment rate adaptation algorithm, that is a member of variable sampling group methods [12]. The majority of this group of methods apply the control in an event driven mode (see [13] and references therein), i.e. when a control packet arrives at the local system. Such an approach usually violates economic requirements (the use of low-end local computers) due to the necessity of performing sophisticated computations. The developed algorithm can be implemented on low-end machines, ordinary PC, or even a virtual one. In contrast to the earlier studies, e.g. [13], primarily limited to computer simulations, the algorithm is tested within a consistent experimental platform. Since the tests involve a structurally unstable plant, low-end devices, and general-purpose networks, the conclusions are applicable to a relatively broad class of customary systems (not just high-end industrial solutions).

2   System Model

Consider the system depicted in Fig. 1. The plant dynamics are modeled as

$$\dot{\mathbf{x}}_{\mathbf{p}}(t) = \mathbf{F_c}\mathbf{x_p}(t) + \mathbf{G_c}\mathbf{u_p}(t) + \mathbf{d_p}(t), \tag{1}$$

where t is a continuous variable denoting the evolution of time, $\mathbf{x_p}(t) \in \mathbb{R}^n$ is the plant state vector, $\mathbf{u_p}(t) \in \mathbb{R}^r$ is the input applied at the plant, $\mathbf{F_c} \in \mathbb{R}^{n \times n}$ and $\mathbf{G_c} \in \mathbb{R}^{n \times r}$ where r, n are positive integers, and $\mathbf{d_p}(t)$ represents the cumulative impact of model uncertainty and external disturbances.

The plant is to be controlled remotely according to a suitably chosen regulation strategy. The signals between the controller and the plant are exchanged through a communication network. Such configuration, while flexible and cost efficient, introduces additional phenomena that need to be accounted for in

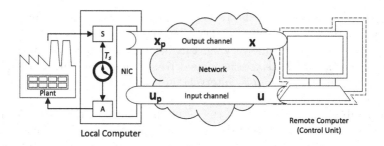

Fig. 1. Remote plant control. S – Sensor, A – Actuator, NIC – Network Interface Controller.

proper control system design. In particular, one needs to answer the challenges related to network-induced effects.

First of all, handling events at the network interfaces introduces an extra latency that limits the achievable sampling rate. At the input side of a communicating party, this latency is associated with capturing the arriving packets, extracting the relevant information, and applying the retrieved information for the control purposes. Similarly, at the output side, finite time is needed to form and transmit packets through the networking device. Additional time is necessary to process the data at the remote computer. Therefore, unless the dominant time constant of the controlled process significantly exceeds the sampling period, in a networked-based implementation of control tasks the effects of finite sampling rate should be addressed. In the considered setting, it is assumed that the plant control system is time driven with period T_s and the input is applied through zero-order hold (ZOH), i.e. $\mathbf{u_p}(t) = \mathbf{u_p}(kT_s)$ for $t \in [kT_s, (k+1)T_s)$, with $k = 0, 1, 2, \ldots$ For brevity, in a latter part of the text kT_s will be written shortly as k. In discrete time domain, the nominal plant dynamics evolve as

$$\mathbf{x_p}(k+1) = \mathbf{F}\mathbf{x_p}(k) + \mathbf{G}\mathbf{u_p}(k), \tag{2}$$

where $\mathbf{F} = e^{\mathbf{F_c}T_s}$ and $\mathbf{G} = \mathbf{G_c} \int_0^{T_s} e^{\mathbf{F_c}(T_s - \lambda)} \mathrm{d}\lambda$.

The other group of effects in remote control systems is related to passing the data between the communicating parties through the network channels. This process is never instantaneous. Moreover, the common usage of shared communication channels in NCSs (especially the wireless ones) results in delay fluctuations arising from various media access procedures, link layer retransmissions, queuing at the network devices, etc. Let $\mathbf{x}$ denote the information about plant state delivered to the remote controller and $\mathbf{u}$ – the command established by that controller. It is assumed that the controller operates in an event-based mode, i.e. the input computation is triggered by the reception of a packet carrying the information about the plant state, and the clock synchronization with the plant is not required. The controller command is received at the plant with forward (input) delay $n_f(t)$ and the state measurement is delivered with backward (output) delay $n_b(t)$. The remote computer processes the data by $n_r(t)$. Unless only local signal exchange is considered with the transmission delay incorporated into the discretization period, the overall loop delay $\tau(t) = n_f(t) + n_b(t) + n_r(t) \geq T_s > 0$. Although uncertain and time-varying, this delay may be assumed to comprise an integer multiple of T_s, since the time-driven actuator operation permits the input changes at the plant in T_s intervals only. Thus, $\tau(t) = h(k)T_s$, where the uncertain $h(k) \leq H$, $H < 0$.

When subject to linear control the actual placement of the controller in the loop does not affect the system dynamic description. Therefore, following this common convention in modeling the phenomena encountered in remote control systems [12], the overall system dynamics may be represented as

$$\mathbf{x_p}(k+1) = \mathbf{F}\mathbf{x_p}(k) + \mathbf{G}\mathbf{u}(k - h(k)) + \mathbf{d_p}(k). \tag{3}$$

On the other hand, with the nominal delay $\tau = hT_s$, the system dynamics become

$$\mathbf{x}(k+1) = \mathbf{F}\mathbf{x}(k) + \mathbf{G}\mathbf{u}(k - h) + \mathbf{d}(k), \tag{4}$$

where $\mathbf{d}(k) \in \mathbb{R}^n$ captures all the effects related to the plant and communication process uncertainty.

Remark 1. In addition to experiencing variable, uncertain delay, the information transmitted in the network may be lost or corrupted, thus increasing the overall system uncertainty. Since the effect of lost and late packets is essentially the same from the perspective of preserving the continuity of the control process, the case of a lost packet is treated here as transmission with infinite delay thus confining the discussion to the system with uncertain, time-varying delay and lossless transfer.

Remark 2. In general the overall system delay could be decreased by shortening T_s. However, such an approach is counterproductive in NCSs because of the physical limitations of communicating channels and risk of collapse when non-expensive shared channels are used. Additionally, as a side effect, decreasing T_s increases the error in reconstructing unmeasured variables [5].

3   Control Strategy Selection

The presence of delay enforces the use of appropriate mechanisms to counteract its potentially destabilizing effect on the closed-loop system [14]. One of the popular approaches concentrates on finding a suitable control gain $\mathbf{K}$ so that the closed-loop system remains stable irrespective of the delay and perturbations. Thus considered control,

$$\mathbf{u}(k) = -\mathbf{K}\mathbf{x}(k), \tag{5}$$

while straightforward to implement and convenient for the analytical studies, introduces operational conservatism in the effort to preserve stability under worst-case delay (and disturbance) conditions.

In order to avoid throttling the dynamics, an attractive alternative is the application of dead-time compensators (DTC) [15]. By introducing the variable

$$\mathbf{v}(k) = \mathbf{F}^h \mathbf{x_p}(k) + \sum_{i=1}^{h} \mathbf{F}^{i-1} \mathbf{G} \mathbf{u}(k - i), \quad \mathbf{v} \in \mathbb{R}^n, \tag{6}$$

closed-loop dynamics (4) become

$$\mathbf{v}(k + 1) = \mathbf{F} \mathbf{v}(k) + \mathbf{G} \mathbf{u}(k) + \mathbf{F}^h \mathbf{d}(k). \tag{7}$$

With the use of DTC (6), the gain selection for system (7) may proceed using the classical state feedback design for undelayed processes with the control of the form

$$\mathbf{u}(k) = -\mathbf{K} \mathbf{v}(k). \tag{8}$$

As long as the delay used in the DTC matches the loop delay gain $\mathbf{K}$ may be chosen independent of h, which, in principle, implies the possibility of obtaining arbitrarily fast dynamics. The performance is then tuned with respect to e.g. disturbance attenuation requirements. The drawback of this method is the need for recreating an accurate model of the physical process inside the DTC. If the model is imprecise, in particular, if the delay used for the internal state evaluation (6) differs from the one that actually exists in the control loop, the system is susceptible to stability issues. In the paper, the system is assumed to be governed by a linear controller, in the form given by (8) (rather than (5)), with the gain pre-assigned according to certain design requirements. The gain selection may be realized in any way. The purpose of the presented study is to evaluate the methods of reducing the influence of network-induced uncertainty, in particular, random delay variations, on the control system performance.

4   Reducing Influence of Network-Induced Uncertainty

While potentially problematic, owing to the extra uncertainty source, the control system implementation in a networked environment offers also additional means of reducing that uncertainty influence.

Among the methods commonly used to handle network-induced perturbations in NCSs (see e.g. [3] for an excellent review of current trends), we have selected two most willingly used, similar in implementation but different in a way subsequent inputs are evaluated.

Both approaches are based on the observation that instead of issuing a single control command (calculated for the nominal delay, for instance) the networked implementation of a control strategy permits simultaneous transmission of multiple commands. In such a case, the controller determines multiple input values and sends them in the control packet for the actuator. Then, based on timestamp comparison, the most appropriate command corresponding to the current delay is chosen and supplied to the actuator. As a result, irrespective of the pattern of delay variation, the loop delay is compensated.

Remark 3. The associated increase of packet payload only slightly increases the overall delay because the enlarged transmission time is comparable to media access time. The computation time $(n_r(t))$ increases approximately linearly with H, but is still shorter than $n_b(t)$ and $n_f(t)$.

The investigated methods (in different variants) were the subject of many analytical and numerical studies in recent years [16,17]. However, their assessment when acting in the actual networked environment (where the plant and network-induced perturbations influence the control process concurrently) has not been given sufficient attention. In this study, the practical rather than formal perspective is taken, and comparative, experimental evaluation of the efficiency in mitigating the impact of delay uncertainty on remote control system performance is provided. The method variants are selected so that the protection against instability is achieved without much deterioration of the response speed or excessive implementation and operational burden owing to large level of sophistication. They are discussed in details in the subsequent sections. In each case, the network interface of the local computer creates a packet every T_s containing the current state measurement, the history of previously applied inputs $\mathbf{u_p}(k-1)$, $\mathbf{u_p}(k-2)$, ..., $\mathbf{u_p}(k-H)$, and a timestamp that uniquely describes time instant k. The controller copies the timestamp, computes a set of control values and sends them to the actuator. Since the controller operates in the event-based mode, the time synchronization with the plant is not required.

The information exchange process, illustrated in Fig. 2, proceeds as follows. Every T_s, the sensor interface issues a timestamped packet containing the plant state measurement and the history of inputs applied to the actuator in the last H periods.

Upon reception (if the timestamp comparison indicates the packet is not outdated, e.g. due to reordering) the entire set of H commands is stored in the local buffer. At sampling instant $k + j$ the value corresponding to loop delay j, $\mathbf{u}_j(k)$, is taken from the buffer and applied to the actuator. If no new packet arrives before the next sampling instant the loop delay is perceived to have increased by 1, so $\mathbf{u}_{j+1}(k)$ is chosen as the actuator input. Otherwise, if a new packet does arrive and the buffer contents are updated, $\mathbf{u}_j(k + 1)$ is selected as the plant input. The history of H actually applied inputs is recorded in the measurement packet sent towards the controller and the process repeats itself.

4.1   Predictive Based Method (PB)

In this method the controller extracts the received data and calculates H command proposals

$$\mathbf{u}_j = -\mathbf{K}\mathbf{v}_j, \mathbf{v}_j = \mathbf{F}^j \mathbf{x_p} + \sum_{i=1}^{j} \mathbf{F}^{i-1} \mathbf{G} \mathbf{u_p}(j-i), \quad j = 1, \ldots, H, \qquad (9)$$

that are all sent in a single packet towards the plant. The operational details of this method were discussed in [18].

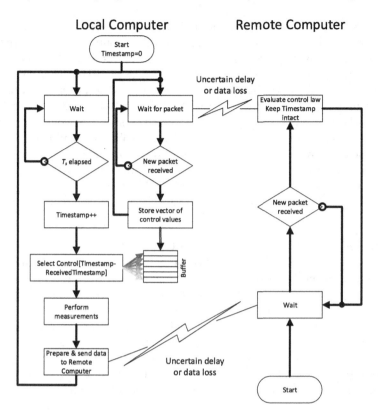

Fig. 2. The local computer operating in a time-driven mode performs the measurements, and sends them to the remote controller at discrete instants kT_s

4.2   Sampling Rate Adjustment (ASR)

The second group of methods investigated in this paper is based on the dynamic adjustment of the sampling rate. A drawback of the PB method, which incorporates the set of values computed for the loops with different delay, is decreased accuracy if the plant model used to establish the compensator signal is imprecise. The error may propagate in the subsequent powers of state matrix $\mathbf{F}$ (formulae (7) and (9)), leading to performance deterioration for loops with long delay. In order to address this issue, instead of using one model with matrix $\mathbf{F}$ determined for the continuous-time process discretized with period T_s, one can construct a set of discrete-time models, each corresponding to a different sampling period. Then, if the delay equals τ and the input calculated for the model sampled with period τ is supplied to the actuator for the duration of τ, a closed-loop system with a unity delay is obtained (6). The influence of uncertain delay variation is thus reduced. In order to cover the full spectrum of possible delay values an infinite set of models would be required. In the practical implementation, the set of delays is restricted to integer multiples of fundamental sampling period T_s.

The proposed algorithm works in the following way: every T_s the sensor interface sends a packet towards the remote controller containing the current and previous state measurements and the history of inputs applied to the plant in the last H fundamental sampling periods T_s. Unlike the PB method, in addition to the input also the state history needs to be provided for the reconstruction of unmeasured variables in the interval $(k - H, k]$. This task should not be delegated to the controller since packet drops in the sensor-to-controller channel would lead to erroneous view of the plant state evolution should the controller retrieve the data by itself. Upon the packet reception, the controller establishes H commands, each corresponding to a system sampled with period iT_s. Assuming a unity loop delay, the commands are calculated as

$$\mathbf{u}_i = -\mathbf{K}_i \left[\mathbf{F}_i \mathbf{x} + \mathbf{G}_i \mathbf{u_p}(k - i) \right], \tag{10}$$

where the matrices $\mathbf{F}_i = e^{\mathbf{F_c} iT_s}$ and $\mathbf{G}_i = \mathbf{G_c} \int_0^{iT_s} e^{\mathbf{F_c}(iT_s - \lambda)} \mathrm{d}\lambda$, $i = 1, 2, \ldots, H$, represent the discrete-time models of continuous-time system (1) sampled with period iT_s. $\mathbf{K}_i$ is the feedback matrix for model $(\mathbf{F}_i, \mathbf{G}_i)$, determined so that the closed-loop system under control (10) is stable. All H input proposals are placed in the packet sent to the plant. The actuator part of the system is exactly the same as in the PB method. Upon the packet reception at the plant, the local buffer is updated, and a new control input is applied at the next T_s sampling instant according to the current loop delay. The control packet reception at the plant right before instant $k + i$, with the command calculated for the sensor measurement sent at instant k, indicates input $\mathbf{u}_i$ should be applied to the actuator at $k + i$. Instead of keeping this input at the ZOH for the duration of i, however, the proposed variant operates with more frequent input updates. Namely, the check whether a new input should be latched in the ZOH is performed with period T_s rather than iT_s. In particular, if between instants $k + i$ and $k + i + 1$ no new control packet is received, it means the loop delay has increased by 1. Therefore, instead of keeping $\mathbf{u}_i$ at the ZOH at instant $k + i + 1$, it is more appropriate to apply command $\mathbf{u}_{i+1}$ from the buffer, corresponding to the discrete subsystem sampled with period $i + 1$. In this way, a subsystem with the control based on more recent data is triggered. In addition to obtaining higher responsiveness, the overall control quality may also improve since the obtained finer sampling period leads to more accurate representation of the physical process (continuous-time and discrete-time system representation coincides at the sampling instances). On the other hand, if between instants $k + i$ and $k + i + 1$ a new packet arrives and the buffer contents are updated, the timestamp comparison provides the value of the current loop delay that allows one to choose the corresponding command directly from the updated buffer.

4.3   Comments

The efficiency of the considered methods in reducing the adverse effects of channel uncertainty and the their impact on the process control quality are evaluated in the experimental part reported in Sect. 5. In this section, the methods are compared in terms of the computational overhead and resource requirements imposed on the plant, remote controller, and network channels.

Computational Overhead. Since the command for the plant is determined by the remote controller, the computational overhead experienced by the local computer (at the plant) is nearly the same, except preparation of the packet to be sent to the controller. Basically, it reduces to choosing the actuator input every T_s. Similarly, the on-line computations performed by the controller are the same except for the necessity of retrieving the unmeasured state variables in the ASR method. In both cases, the controller performs H matrix computations, as the appropriate data should be prepared off-line.

Resource Usage. The plant control system maintains only a few local variables of integer type. The corresponding floating point variables are handled at the remote controller, so the amount of memory consumed by the methods is insubstantial. Both methods require a buffer at the plant to hold last H plant input values and input proposals, and in the case of the ASR method also the plant state history (H values for each state variable).

Network Utilization. The network utilization depends mainly on T_s and the size of transfered variables. For instance, assuming 16-bit long variables, system order $n = 4$, number of measured outputs $l = 2$, sampling time $T_s = 10\,\text{ms}$ and history size $H = 10$, maximum throughput of controller-actuator link for both methods is $(H + 1) * 16/T_s = 17.6\,\text{kbps}$, and maximum throughput for sensor-controller link is $(H + n + 1) * 16/T_s = 24\,\text{kbps}$ for PB method and $(H + Hl + 1) * 16/T_s = 49.6\,\text{kbps}$ for the ASR one.

5   Experiments

The method performance is studied in the networking environment with the physical object controlled remotely over a real communication network. The constructed laboratory setup, involving the object, sensors, and actuator (motor) connected to a microcontroller unit (MCU) equipped with a networking interface is shown in Fig. 3.

Plant. The controlled plant reflects a structurally unstable 4th-order inverted pendulum-on-a-cart system. The plant parameters are as follows: mass of the cart $M = 0.768\,[\text{kg}]$, mass of the pendulum $m = 0.064\,[\text{kg}]$, moment of inertia (around the center of gravity) $J = 0.00231\,[\text{kg}\,\text{m}^2]$, and distance between the pendulum center of gravity and the shaft $l_p = 0.205\,[\text{m}]$. For the purpose of controller design a linearized plant model is considered – the neglected friction, nonlinearities, and actuator dynamics constitute the plant uncertainty. Thus obtained nominal plant dynamics are given by

$$\dot{\mathbf{x}} = \begin{bmatrix} v \\ a \\ \omega \\ \varepsilon \end{bmatrix} = \begin{bmatrix} 0 & 1 & 0 & 0 \\ 0 & 0 & 0.2909 & 0 \\ 0 & 0 & 0 & 1 \\ 0 & 0 & 27.9842 & 0 \end{bmatrix} \begin{bmatrix} s \\ v \\ \alpha \\ \omega \end{bmatrix} + \begin{bmatrix} 0 \\ 1.1663 \\ 0 \\ -3.4286 \end{bmatrix} u, \tag{11}$$

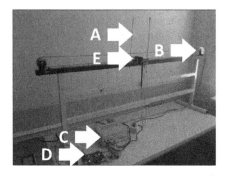

Fig. 3. Laboratory setup. Inverted pendulum (A) is mounted on cart (E) whose horizontal displacement is regulated by motor (B). The motor input is supplied through signal manipulation device (C). The device is connected to microcontroller unit (D) that provides the networking interface. The remote computer is not shown in the picture.

where v is the cart velocity, s – cart position, a – cart acceleration, ω – pendulum angular velocity, α – pendulum angular position, and ε – pendulum angular acceleration. Input u is the motor driving force.

The motor speed is regulated through a PWM wave generated from the MCU (in our case Phillips LPC1768). By design, the MCU collects the data from the sensors, selects an appropriate PWM duty cycle for the actuator, and handles the network communication with the remote controller. The position of the cart and pendulum is obtained from the incremental encoders with 1024 impulses per rotation. The remaining state variables – the cart and pendulum velocities – are determined from the position measurements using a differentiating filter with coefficients $[1, -1]$. The calculations are performed on the local computer (the PB method) or on the remote one (the ASR method). The MCU works in a time-driven fashion. It adjusts the actuator signal, collects and sends the data in $T_s = 10$ ms intervals. The MCU, equipped with an Ethernet interface, communicates with the remote computer using a general-purpose IP-based network. The internal buffer size is adjusted to accommodate $H = 10$ input and measurement values. Consequently, the maximum loop delay expected in the system is $HT_s = 100$ ms. If more than H subsequent packets are lost, the system switches to a fail-safe state (cuts off the supply power), and restarts after the connectivity is reestablished.

Networking Environment. The information between the plant and the controller is exchanged within the standard UDP transmission without acknowledgments. The data passes through three layer 2 and layer 3 switches. During all the conducted experiments the control information is transferred in the presence of regular traffic so that the opportune conditions of an isolated networking environment are avoided. As soon as the remote computer receives a packet, the relevant information is extracted and used to determine the control signal. The established controller command is placed in the packet directed to the MCU.

The MCU places the data retrieved from the incoming packet in the internal buffer and at the next sampling instant it applies the most appropriate value to the actuator. The controller uses the algorithm presented in Sect. 4 to determine the set of control values for the plant. Special real-time considerations (e.g. synchronization) need to be applied neither to the remote computer, nor the network connectivity plane. Only the local computer – the MCU – tracks the evolution of time.

Controller. Since the discussed methods of counteracting negative impact of network-induced perturbations operate independently of the control law selection, basically any stabilizing controller **K** can be used. In the tests, static-gain LQ-optimal control was considered with the state weighting matrix **Q** = diag$\{60, 4, 40, 1\}$ and input weighting matrix **R** = 1. For the minimum (constant) delay in a given range both methods operate in the same way. Therefore, the experiments conducted for the connectivity with minimum delay is used as a baseline for comparison.

Experiments. Three series of experiments are conducted. In each case, the initial state is set as $[0\ 0\ \pi/6\ 0]^T$, which reflects the situation of the pendulum (and cart) originally at rest with the angular position shifted from the upward (unstable) equilibrium by 30°. The objective is to stabilize the pendulum in the upward position despite the plant model uncertainty and network related perturbations by horizontal cart displacement. Note that setting the initial pendulum position at 30° with respect to the vertical axis violates the small signal approximation of the linear model and further strains the tests. The network perturbations are artificially injected into the network channels to establish a consistent comparison plane. The background traffic, in turn, is uncontrolled – it depends solely on the activity of other users. The numerical data gathered in each of the corresponding tests constitute an average of 10 experiment runs.

Tests 1–3. The purpose of the first group of tests is to assess the control system sensitivity to the parametric uncertainties and unmodelled dynamics neglected in the control gain selection. The network transmission proceeds without errors, the channels are reliable (no data loss is experienced), and packets are delivered with constant delay. Consequently, this group of tests forms the basis for evaluating the method performance when the network related perturbations actually disturb the communication process in Tests 4–9. Results of these experiments are presented in Fig. 4.

Tests 4–6. In the second group of tests, the communication channels are reliable – no error or statistical data loss occurs – but the limited bandwidth does not allow to transmit all the required information. These experiments show how a remote control system behaves when the dedicated channel connectivity is used yet an expensive Commitment Information Rate is too low to transmit the packets generated at each sampling period. The reduced throughput may also result from bottlenecks on the data path, or regular (as opposed to burst)

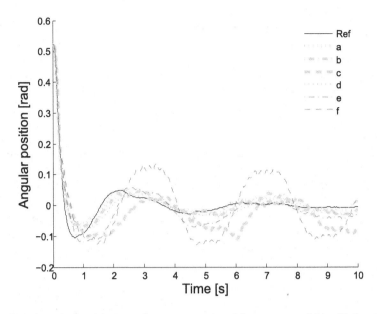

Fig. 4. Results of algorithm performance tests with constant delay. Ref: minimum delay, a, d: 20 ms delay, b, e: 30 ms delay, c, f: 40 ms delay. a, b, c: PB method, d, e, f: ASR method.

packet dropouts. The successful transmission rate amounts to 1/3 in all three experiments, i.e. only one packet out of three arrives at the plant. The results of these experiments are presented in Fig. 5.

Tests 7–9. The last group of tests allows one to assess the method performance in a stochastic environment. The packets are directed through ordinary packet transfer network. The delay variations follow the Poisson process with mean = 30 ms, enlarged by different constant delays. Results of these experiments are presented in Fig. 6.

Result Analysis. Overall better performance (smaller deterioration of the control quality as measured by the LQ performance index) presents the PB method, while the ASR one results in smaller overshoots. The quality of the ASR method rapidly degrades with delay increase. This effect is caused by a larger error in discrete-time system representation for longer sampling periods of the linearized, continuous-time plant model. Both methods follow closely the baseline for small delays with the stabilization effectiveness degrading for longer delays. The nonlinearities and plant uncertainty neglected in the controller design do not allow the error to be eliminated. Instead, a limit cycle, stable in the Lypaunov sense, can be observed. The ASR method results in the larger limit cycle (measured by the amplitude of oscillations).

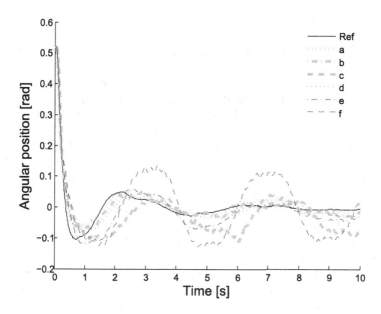

Fig. 5. Results of algorithm performance tests with throughput limitation: 1/3 of packets arrive at the actuator. Ref: minimum delay, no drops, a, d: no additional delay, b, e: additional 10 ms delay, c, f: additional 20 ms delay. a, b, c: PB method. d, e, f: ASR method.

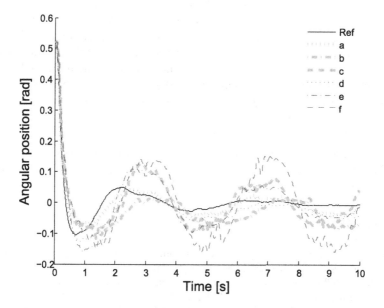

Fig. 6. Results of algorithm performance tests with Pareto distributed delays with mean 30 ms. Ref: minimum delay, no drops, a, d: no additional delay, b, e: additional 10 ms delay, c, f: additional 20 ms delay. a, b, c: PB method. d, e, f: ASR method.

6   Conclusions

The sensitivity of linear controllers to uncertainties enforces conservative designs which may discourage their implementation in practical systems. On the other hand, robust, frequently nonlinear, control strategies tend to address stability issues at the expense of increased complexity and implementation difficulties (e.g. the chattering problem in sliding-mode control systems). In the paper, two groups of algorithmic methods accessible in remote control systems for reducing the impact of network-induced perturbations were examined. For the purpose of comparison, a series of experiments were conducted involving a physical plant and a real communication network thus allowing for consistent empirical investigation in the presence of model imperfections as well as network-related ones.

Two, similar in implementation but different in design, classes of methods were identified and subjected to experimental evaluation: predictor based and adaptable sampling rate methods. The PB method generally outperforms the ASR one in terms of reducing the output error. The observed performance degradation of the ASR method originates from two successive approximations, one in the linearization of nonlinear process dynamics and another in the discretization with longer sampling period, that reinforce the plant model uncertainty. It occurs that the (potential) error propagation in determining the compensator signal due to imprecise plant model in PB method is less severe for the control process quality in networked environment than applying a longer sampling period.

The investigated methods support the control law in decreasing the influence of network-induced perturbations on the closed-loop system. Their performance, however, depends on the quality of the model of the regulated process. Both methods benefit from more accurate representation of the physical system. Therefore, for achieving best performance, the effort of reducing the influence of network-induced phenomena should be assisted by decreasing the plant-related uncertainty. In the case of highly uncertain processes one needs to recur to inherently robust control algorithms.

Acknowledgments. This work has been performed in the framework of project no. 0156/IP2/2015/73, 2015–2017, under "Iuventus Plus" program of the Polish Ministry of Science and Higher Education.

References

1. Baillieul, J., Antsaklis, P.: Control and communication chellenges in networked real-time systems. Proc. IEEE **95**(1), 9–28 (2007)
2. Gupta, R., Chow, M.: Networked control systems: overview and research trends. IEEE Trans. Industr. Electron. **57**(7), 2527–2538 (2010)
3. Zhang, L., Gao, H., Kaynak, O.: Network-induced constraints in networked control systems - a survey. IEEE Trans. Industr. Inf. **9**(1), 403–416 (2013)
4. Willig, A.: Recent and emerging topics in wireless industrial communications: a selection. IEEE Trans. Industr. Inf. **4**(2), 102–124 (2008)

5. Heemels, W., Teel, A., van de Wouw, N., Nešić, D.: Networked control systems with communication constraints: tradoff between transmission intervals, delays and performance. IEEE Trans. Autom. Control **55**(8), 1781–1796 (2010)
6. Ignaciuk, P., Bartoszewicz, A.: Discrete-time sliding-mode congestion control in multisource communication networks with time-varying delay. IEEE Trans. Control Syst. Technol. **19**(4), 852–867 (2011)
7. Ignaciuk, P.: Nonlinear inventory control with discrete sliding modes in systems with uncertain delay. IEEE Trans. Industr. Inf. **10**(1), 559–568 (2014)
8. Ignaciuk, P.: Discrete-time control of production-inventory systems with deteriorating stock and unreliable supplies. IEEE Trans. Syst. Man Cybern. Syst. **45**(2), 338–348 (2015)
9. Zhang, J., Lam, J.: A probabilistic approach to stability and stabilization of networked control systems. Int. J. Adapt. Control Signal Process. **29**(7), 925–938 (2015)
10. Gao, H., Meng, X., Chen, T.: Stabilization of networked control systems with a new delay characterization. IEEE Trans. Autom. Control **53**(9), 2142–2148 (2008)
11. Natori, K., Tsui, T., Ohnishi, K., Hace, A., Jezernik, K.: Time-delay compensation by communication disturbance observer for bilateral teleoperation under time-varying delay. IEEE Trans. Industr. Electron. **57**(3), 1050–1062 (2010)
12. Peng, C., Yue, D., Han, Q.L.: Communication and Control for Networked Control Systems. Springer, Heidelberg (2015)
13. Truong, D.Q., Ahn, K.K.: Robust variable sampling period control for networked control systems. IEEE Trans. Industr. Electron. **62**(9), 5630–5643 (2015)
14. Sipahi, R., Niculescu, S.I., Abdallah, C., Michiels, W., Gu, K.: Stability and stabilization of systems with time delay. IEEE Control Syst. Mag. **31**(1), 38–65 (2011)
15. Normey-Rico, J., Camacho, E.: Dead-time compenstarors: A survey. Control Eng. Pract. **16**(4), 407–428 (2008)
16. Zhang, J., Xia, Y., Shi, P.: Design and stability analysis of networked predictive control systems. IEEE Trans. Control Syst. Technol. **21**(4), 1495–1501 (2013)
17. Zhao, Y., Liu, G., Rees, D.: Stability and stabilization of discrete-time networked control systems: a new time delay system approach. IET Control Theory Appl. **4**(9), 1859–1866 (2010)
18. Ignaciuk, P., Morawski, M.: Mitigating the impact of delay uncertainty on the performance of LQ optimal remote control systems. In: Proceedings of 12th IFAC Workshop on Time Delay Systems, Ann Arbor, MI, USA, pp. 452–457, June 2015

A SIP-Based Home Gateway
for Domotics Systems:
From the Architecture to the Prototype

Rosario G. Garroppo, Loris Gazzarrini, Stefano Giordano,
Michele Pagano$^{(\boxtimes)}$, and Luca Tavanti

Department of Information Engineering,
University of Pisa, Via Caruso 16, 56122 Pisa, Italy
{r.garroppo,l.gazzarrini,s.giordano,m.pagano,l.tavanti}@iet.unipi.it

Abstract. The integration of the various home devices into a single, multi-service, and user-friendly platform is still an area of active research. In this scenario, we propose a domotics framework based on the Session Initiation Protocol (SIP) and on a SIP-based Home Gateway (SHG). The SHG retains the compatibility with the existing SIP infrastructure, allowing the user to control all domotics devices through his usual SIP client. Particular attention has been paid to the usability and scalability of the system, which brought us to define a functional addressing and control paradigm. A working prototype of the SHG and a customized SIP event package have been used to provide a proof-of-concept of our architecture, in which the SHG has been interfaced with ZigBee and Bluetooth networks.

Keywords: Domotics systems · SIP-based Home Gateway · ZigBee

1 Introduction

In the domotics archetype, all subsystems and devices will be able to talk to each other and interact in a seamless manner, realizing an intelligent structure that improves the quality of life, reduces the costs, and achieves energy savings, allowing the user to manage them from heterogeneous devices either at home or remotely. To put this paradigm into practice, the communication among the single home devices and between the various domotics subsystems is the fundamental operation. The set of sensors and actuators usually comes in the form of one or more networks (Domotics Sensors and Actuators Networks, DSANs), backed either by a single technology or by different ones. Often, however, the specifications do not define a common control plane to contemporarily manage devices belonging not only to different standards, but even to different application profiles (e.g. ZigBee's Home Automation, and Smart Energy, or KNX's Lighting, and Heating – just to cite the most appealing ones). As a result, the burden of coordinating and making devices interoperate is left entirely to the system implementer. From the user perspective, the devices belonging to the diverse

P. Gaj et al. (Eds.): CN 2016, CCIS 608, pp. 344–359, 2016.
DOI: 10.1007/978-3-319-39207-3_30

subsystems can be typically controlled through dedicated appliances located in the house (e.g. a touch panel, a smart telephone, a TV remote). However, this paradigm no longer holds for remote control operations that occur when the user is far from home. In this case he/she would normally have a single device at hand, such as a tablet or a smartphone, by means of which he/she would like to control any device in the home, possibly without profile- or technology-specific configuration procedures.

In this scenario, we designed a domotics architecture aimed at gaining interoperability among devices belonging to different technologies and profiles. In our vision, the common control plane is realized through the Session Initiation Protocol (SIP) [1]. Indeed, SIP is able to meet many of the requirements that a domotics system may impose, such as various communication paradigms (command-based, event-based, session-based), user mobility, integration with existing services (e.g. telephony), fast development times, and security[1]. A *SIP-based Home Gateway* (SHG), the major enabler of the envisioned system, is devised to translate the user commands from and to the specific DSAN languages and to retain the compatibility with the existing SIP infrastructure and deployed SIP clients, which can thus be exploited in full. In addition, we defined a *functional addressing and control scheme* (for short, FACS) by means of which the user can "talk" with the domotics network using just mnemonic names and commands. These are interpreted by the SHG and managed transparently to the user. We implemented a working SHG prototype, used to build a complete proof-of-concept of our SIP-based domotics architecture. A customized SIP event package and a notification server have also been developed to validate a possible extension to new services. On the user side, the SHG was interfaced with both a common SIP client and a client we developed with some "extended" features. On the DSAN side, the SHG was interfaced with an actual ZigBee network and a Bluetooth PAN. The mapping of the SIP methods to the domotics architecture has also been discussed, and the synergies between SIP and the ZigBee DSAN have been identified and exploited for a more efficient implementation.

2   State of the Art

Some authors approached the integration between wireless sensor networks and control plane protocols by bringing customized or reduced versions of SIP or REST on the sensor nodes [3,4]. This kind of approach suffers from a series of drawbacks. Since the devices are resource constrained, the protocols must be stripped of many functions. Then, due to the particular operating system running on the sensor nodes, the development times might be non-negligible. However, given the high heterogeneity of the devices, it might be necessary to repeat and modify the customization and development steps for every technology that is going to be integrated into the system. Finally, compatibility with deployed

---

[1] The domotics system can smoothly exploit the SIP security options. Details about these options and their impact on the system performance can be found in [2].

hardware and software is not retained. Alia et al. follows the philosophy of making an open and flexible service platform based on a two-tier middleware [5]. Yet, the main focus of this proposal is on the software and middleware aspects of the system, with primary target the multimedia home entertainment area. As a result this work cannot be easily extended to other domotics sub-systems. The definition of an abstraction layer to be located on top of various MAC layers is the focus of the IEEE P1905.1 working group [6]. The purpose is to make applications and upper layer protocols agnostic of the underlying networking technologies. A preliminary implementation of this concept has been presented by Nowak et al. [7]. The major drawback of this approach is the need to tailor the abstraction layer to each MAC technology and implement it on all home devices. The work closest to ours is perhaps the one by Bertran et al. [8], who tested SIP as a universal communication bus for home automation environments. A SIP gateway and a series of SIP adapters and interpreters have been implemented and deployed to make all devices SIP compliant. However, there are some aspects that may put our framework one step ahead. Bertran et al. did not consider the issues with addressing and accessibility of the single DSAN devices. Conversely, the FACS greatly simplifies the user interaction and does not require the DSAN nodes to register to any SIP server or other additional entities. Then, we devised a way of keeping the compatibility also with the already deployed user terminals, providing the users with functionalities that are not natively supported by their devices. Bertran et al. did not pay much attention to this aspect. A third distinguishing point is in the adaptation between the SIP and the DSAN worlds. While Bertran et al. design a single software module to be put in the gateway, we perform this operation in two steps. This allows decoupling the implementation of the two domains, making the system more flexible. Finally, we studied in much more detail the integration with two possible DSANs, namely ZigBee and Bluetooth, and showed how it is possible to exploit their features to simplify the integration into the system. The main focus of Bertran's experimental platform [8] was on the performance figures of the gateway, which, if we consider the current hardware technology, might not be the most relevant hurdle. Recent research activities have been focused on the realization of REST (Representational State Transfer) architecture in a suitable form for the most constrained nodes, in order to permit the development of Internet of Thing (IoT) services [9]. Similarly, a constrained version of the SIP, named "CoSIP", has been designed to allow constrained devices to instantiate communication sessions in a lightweight and standard fashion [10]. A major disadvantage that is common to proposals like Alia et al. [5], Bertran et al. [8], and, to a smaller extent, TinySIP [4] and CoSIP [10], is the need for every home device to register with its own identifier (a.k.a. URI). When the number of devices increases (heavily monitored and automated buildings may have hundreds of nodes), the user capability of handling them through their URIs is clearly hampered. On the other hand, in our system the sole SHG must perform the SIP registration, while we mask the multitude of DSAN nodes by means of the FACS.

3   System Architecture and Components

Figure 1 shows the general architecture of the conceived domotics system, which is composed by four major physical elements: the Clients, the SIP Servers, the SHG, and the DSANs. The SIP-based Home Gateway (SHG) is the key element of the system. It enables the remote control of the various DSANs by interpreting and mapping user commands to device-specific actions. It is in charge of using the proper set of SIP methods in order to ensure backward compatibility with Legacy Clients and full functioning to Enhanced Clients. It performs "intelligent" operations, such as piloting devices of a DSAN in response to events from another DSAN. An intermediate entity, named Domotics Facility Abstraction (DFA), has been introduced with a double goal: disjoin the implementation of the two domains of the system (i.e. SIP and the DSANs), and create a single and user-friendly service abstraction. The former goal is meant to ease the development of the SIP and the DSAN interfaces, which may be carried on separately. The latter goal is meant to be achieved through the definition of the *functional addressing and control scheme* (FACS) employed for the user interface. In practical terms, this abstraction lets the user deal with mnemonic names for the domotics services (such as the room names and the device functions), keeping SIP and the specific DSAN technologies hidden to the user. A pictorial description of the framework can be seen in Fig. 2.

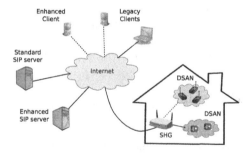

Fig. 1. Reference architecture of the SIP-based domotics system

The functional service abstraction is implemented through the user interface on the Client (including any legacy SIP application), and is understood and processed at the SHG. The DFA is realized for the most part in the SHG, which stores the set of actions and performs the necessary tasks to accomplish the user's directives. The client is any device at the user's disposal supporting a minimum set of SIP methods. All widely deployed SIP clients fit this description, without the need for any customization or added feature. However, in most of the deployed SIP-clients, very few SIP functionalities are available. Hence, for the full exploitation of the domotics features envisioned in our system, it might be necessary to enhance the client with further capabilities and SIP methods. Therefore we can distinguish two types of client: a Legacy Client,

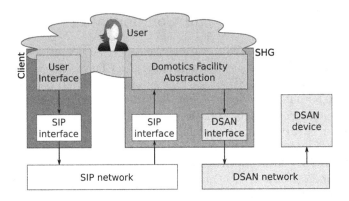

Fig. 2. The functional paradigm implemented by means of the DFA

supporting only the basic SIP methods, and an Enhanced Client, which supports domotics-specific operations. A SIP server is necessary for the registration of the devices and to support the Publish/Subscribe-Notify semantic. This server can be any server on the SIP infrastructure providing such functions. The distinction between "legacy" and "enhanced" applies to the servers too. Typically, existing servers are unaware of the domotics functions; hence only standard SIP features are available. However, as the domotics paradigm gains ground, new domotics-aware servers can be expected to appear. The sensors and actuators deployed in the house are usually connected through one or more domotics networks (i.e. the DSANs). Each DSAN has its own transport technology and can use either wired or wireless connections (or both). All DSANs are then attached to the SHG.

4   Selected SIP Methods

In this section we provide a brief description of the SIP methods we used and, above all, how we integrated them into the domotics system. Note that, since the integration mode is not univocal, particular attention is paid to the reasons that drove our choice, how these methods have been exploited, and how they interact with the various elements of the system. The SIP *message* method (defined in RFC 3428) is used to supply the real-time dispatch of short text messages. Each *message* transaction[2] is self-contained (i.e. each text message is independent from the others) and requires no session set-up. We used this real-time low-overhead method to implement the request/response paradigm. In detail, the *request* is mapped to a first *message* transaction, and the *response* is mapped to a second *message* transaction. Therefore, four SIP messages are used to realize this paradigm. It should be noted that, since the SIP acknowledgment message (such as *200 ok*) could in theory include application data,

---

[2] A "transaction" consists of a message that invokes a particular method on the server and at least one response message.

a single *message* transaction could have been used. Yet, by means of two separate transactions we achieve a semantic separation too. The first "low-level" SIP acknowledgment (i.e. a *200 ok* message) is used to confirm just the correct reception of the *request message*, whereas the second "high-level" *response message* is sent in an asynchronous way to inform the user about the outcome of the requested action (which may occur several seconds, or even minutes, later). A very important aspect of the *message* method is its compatibility with all existing SIP clients. Since every SIP client must support this method, this ensures that the basic managing functions of our system are also supported. The SIP Event Notification Framework (ENF) (RFC 3265) provides a way for SIP elements to learn when "something interesting" has happened somewhere in the network. Briefly, an initial *subscribe* message is sent by the user that is interested in the event (subscriber) to the node that is first aware of it (notifier). The events are then reported from the notifier to the subscriber by means of the *notify* method. RFC 3903 provides a framework for the publication of event states on a notification server, called Event State Compositor (ESC). This task is accomplished using the *publish* method. The ESC is responsible for managing and distributing this information to the interested parties through the ENF. Note that the ESC is a logical entity, which can physically reside in diverse parts of the domotics system. In our prototype we arranged for the ESC functions to be provided by the SHG. In particular, the SHG is the only entity that publishes the events. DSAN devices are thus preserved from knowing anything about the SIP existence. In addition, the SHG can filter and compose events that are not available in the single DSAN domains. The *register* method is used by SIP to create bindings that associate a user's URI with one or more "physical" locations of the user (e.g. the IP address). SIP network elements (such as the User Agents (UAs) and the various servers) handle the URI-location binding on behalf of the user. In the proposed architecture just two elements must be registered: the user and the SHG. The user shall register every time he/she changes the point of attachment to the network (in fact, this is performed automatically by the UA). Since the SIP infrastructure will automatically take care of keeping the connection with the SHG alive, the user mobility is made transparent to both the user and the SHG. On the other side, given that all sensors and actuators are managed by the SHG via the specific DSAN interfaces, they can be completely unaware of the SIP control plane. Yet, the user can interact with them by knowing the SIP URI of the SHG and referring to the DSAN devices through the FACS. This makes the system extremely user-friendly and also highly scalable. No matter how many devices are in the house, the user can control them from anywhere he/she is and through invariably the same URI (the SHG one). Note that this single-URI approach is neither an intrinsic feature of SIP nor of the domotics concept itself. Rather, it is a "plus" of the way we built our architecture and the SHG.

5   Proof of Concept

To put the ideas expressed in the previous sections into practice, we have realized
a small testbed involving all the elements of the architecture. The SHG, being the
core and most innovative element, has been built from scratch. Two DSANs have
been implemented using two sets of ZigBee and Bluetooth devices. An Enhanced
SIP Client has also been developed. Finally, we have defined and implemented
a specific SIP event package for the domotics framework. To build the SHG we
used a generic single board computer (SBC) with a processor running at 720 MHz
with 256 MB of DRAM and 256 MB of NAND flash memory. To give the SHG
the physical interfaces towards the two DSANs, a ZigBee module was embedded
into the board and connected to the main processor via a serial interface, and
a Bluetooth adapter was inserted into a USB port. Figure 3 shows the prototype
SHG. The SHG software was built on top of GNU/Linux (with kernel 2.6.36),
which provides the necessary support and development tools (e.g. a SIP library,
the interface drivers). The software that implements the SHG functionalities has
been written from scratch using the C++ language. The prototype is provided
of a ZigBee and a Bluetooth DSAN. The nodes of the ZigBee sensor network
are based on a 32-bit micro-controller unit and a low power 2.4 GHz transceiver.
A fully compliant ZigBee stack was installed on the nodes. A custom application
that supports environmental data collection (temperature and pressure) and
remote light control has been developed on top of the ZigBee stack by means of
the ZigBee Cluster Library (ZCL) functions[3].

Fig. 3. The prototype SHG, with the ZigBee and Bluetooth modules

The APS ACK feature (an end-to-end acknowledgment mechanism) was
enabled to make the ZigBee transmissions reliable. Ambient data are retrieved
both on regular time basis and on-demand. Both approaches are available to
the user, who can either subscribe to this event or ask the SHG to check a specific sensor value. As for remote light control, the sensor boards are equipped

---

[3] The ZCL is a set of attributes, data, and commands used to speed up the application
development.

with an array of LEDs, which was used to mimic a multi-level light. For both ambient data collection and light control we defined a set of textual commands (see Table 1). Combining them with the name of a room allows the user to set the desired light level or retrieve the sensor reading. An important aspect of the ZigBee system is that it provides for a mechanism, known as binding, to connect endpoints[4] on different nodes. Binding creates logical links between endpoints and maintains this information in a binding table. The binding table also has information about the services offered by the devices on the network. A notable advantage of this structure is that it allows the implementation of the service discovery procedure via bindings. The services available inside the ZigBee network can thus be discovered directly within the ZigBee domain, without resorting to any additional software or external entities. With specific reference to the SIP control plane, this means that there is no need to port the SIP registration procedure to the ZigBee network, since this would be a duplication of the ZigBee service discovery. The Bluetooth adapter connected to the SHG board embeds a version 2.0 compliant chipset. To operate this device we took advantage of the Bluetooth Linux stack (BlueZ), which gives support for basic operations such as scanning and pairing. On top of these basic functionalities we built the Bluetooth interface, which is capable of listing and managing the connected devices. For the purpose of validating the domotics system, we set up an audio streaming test. A Bluetooth-enabled headset was used as the client device. The audio streaming was handled directly by the BlueZ (on the SHG side) and the headset, by means of the A2DP profile. On the SIP-based control plane, we defined and implemented some simple commands, such as listing the available content and playing an audio stream on the Bluetooth headset (see Table 1). The SIP ENF provides just the procedures that enable notification of events, but does not define any specific event package. A few packages for the SIP ENF have currently been ratified. Among them, the Presence package (RFC 3856) is supported by some widely available SIP clients. However, this package provides just a single functionality (the presence of a user) and refers to the bare ENF, without taking advantage of the *publish* method. To test our domotics system with a complete and flexible Publish/Subscribe-Notify paradigm, we built a new package, named "Home Automation". The basic features of Home Automation are similar to the ones of Presence, but our package employs the *publish* method too. We designed it to embed a customized XML text, which contains domotics specific data (such as the values read by some ambient sensors). Clearly, this XML scheme can be replaced with any other kind of text format, like REST or SOAP. To verify the domotics-specific features of our framework, we developed a prototype Enhanced Client that supports the full Publish/Subscribe-Notify paradigm and the Home Automation package described before. One such client is in all aspects a SIP-compliant software, but with the extra feature that it can control the DSAN with its native semantic. A popular smartphone with SIP support has been used as a sample Legacy Client. In our proof-of-concept we employed three different

---

[4] An endpoint is a logical wire connecting distributed applications residing on different nodes.

servers, which represent the three possible situations for our domotics system. An external registrar and proxy server were used as a sample of a pre-existing SIP network element. This server is compliant to existing SIP standards, and is completely unaware of the nature of our domotics testbed. Specifically, we selected a server provided by *iptel.org*, the well known free IP-telephony service. A customized SIP server, called Enhanced Notification Server (ENS), was built to support the Home Automation event package. Hence, this server is representative of a domotics-aware element in the SIP infrastructure. Note that the ENS also supports the registration and proxy procedures. Finally, a SIP server was set up within the SHG. This solution is the archetype for a self-contained domotics system, in which the gateway provides all the necessary SIP functionalities.

6   Functional Tests

To check the correct functioning of our prototype we performed a series of tests, which cover the three major communication paradigms: the request/response paradigm for remote control, the Publish/Subscribe-Notify semantic for event management, and the session management to support an audio streaming session. For these tests, the messages exchanged on the various interfaces have been captured by means of a network analyzer and synthetically reported in the form of diagrams.

6.1   Network Topology

All tests have been carried out within the premises of our Department. The realized testbed is made of five ZigBee sensor nodes, including the ZigBee Coordinator (ZC), which is embedded into the SHG board, as already shown in Fig. 3. The physical location of the nodes is illustrated in Fig. 4. Due to the indoor environment, the nodes *Stairs*, *Office*, and *Corridor* use a multi-hop path to reach the ZC. The Bluetooth headphones (*bths*) are placed in a room adjacent to the one with the SHG.

6.2   Remote Control

The remote control test consists in a user controlling a light from a remote location. In this test, the Client can be either Legacy or Enhanced, since all operations are performed through the *message* method. The user's terminal and the SHG are located on two different networks (though both belonging to our University). The registration procedures made use of the external SIP server. The user can make use of a set of commands to pilot the light, as detailed in Table 1, and can address the end nodes with the name of the room where they are located (the mnemonic names of the rooms coincide with those of the ZigBee nodes). The relationship between the SIP messages (expunged of the registration and authentication procedures, as well as of the *200 ok* answers), the SHG translations, and the ZigBee frames is schematically illustrated in Fig. 5.

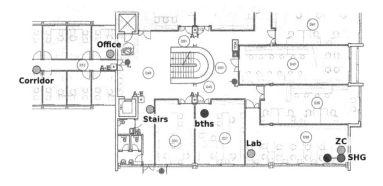

Fig. 4. Experimental scenario with the position of the SHG, the ZigBee and the Bluetooth nodes

Table 1. Some of the commands implemented in the testbed

Command	Effect
switch_on_light	Turn on the light at the maximum level
switch_off_light	Turn off the light
set_light_level N	Set the light level to N
subscribe home_auto	Subscribe to the Home Automation events in backward compatibility mode
list_audio	List available audio files
play_audio T on D	Play audio track T on device D

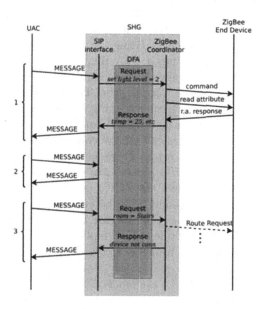

Fig. 5. Messages on the SHG and ZigBee domains exchanged during the remote control tests. Only the SHG interfaces are shown.

Fig. 6. Screenshot of the chat window taken during the remote control test

After the registration, the SHG is reachable through its mnemonic address (*sip.home.gateway@iptel.org*[5]). Then, the test can be split in three phases (as indicated on the left of Fig. 5). The first phase refers to a successful user interaction. To control a specific device the user types one of the previously listed commands preceded by the room name. For example, assuming the user wishes to turn on the *Office* light at an intermediate level, the resulting text would be something like "Office set_light_level 2". Figure 6 shows the text messages that appear on the Client interface. The SIP UA encapsulates this text into a *message* and sends it to the SHG.

At the reception of the *message*, the SHG SIP interface performs a syntax check. Since the syntax is correct, a *200 ok* is sent back to the Client via the *iptel.org* server. Then, another verification is made by the translation module to ensure that a valid request is received. Being the command and room correct, the SHG transforms the command into a DFA Request. This is in turn translated into a ZigBee command, containing the MAC address of the *Office node*, which is sent to the embedded ZigBee Coordinator. The ZC builds a proper ZigBee frame (the ZCL "command" frame shown in Fig. 5) and casts it over the air. The specified sensor node is thus reached by the command, possibly through some intermediate router nodes. It verifies that the command is correct and executes the action (i.e. the light level is set to 2). Since no explicit acknowledgment is sent by the sensor node[6], the SHG proceeds to check for the correct execution of the command. It thus instructs the ZC to ask for an update of the sensor status, which, in our example, comprises the current light level, temperature, and pressure values. The ZigBee Coordinator builds and sends a ZCL "read attribute" frame to the sensor node, which replies with a ZCL "read attribute response" frame. The response frame is received by the ZigBee Coordinator,

---

[5] This is a fictitious address valid for the proof-of-concept only.

[6] According to the ZCL specifications, the sensor is not required to confirm neither the reception nor the execution of the command. A MAC layer acknowledgment frame is indeed sent back, but obviously this does not carry any application layer information.

which informs the SHG (via the ZigBee Manager module). The SHG can thus build a new *message* packet, with the new light level and sensor readings, and send it to the user. The user terminal will then display a confirmation message, as shown in Fig. 6, part 1.

To verify and give an example of the behavior in case of a *message* containing a wrong command, we let the user issue a non-existent command ("set_water_temp"). The SHG replies at first with a *200 ok*, because the syntax of the *message* element is correct, but after parsing the user message, it detects a wrong command, and therefore builds a *message* to inform the user that the typed command is wrong (Fig. 6, part 2). Note that no message translation occurs and no traffic is sent on the ZigBee network (Fig. 5, part 2).

Finally, we report a test in which a user is trying to control a node that is not connected to the network (part 3 of Figs. 5 and 6). Assuming that the command is correct, the SHG passes the request to the DSAN. However, when the ZC looks for the ZigBee network address of the device in its binding table, nothing is found. The ZC may then try to find this device, e.g. by means of a discovery procedure, and, after a specified number of failed attempts, an error message is sent from the ZC to the SHG and then from the SHG to the Client.

6.3   Event Notification

The event notification test involves the use of the *publish, subscribe*, and *notify* methods together with the newly defined Home Automation event package. Thus we also employed the ENS. The *iptel.org* server is still used as both registrar and SIP proxy for the Clients. The test consists in a remote user interested in following the variations of the ambient values of a specified room, in our case the *Lab*. As already outlined, we have two different possibilities, depending on the client employed by the user (either Legacy or Enhanced). Both of them are shown in Fig. 7, which reports the SIP transactions occurred during the test.

To keep the figure clean, we do not show the acknowledgment messages, the Client registration procedures, and the ZigBee devices, and the *iptel.org* server. The messages exchanged on the ZigBee side of the system, however, are quite essential: every time an event is generated by a ZigBee node, a frame is sent to the ZC and hence to the SHG. The reception at the SHG is indicated by a disc on the SHG line. In our testbed, events are generated on a regular time basis. At startup (phase 0), the SHG registers with the ENS, so that it can publish the Home Automation events. A *publish* is sent immediately after the reception of the *200 ok* message to perform a first update of the DSAN data. Both clients register instead with the *iptel.org* server. The interaction with the Enhanced Client embraces phases 1 and 2. When the user wants to monitor the variations in the ambient values, he/she types the mnemonic address of the device (e.g. "Lab") and the duration of the subscription (e.g. 100 s). A subscription to the Home Automation event regarding the *Lab* room is sent towards the ENS by the UA. Note that this message traverses the "legacy" SIP network (dotted line in Fig. 7). If the command is correct, the server accepts the subscription and provides the client with the current status of the *Lab* readings via a *notify* message. The shadow on the Enhanced Client line indicates that the notification service is active for this device. Therefore, when the ZigBee network sends an event to the SHG, this casts a *publish* to the ENS which in turn notifies the Enhanced Client with a *notify* message. Figure 8 shows a screenshot of our Enhanced Client. When the user wants to monitor the ambient values from a Legacy Client (e.g. a smart-phone), he/she performs the subscription by typing the "subscribe home_auto" command, followed by the room name and the duration. The UA then encapsulates this text into a *message* and sends it to the SHG (phase 3). Indeed, from the analysis of the traces captured on the SIP network, we saw that the message reaches the SHG via *iptel.org*. The mechanism is therefore transparent to the SIP network. If the command is correct, the SHG replies with another *message* in which the user is notified of the acceptance of the subscription. A further *message* with the observed values immediately follows (to mimic the *notify* message sent to the Enhanced Client). The user can now take advantage of the notification service managed directly by the SHG. Note that the notification setup procedure for the Legacy Client involves three transactions (in place of the two for the standard procedure). When an event occurs (phase 4), the SHG sends a publish to the ENS and also a *message* to the Legacy Client. Note that all *message* messages are sent from the SHG directly to the client. The outcome of the event notification service on the chat window is shown in Fig. 9 (where two events on two different rooms have been subscribed).

6.4 Audio Streaming

In the scenario conceived for this test, the SHG acts as a Media Center capable to share contents with other devices inside the domotics network[7]. In the specific case, the content is a set of mp3 audio files. The user activates the streaming by means of its SIP-enabled terminal – in our proof-of-concept the same smart-phone of the event notification test. The messages exchanged between the user terminal, the SHG, and the headset are summarized in Fig. 10. At first, the user asks for the available audio tracks (by means of the "list_audio" command). Since the tracks are stored on the SHG, the request and response commands are handled directly by the SHG engine. In case an external repository was present, they would have been forwarded towards the proper interface.

Then, the user asks for playing an audio track (e.g. "play_audio track01.mp3 on bths"). The SHG receives the SIP message, interprets the embedded request and (i) instructs the Bluetooth manager to set up a connection towards the *bths* device (if not active already); (ii) starts a software to play "track01.mp3"; (iii) redirects the audio stream towards the Bluetooth manager, which in turn sends it to the Bluetooth interface and hence to *bths*. An acknowledgment *message* is sent to the client to confirm that the playback is about to start. On the Bluetooth side, we report the setup phase and the beginning of the data exchange phase

---

[7] Clearly, in a realistic domotics environment, the content would be stored on a dedicated streaming server, or come from a remote repository or service provider, but for the purposes of the test this does not make much difference.

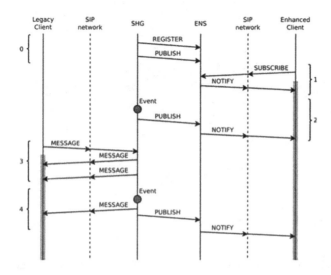

Fig. 7. Messages on the SIP domain exchanged during the event notification test

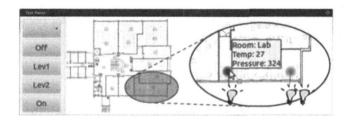

Fig. 8. Screenshot of the Enhanced Client taken during the event notification test

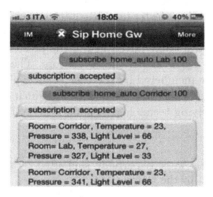

Fig. 9. Screenshot of the chat window taken during the event notification test

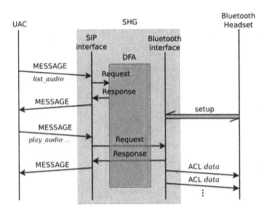

Fig. 10. Messages exchanged during the audio streaming test

(via the ACL commands, since the audio has been treated as generic data), whereas the one-shot preliminary pairing phase is not shown. The setup phase, which comprises several messages, is not necessarily synchronous with the user-side transactions. The only requisite is that, when the user asks for playback, the setup must be completed. The SHG checks this before replying to the user "play_audio" command. For convenience we assumed the user turns on his/her headset after having received the list of audio tracks.

7   Conclusion

The presented architecture and SIP-based home gateway, with the related prototype and proof-of-concept, showed that building a domotics system that integrates diverse home networking technologies, and at the same time provides the user with a uniform and easy-to-use interface, is indeed feasible by means of the technologies currently available. Moving the development effort to a single "high-end" device (the SHG), allows faster implementation times and full compatibility with both existing sensor and actuator networks and also with the deployed SIP terminals. A further and notable advantage of the presented system is that the sole SHG must register with the SIP infrastructure. The SHG URI is thus the only one that must be remembered by the user. This expedient, joined with the functional addressing and control scheme (FACS), which exempts the user from the jargon of the underlying technologies, makes the system extremely scalable and user-friendly.

Acknowledgement. This work is partially supported by the PRA 2016 research project 5GIOTTO funded by the University of Pisa.

References

1. Rosenberg, J., et al.: SIP: Session Initiation Protocol. IETF RFC 3261, June 2002
2. Callegari, C., Garroppo, R.G., Giordano, S., Pagano, M.: Security and delay issues in SIP systems. Int. J. Commun. Syst. **22**(8), 1023–1044 (2009)
3. Luckenbach, T., Gober, P., Arbanowski, S., Kotsopoulos, A., Kim, K.: TinyREST: a protocol for integrating sensor networks into the internet. In: Workshop on Real-World Wireless Sensor Networks (REALWSN), June 2005
4. Krishnamurthy, S.: TinySIP: providing seamless access to sensor-based services. In: Annual International Conference on Mobile and Ubiquitous Systems, July 2006
5. Alia, M., Bottaro, A., Camara, F., Hardouin, B.: On the design of a SIP-based binding middleware for next generation home network services. In: Meersman, R., Tari, Z. (eds.) OTM 2008, Part I. LNCS, vol. 5331, pp. 497–514. Springer, Heidelberg (2008)
6. IEEE P1905.1: Draft Standard for a Convergent Digital Home Network for Heterogeneous Technologies
7. Nowak, S., Schaefer, F., Brzozowski, M., Kraemer, R., Kays, R.: Towards a convergent digital home network infrastructure. IEEE Trans. Consum. Electron. **57**(4), 1695–1703 (2011)
8. Bertran, B., Consel, C., Jouve, W., Guan, H., Kadionik, P.: SIP as a universal communication bus: a methodology and an experimental study. In: IEEE International Conference on Communications (ICC), May 2010
9. Shelby, Z., Hartke, K., Borman, C.: The Constrained Application Protocol (CoAP). IETF RFC 7252, June 2014
10. Cirani, S., Picone, M., Veltri, L.: CoSIP: a constrained session initiation protocol for the internet of things. In: Canal, C., Villari, M. (eds.) ESOCC 2013. CCIS, vol. 393, pp. 13–24. Springer, Heidelberg (2013)

The Cloud Computing Stream Analysis System for Road Artefacts Detection

Marcin Badurowicz[1]([✉]), Tomasz Cieplak[2], and Jerzy Montusiewicz[1]

[1] Institute of Computer Science, Electrical Engineering and Computer Science
Faculty, Lublin University of Technology,
Nadbystrzycka Street 38D, 20-618 Lublin, Poland
{m.badurowicz,j.montusiewicz}@pollub.pl
[2] Management Faculty, Lublin University of Technology,
Nadbystrzycka Street 38A, 20-618 Lublin, Poland
t.cieplak@pollub.pl
http://cs.pollub.pl

Abstract. The paper presents the cloud computing system designed for monitoring the state of roads by processing data packages covering such data as the car's acceleration and position acquired by mobile devices (smartphones and tablets) mounted in cars and implemented on IBM BlueMix platform. Such data are being directly sent to the cloud system, where they are being saved and processed online. The system is capable of performing the authors' pothole detection algorithm which is characterized by high detection rate, but on continuously arriving data. Finally, the system is presenting its results in the form of website and data packages sent back to mobile devices.

Keywords: Potholes · Accelerometer · GPS · Cloud computing · Internet of things · Bluemix

1 Introduction

The problem of road quality analysis is not a sparking discussion in the topic of comfort of the road users – drivers and passengers in private cars and public transport, which is a serious issue in many areas. When travelling by car for a long time, both the driver and passenger need to feel comfortable instead of trying to avoid all the possible potholes, ruts or other elements negatively impacting user comfort, which the authors dubbed road artefacts. But the concept is not only limited to that – evidence of possible road artefacts is also a very important aspect of medical transport, where patients should not be transported on low-quality roads due to their medical condition.

The goal of this research is to provide an automatic computer system gathering data about road quality and providing end users with visual cues on it. The system must be automated as much as possible. The achieve the ubiquity of possible data sources for such a system, the authors decided to use smartphones, which are popular devices, available to a wide spectrum of users, as well

© Springer International Publishing Switzerland 2016
P. Gaj et al. (Eds.): CN 2016, CCIS 608, pp. 360–369, 2016.
DOI: 10.1007/978-3-319-39207-3_31

as equipped with multiple environmental sensors which may provide information about road quality. The smartphone may now be viewed as a representation of data coming from a car itself, therefore diving into the concept of the Internet of Things, treating the car as a Thing connected to the global network.

The solution presented in this paper is to provide a never-ending, continuous, acquisition of streams of data from multiple devices and process them on-the-fly to find road artefacts as soon as they are detected by the automated system and share them in a graphical form.

2   Previous Works and Motivations

The base of the concept the authors are presenting in this paper is the use of accelerometer in the smartphone – if a smartphone is stably mounted inside a car, there is a direct correlation between the car's motion and the phone's motion. Thus, acceleration data acquired by smartphone will directly correlate with potential road artefacts the car is meeting on the road. Because the smartphone is also equipped with a GPS (Global Positioning System) receiver, all acceleration data are paired with a current location, which allows to place detected anomalies and performed assessments on maps. Acceleration can be measured using accelerometers, cheap devices included across virtually every price range of smartphones currently available. Data from the accelerometer are available from the proper software interfaces to 3rd party developers of smartphone applications.

This subject has been studied by only a few researchers in recent years. The authors published their previous results [1] in 2015, where a basic proposal for this kind of system was presented, and real life experiments were performed for checking the algorithm's correctness, resulting in 100 % detection rate, however with a number of false positives. The Nericell and TrafficSense systems [2] focused on traffic monitoring. The researchers were using three discrete devices – a smartphone, a GPS receiver, and a standalone accelerometer. In contrast, in the proposed solution everything is built into one device, with a possible loss of GPS signal or acceleration accuracy, however with a positive impact on battery life and overall mobility of measuring equipment.

In study [3], the mobile anomaly-sensing systems are discussed where data from mobile agents were processed on a central server after data acquisition process, the acceleration of vertical pulse was measured, and the vertical axis was determined using Euler angles. In the solution presented here a real world vertical axis is calculated by using the translation from the local to the global coordinate system by means of the rotation matrix and the data points are processed in real-time in the cloud computing system.

The authors of [4] collected accelerometer readings with the frequency of 38 Hz along with GPS location, and the noises were nullified by using the Kalman filter. For final post-processing, Fast Fourier Transform (FFT) was used, and the data were verified in comparison to a video camera. In a previous study automatically detected road artefacts by the software were compared with a human operator pressing the button on every road artefact traversed.

In [5], detection with smartphones with a 3-axis accelerometer and GPS receiver during 60 s segments was also discussed, with correct identification of road artefacts ranging from 45 % to 70 %. Better results were obtained using an analysis of three elements of the acceleration vector in the SWAS system [6], with correct identification up to 90 %. The authors' study mentioned earlier, as well as subsequent experiments, achieved similar values of positive identification with analysis of only Z axis acceleration – in the authors' approach, the most important data aspect is acceleration on the Z axis, perpendicular to the Earth's surface, instead of data from all three axes.

The solution is based on detecting anomalies and grading the current state of road surface based only on these data. A previous attempt of event detection using standard deviation of Z-axis acceleration over a defined threshold was proposed in [7], where this method had the lowest false positive, but the system presented there lacked real-time data analysis. The next study from the same team [8] introduced real-time monitoring, however on the device itself, which is constrained by limited hardware and software resources. Our study is also using a very basic detection algorithm similar to Z-TRESH, however modified to adapt to overall road quality, which improved detection and reduced the time of fine-tuning. That study was also limited by strictly controlled device orientation. The Wolverine system [9] uses the same rotation matrix technique to achieve independence from the device's actual orientation, however the general objective of the research was concentrated on traffic smoothness, not road quality.

The need to use a cloud computing solution comes from the basic data amount problem – to learn about road quality we need many sensors describing acceleration data, sent from many units. If the analysis is based on a huge geographical solution, we soon need a solution to store and analyse megabytes of data every second. Cloud computing allows to gather lots of data in real time, scaling them appropriately to the needs. When more and more devices will be added to the system, scalability can be achieved by increasing the quantity of the processing units, which is hardly achievable in regular data analysis systems.

Since continuous stream analytics is a basic problem for whole Internet of Things scenarios, where multiple devices are sending their data to the processing systems, the issue is heavily discussed, but no solution like the one proposed exists at the moment.

The IoT allows connecting each device being part of e.g. a wireless sensor network (WSN) with user nets by using software, services and networking technologies. As shown in Fig. 1 above, three models of IoT implementation are mainly used. In the first model all actors of the IoT are located in closed vertical silos [10]. This model is very hermetic and does not allow data to be exchanged with other systems. Two other IoT deployment models are based on cloud computing services and it is mainly these that are depicted in the present paper.

The important factor is that not only one-time road artefact detection for one fragment of a road is sufficient, but data need to be processed continuously because of the dynamic character of road quality, especially due to the current funds from the European Union and local governments for improving the situation.

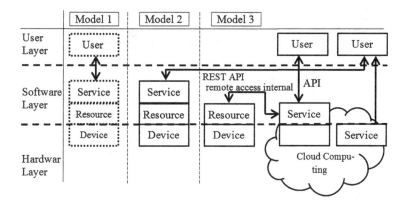

Fig. 1. Devices, resources and services composing the Internet of Things [11]

3   Data Acquisition and Server Communication

The low speed areas created by law (traffic-calming zones, school zones) are lowering effective car speed to the range of 30–40 km/h, where road artefacts may be the most problematic for the road user, and are also full of traffic-calming means like speed bumps. Such traffic zones were chosen as the base for the research. To detect artefacts of about 50 cm in length, data are being acquired from the device with the frequency of 10 Hz, which the authors previously [1] described as sufficient.

Each data package consists of acceleration in three dimensions, as well as such acceleration translated to the global coordinate system, current location, speed and course taken from a GPS module integrated in the smartphone, topped up by GPS accuracy and current time previously synchronized with an NTP (Network Time Protocol) server to ensure correct timings.

The data are not only saved on a device but, crucially, are processed in the cloud. To achieve that, the device is sending all measurements as MQTT protocol packages, published under the topic of the unique device identifier. Because of the fact that, in view of its relativity, every data package (acceleration, time and location) is useful for further processing only as a whole package, the packages are sent as one, rather than being published separately in different MQTT topics.

The MQTT protocol uses the publish/subscribe principle, as presented in Fig. 2. It is a lightweight messaging protocol useful with low power sensors and lossy networks. MQTT is TCP-less and does not require the IP protocol layer [12], which is used in the proposed solution anyway. The MQTT protocol does not define data format, and for the sake of easiness of implementation data are sent as JSON (JavaScript Object Notation).

These data are being continuously received by cloud computing system subscribing to the MQTT broker and saved in the document database system, stamped with an unique identifier, and accessible to the processing units. At this moment, data are not being removed from the document database after

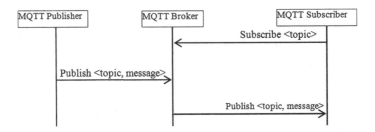

Fig. 2. MQTT publish-subscribe mechanism [12]

post-processing, however this aspect will be introduced if the amount of data after introduction of multiple devices will be too large.

4   Proposed Detection Algorithm

For detection of road artefacts the authors are willing to reuse a modified version of the algorithm called Z-TRESH in [8], previously presented in [1]. The authors previously proved that this method is sufficient to detect such road artefacts as potholes, speed bumps or overall road surface degradation.

The base concept is that when riding through a road artefact the whole car is subjected to vibrations, which may be measured using an accelerometer. For example, when the car is entering a speed bump, it is going "up" the entering slope, and then going "down" the exiting slope, and the same thing happens for the second pair of wheels. This means that to find a road artefact's value of acceleration the Z axis was used. An example of readings of acceleration on the Z axis in a device during riding through a pothole are presented in Fig. 3 below.

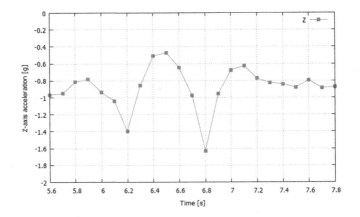

Fig. 3. Acceleration data when driving through a road artefact

Then, the value called Z_s is calculated using the formula below:

$$Z_s = |Z - \sigma|. \tag{1}$$

Z is an acceleration value for the Z axis and σ is standard deviation of Z axis acceleration values for the current road segment.

If the value for Z_s for a current data package was greater than a certain threshold, that data package was marked as a possible artefact. A threshold was previously compared to positions of road artefacts entered by the human operator. Choosing a proper threshold value is a key factor to an algorithm's effectiveness. By contrast to the original Z-THRESH method, the authors are representing the threshold value also as relative to overall road quality. While relative road quality is represented by standard deviation (σ), the defined threshold is a multiplication of that value. That also gives a possibility to detect fluctuations both beyond and above a certain threshold without its modifications.

In previous research it was calculated that best results are achieved with a threshold of 4.3 times the standard deviation (4.3σ). In Fig. 4 below there is a standard deviation times threshold presented over the calculated Z_s data to show that three data points (at 6.2, 6.5 and 6.8 s, respectively) are above the defined threshold and will be recognised as road artefacts.

This causes two issues. First, it is necessary to determine how acceleration data are represented by the device itself – because when lying on the back of the Z axis will be the information the authors were seeking for – acceleration "up" from the Earth's surface, but when the device is mounted in a holder, this may not be true. To fix that, the global coordinate system is used, where data for the Z axis are always "up" from the Earth's surface. To calculate values from the local device coordinate system to the global one, a rotation matrix is used, which is derived from other sensors of the device (magnetometer and gyroscope, respectively) and easily accessible from the device's APIs (Application Programming Interface). In contrast, most of the previously presented relevant work was

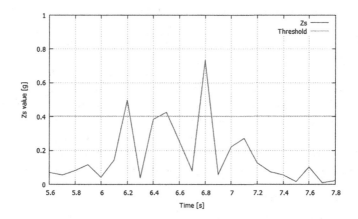

Fig. 4. Calculated Z_s in relation to the threshold value

based on carefully controlled device orientation. With data of the course of the car there is also a possibility to calculate device orientation in relation to the car's movement, positioning the X-axis always "in front" of a car. These calculations were not used in the proposed solution, but the authors are willing to try them in the future.

Second, because the algorithm may detect two data points each as an artefact, while they are parts of only one real artefact (but acceleration change occures when entering and exiting the speed bump or pothole). To cope with that, a very basic concept of data grouping is used. Because of GPS accuracy never surpassing 4 meters, it is impossible to distinguish data points which are located within a distance below this range. And because of road artefacts being much smaller than 4 meters, the authors chose to group all the data points in the distance of twice the current GPS accuracy value and mark such a group as a road artefact in contrast to the original method, where all such points were classified as road artefacts, resulting in a large number of artefacts representing only one physical one.

In addition, while the previously used algorithm was based on a whole road segment from start point to end point, as such it cannot be used in continuous data acquisition and the processing scheme presented in this paper. To solve this problem, standard deviation is not being calculated from the whole road between start and stop points, but using a sliding window of the length of 1000 data packets, which are representing 10 s of the data stream. Detection is performed using the algorithm presented in this sliding window.

5 Implementation of the Cloud Computing System

To simplify the IoT model for the presented research, a diagram of the measurement system in the cloud is shown in Fig. 5. The basis of the system is the measuring equipment included in the mobile device. The mobile device and the car are treated like one object and a source of physical phenomena. Devices of this type form the sensor layer. Next, the communication layer gathers all sorts of devices that allow retrieving basic signals from the meters and detectors. In case of this research, a mobile device acts as communication as well as integration of signals to one data stream. Then, the data goes to an intermediate layer. The aim of this layer is to prepare the data in such a way that they can be transferred to a database system, e.g. localised in the cloud service. The intermediate layer registers the measurement system as virtual devices with a set of measured parameters. This layer keeps rules of security and device access to the system. In the next step, the data are stored in the document database.

Based on the architecture depicted above was the implemented software system. The application was built by means of Node-RED, an open-source tool for building solutions for the Internet of Things [13]. For the purpose of development and implementation of the system, the innovative IBM BlueMix platform and the runtime environment based on Cloud Foundry were used.

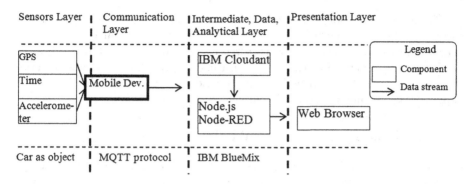

Fig. 5. Diagram of the testbed setup

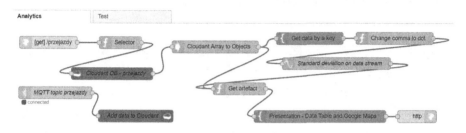

Fig. 6. Screenshot of the application's Node-RED flow

The Node-RED flow shown in Fig. 6 is an implementation of code written in Python language and used in the previous research [1]. The original algorithm was extended by the connectivity and database specific nodes for needs of implementation in the Cloud Computing model.

Listing 1.1. A JSON package as received by the server

```
{
  "_id": "181bd48bfdffb0ae85962b5f32ec7039",
  "_rev": "1-e2f30de277c309c015efd8cb59adfcf6",
  "X": "0,0189453125",
  "Y": "0,0234375",
  "Z": "0,1380859375",
  "N": "0,0389922831469448",
  "E": "0,0309265453470289",
  "Z2": "0,132284240551235",
  "Latitude": "51,2622323445976",
  "Longitude": "22,5416084378958",
  "Time": "2015-07-30 00:00:33,531",
  "Speed": "11,5",
  "Course": "6,6",
  "Accuracy": 13
}
```

The connection node (input node) allows to subscribe to the MQTT topics and then it is connected with a storage node. In this particular case the information taken from MQTT stream is stored in the Cloudant NoSQL database in the format represented in Listing 1.1. On the other hand, the stored data were the source for the function nodes. The function nodes contain the JavaScript implementation similar to the original one.

6 Conclusions and Future Work

The output node in the presented Node-RED system was the HTTP response, which returns the data to the user's web browser. The response result that was returned to the browser was a web page with a data table and a map showing the detected artefacts as well, as shown in Fig. 7.

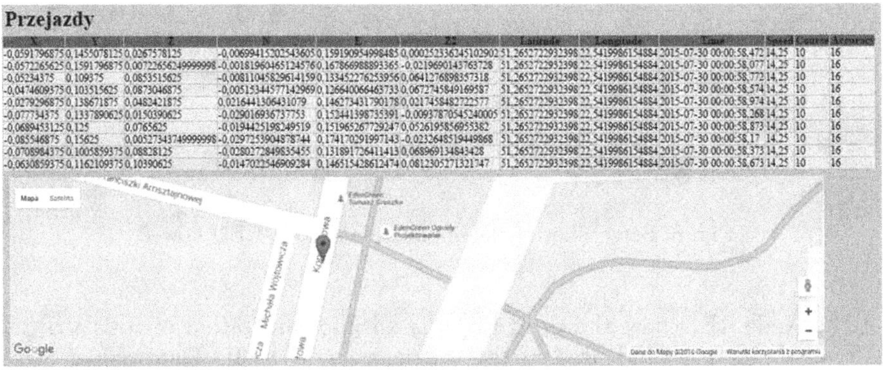

Fig. 7. Generated website with the road artefact positioned

Another possible method of output, not presented here, is a JSON-encoded list of previously detected road artefacts near the user, which may be sent to the mobile device for the sake of presentation to the user – either as a map or a warning messages about near road artefacts. Again, the cloud computing system with its possible scaleability is a wonderful solution for the eventual growth of users.

While the present results are still in the testbed phase, the proposed system is on the way to be a community-driven system for road artefact detection and positioning. By usage of open communication protocols creation of client applications for data sensing and sending will be open to anyone to cover as broad as possible spectrum of client devices.

The authors will now concentrate on better detection algorithms, especially in adaptation to the vehicle speed (e.g. in traffic jams), trying to improve detection where the user is not driving straight through potholes (which is a reasonable behaviour) and on general assesment of the road quality, not only detection of possible road artefacts.

References

1. Badurowicz, M., Montusiewicz, J.: Identifying road artefacts with mobile devices. In: Dregvaite, G., Damasevicius, R. (eds.) Information and Software Technologies. CCIS, vol. 538, pp. 503–514. Springer, Switzerland (2015)
2. Mohan, P., Padmanabhan, V.N., Ramjee, R.: Nericell: using mobile smartphones for rich monitoring of road and traffic conditions. In: Proceedings of the 6th ACM Conference on Embedded Network Sensor Systems, SenSys 2008, pp. 357–358. ACM, New York (2008). http://doi.acm.org/10.1145/1460412.1460450
3. Astartita, V., Vaiana, R., Iuele, T., Caruso, M.V., Giofre, V., De Masi, F.: Automated sensing system for monitoring of road surface quality by mobile devices. Procedia Soc. Behav. Sci. **111**, 242–251 (2014)
4. Pertunen, M., et al.: Distributed road surface condition monitoring using mobile phones. In: Hsu, C.-H., Yang, L.T., Ma, J., Zhu, C. (eds.) UIC 2011. LNCS, vol. 6905, pp. 64–78. Springer, Heidelberg (2011)
5. Das, T., Mohan, P., Padmanabhan, V.N., Ramjee, R., Sharma, A.: PRISM: platform for remote sensing using smartphones. In: MobiSys 2010, San Francisco, California, USA, June 2010
6. Jain, M., Singh, A.P., Bali, S., Kaul, S.: Speed-breaker early warning system. In: NSDR 2012, 6th USENIX Conference, Boston (2012)
7. Strazdins, G., Mednis, A., Kanonirs, G., Zviedris, R., Selavo, L.: Towards vehicular sensor networks with android smartphones for road surface monitoring. In: CONET 2011, CPSWeek 2011 (2011)
8. Mednis, A., Strazdins, G., Zviedris, R., Kanonirs, G., Selavo, L.: Real time pothole detecion using android smartphones with accelerometers. In: 2011 7th IEEE International Conference on Distributed Computing in Sensor Systems and Workshops (2011)
9. Bhoraskar, R., Vankadhara, N., Raman, B., Kulkarni, P.: Wolverine: traffic and road condition estimation using smartphone sensors. In: Fourth International Conference on Communication Systems and Networks (COMSNETS) (2012)
10. Desai, P., Sheth, A., Anantharam, P.: Semantic Gateway as a Service architecture for IoT Interoperability. arxiv no. abs/1410.4977 (2014)
11. Internet-of-Things Architecture IoT-A Project Deliverable D1.2 – Initial Architectural Reference Model for IoT (2015)
12. Hunkeler, U., Truong, H., Stanford-Clark, A.: MQTT-S: a publish/subscribe protocol for wireless sensor networks. In: Proceedings of Workshop on Information Assurance for Middleware Communications (IAMCOM) (2008)
13. Node-RED, A visual tool for wiring the Internet-of-Things. http://nodered.org. Accessed 10 Jan 2016

RSSI-Based Localisation of the Radio Waves Source by Mobile Agents

Dariusz Czerwinski[1(✉)], Slawomir Przylucki[1], and Dmitry Mukharsky[2]

[1] Lublin University of Technology, 38A Nadbystrzycka Street, 20-618 Lublin, Poland
{d.czerwinski,s.przylucki}@pollub.pl
[2] Al-Farabi Kazakh National University, Almaty, Kazakhstan
amiddd@rambler.ru

Abstract. The paper presents a practical realisation of the localisation of a radio wave source. The system is based on the RSSI principle and uses a set of blind mobile agents. The main goal of the research was to implement the localisation system of the root node in WSN on the low-cost autonomous mobile embedded platform. This platform has a limited ability for complex mathematical operations, therefore more easy algorithms should be used. The study focused on implementation of movement algorithms and exchanging the knowledge between blind agents. In the solution described the gradient search path algorithm was implemented.

Keywords: Mobile agents · RSSI · 805.14 · XBee · ZigBee

1 Introduction

Wireless sensor networks (WSNs) are no longer exclusively the subject of research and tests. They are finding places in a wide range of real-life systems. Despite the diversity of current implementations of WSNs, most of them require same form of an estimation of sensors position or the relative distances among the sensors [1]. Also, the sensor deployment is the important feature of any WSNs installation [2]. It has the influence on the way the sensors collaborate, collect and process the information of the phenomenon in monitoring area. Both issues, localisation and deployment, are becoming even more complex in the case of using mobile sensors. A crucial issue in these systems is the way of the implementation of the cooperation among the group of so-called blind sensors. One of the promising approaches in this field is the use of the Received Signal Strength (RSS). A direct measure of the RSS is greatly affected by the environment and therefore recently, the WSN communication modules have implemented built-in feature able to provide an estimation of RSS by means of the dependable RSS Indicator (RSSI). For this reason, at present a large number of today's test and commercial solutions based on the processing of this indicator [3,4]. Capriglione et al. in [5] describe an approach for reliably investigating the RSSI features in localisation applications. The authors propose a suitable measurement procedure to identify the intrinsic variability of the RSSI measurements. The effects of external interferences on the RSSI values are also discussed.

© Springer International Publishing Switzerland 2016
P. Gaj et al. (Eds.): CN 2016, CCIS 608, pp. 370–383, 2016.
DOI: 10.1007/978-3-319-39207-3_32

Localising objects on the basis of pattern similarity on a RSSI map is discussed in [6]. The WSN is used to create the RSSI map. The paper introduces the energy-aware localisation method. It allows to acquire the actual RSSI map or broadcast a localisation signal, even if there is not sufficient information to perform the localisation by nearby Access Points.

There are also some related works on localisation or distance estimation in WSN using RSSI. The correlation between the distance and the RSSI values in wireless sensor networks is presented by Adewumi et al. in [7]. The authors introduce the RSSI model that estimates the distance between sensor nodes and makes the indoor and outdoor empirical measurements. The result shows that in an outdoor environment the error of the distance estimation is smaller compared to an indoor environment.

The other approach for indoor localisation system based on ZigBee and the measurements of the RSSI level is discussed in [8]. The proposed system for the localisation consists of static nodes and operates in two phases: calibration and localisation. The calibration phase is run any time a blind node needs to be located. In the localisation phase the central server processes all the information and calculates the blind node's position.

The anchor-less relative localisation algorithm for use in multi-robot teams was developed by Oliveira et al. and presented in [9]. The authors use the RSSI as an estimator of the inverse of the distance between any pair of the WSN. They assume that such estimates provide information of the nodes relative localisation. This information is still suitable for several coordination tasks in multi-robot teams. In their work robots perform the task of mine sweeping and move in the settled order.

The subject of our article is the specific type of the mobility in WSNs. It consists in tracing the root node, the so-called coordinator. The task of the agents is to follow the coordinator. Coordinator is the source of RSSI signal used by a group of blind sensors. These blind sensors collaborate by the exchanging their knowledge (the RSSI values). In practice, this means that only the coordinator must know (must be able to determine) the position in the monitoring environment. All blind sensors must only observe changes the position of a coordinator and act according to them. This greatly simplifies the monitoring and control of such a group of sensors. Moreover, the precision of blind sensors localisation is not crucial so the tracking algorithm can be relatively simple and efficient in terms of energy requirements. The algorithm exclusively based on the processing of RSSI values. The merit of this choice has been confirmed in numerous tests of mobile and static WSNs [10].

The mobile WSN, using the ideas presented above may have a whole range of practical applications. Two of them are worth mentioning as a very promising. The first is a situation that a given area is not monitored on a continuous basis. At the same time, however, there is a need for periodic inspection of the area. The inspection requires a group of sensors. due to e.g. the size of the area or the diversity of monitored parameters. In a such case, the coordinator can be programmed to periodically visit that area and the group of blind sensors will follow him. The second situation, in which the discussed sensor mobility can be useful, refers to the case in which area is monitored but the state of controlled processes (monitored phenomenon) causes the necessity of further or more complex investigation. Then, the group of additional sensors can be send there to enrich the current sensor deployment. At this point it is worth to mention that the analysis of the RSSI allows for the localisation as well as the distance assessment [11]. Thus, the blind sensors can accurately take the place unused in the existing deployments of the WSN.

The article presents the developed algorithm for tracking the coordinator by blind sensors. Its operation and characteristics have been tested in the environment using the ZigBee protocol. It bases on the specification IEEE 802.15.4 [12]. The ZigBee is a universal protocol for the construction of small and fast-to-deploy networks. Devices using this protocol are reliable, inexpensive and do not require a big capacity supplying source. Recently a lot of work has been devoted to studying the possibilities of using the ZigBee protocol for networking and device positioning and localisation. Bedford and Kennedy in [13] examine the possibility of applying ZigBee devices in underground navigation in mines and tunnels following an emergency incident. Low-cost wireless sensory network based on ZigBee can also be used as a system for monitoring an indoor localisation. The implementation of such a system in a home automation application, where the real-time knowledge of the location of personnel, assets and portable instruments is important, is presented in [14].

Our contribution, presented in the article consists of the proposition of the collaborative tracking algorithm for group of blind sensors (agents) based on the RSSI processing and multi-scenario tests in real-life environment based on the protocol ZigBee. The remainder of the paper is organised as follows: Sect. 2 describes the hardware and software configuration of the system consisting of the mobile agents, Sect. 3 describes the experiments and results. Conclusions and some further research ideas are presented in Sect. 4.

2   Mobile Agent System Configuration

The proposed system for localisation of radio wave sources consists of the following elements:

- mobile agents, which are fully autonomous mobile robots, for that purpose the Arexx AAR-04 Arduino robots were used [15],
- radio modules, XBee XB24-Z7WIT-004 modules from Digi, which are XBee 2 mW Series 2 wire antenna modules [16],
- coordinator platform, the Arduino Uno board [17] used for data collection, acts as coordinator node in XBee network.

Fitted on top of Arduino Uno board was the XBee shield with XBee radio. The Arduino board collects the data from XBee module, which is set as coordinator in the network (Fig. 1). At the same time, the coordinator also acts as the radio wave source, which should be reached by the mobile agents.

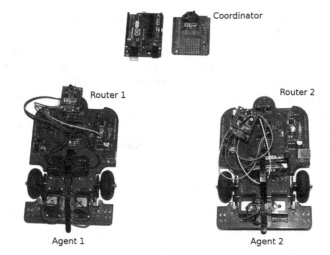

Fig. 1. The architecture of the multi-agent system

Other XBee radios were set up as routers and fitted on the autonomous mobile robots Arexx AAR-04 which is shown in Fig. 1. Mobile robots are based on the embedded platform Arduino Duemilanove. Their features are as follows [15]:

- AVR RISC Processor (ATmega328P) with 32 kB of memory for application code (i.e. sketches) and 2 kB RAM, as well as 1 kB of EEPROM memory for data storing, running at 16 MHz clock frequency,
- optical unit,
- LED indicators,
- two odometers,
- two DC-motors (3 Volt),
- free programmable I/O's,
- USB port for easy PC connection and programming.

The programming of the robots is done in an Arduino open-source development IDE that is based on a simplified C/C++ language [18]. The same software development environment was used for programming the Arduino UNO board for coordinator data storage purposes.

2.1 Hardware Configuration

Three Digi XBee series 2 modules were used to form a wireless sensory network [16]. In all XBee radios the firmware was set to the XB24-ZB family, which enables all the modems to work with ZigBee protocol stack. The Digi XBee modules operate in a 2.4 GHz band at 250 kbps baud rate. The modulation type is QPSK and emission designation is 2M32GXW. The firmware installed in the radio modules supports AT and API commands for their wireless operation. To obtain the RSSI value from the last collected packet the API mode should be set in coordinator and AT modes in routers.

The summary of the XBee modules configuration is shown in Table 1. The personal area network ID in all modules was the same and equal to 123. These settings allow to form the mesh wireless network, which is shown in Fig. 2. The XBee modules configured as routers where connected with adequate pins of AAR-04 mobile robots, i.e.: supply 3.3 V, GND, XBee's DOUT (TX) to pin RXD (Arduino's Software RX), XBee's DIN (RX) do pin TXD (Arduino's Software TX). The coordinator module was connected in a similar way to the Arduino UNO board i.e.:supply 3.3 V, GND, XBee DOUT (TX) to pin 2 (Arduino's Software RX), XBee's DIN (RX) to pin 3 (Arduino's Software TX).

Table 1. Summary of XBee modules settings

Module function/name	Function set	Firmware version
Coordinator/Coord	ZigBee coordinator API	21A7
Router/Agent 1	ZigBee router AT	23A7
Router/Agent 2	ZigBee router AT	23A7

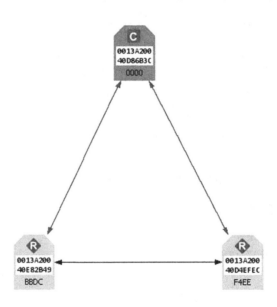

Fig. 2. XBee radio modules mesh connection where: C – coordinator node, R – router node, numbers on white fields – MAC address of the modules

2.2   Software Configuration

To achieve the assumed objectives the Arduino boards need to be programmed. For proper communication with XBee modules the XBee Arduino library

(GNU licence) prepared by Andrew Wrapp has been used [19]. The Arduino Uno board with the XBee radio was set up as coordinator. The software function of the coordinator node was recording the results which comes from the mobile agents. A flowchart for this node is shown in Fig. 3.

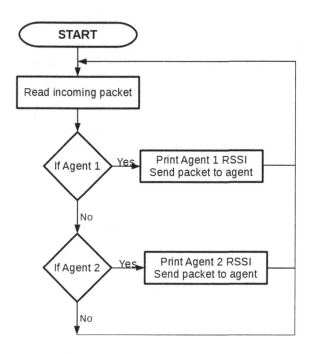

Fig. 3. Flowchart for coordinator node operation

To describe the agent movement additional explanation is needed. Let us assume that there is single radio wave source with reflected signals in the examined area. The intensity distribution of the source in considered area can be described by Eq. (1) [20]:

$$I(X) = \frac{I_m}{(X - X_0)^2} + \sum_{k=1}^{\overline{F}} \frac{I_m(k)}{(X - X_0(k))^2} \tag{1}$$

where $I(X)$ is the intensity value at point X, I_m is the intensity value at point X_0, X_0 is the point coordinate of the source, $\overline{F}$ is the average number of fluctuations distributed in the examined area, $I_m(k)$ is the maximum intensity value for the k fluctuation, $X_0(k)$ is the point coordinate of the k fluctuation, k is the number of fluctuation. If the fluctuations are small, then there is the following condition $I_m \gg I_m(k)$.

Let us consider the algorithm of finding the maximum value of field intensity by blind agent. Agent is moving with constant speed and given speed vector. More formal mathematical description of this algorithm is described below.

Step 1: the initial speed vector $V_i(0)$ and initial coordinates $X_i(0), Y_i(0)$ are given for each i agent; at this point the initial intensity is measured $I_i(0)$.

Step 2: each agent makes move in the direction of vector $V_i(0)$, calculates new coordinates $X_i(t), Y_i(t)$ in the subsequent time points $t = 1, 2, 3, \ldots$:

$$X_i(t) = X_i(t-1) + V_i(t-1). \tag{2}$$

The intensity of the field is determined in each point $I_i(t) = I(X_i(t), Y_i(t))$ and the series of $I_i(0), I_i(1), I_i(2), \ldots, I_i(t)$ is build as long as the Formula (3) is true:

$$I_i(0) < I_i(1) < I_i(2) < \ldots < I_i(t). \tag{3}$$

Step 3: the inequality (3) at some point $t = t+1$ is unfulfilled, than the vector $V_i(0)$ changes direction by an angle α and the Formula (2) is calculated; new intensity $I_i(t+2) = I(X_i(t+2), Y_i(t+2))$ is designated.

Step 4: if $I_i(t+2) < I_i(t+1)$ than Step 3 is repeated until the Formula (3) becomes true.

Now, let us consider the agent, which is in the local maximum generated by reflections as shown in Fig. 4. To escape from this the agent have to move from the centre of local maximum to reasonable distance in finite time strays.

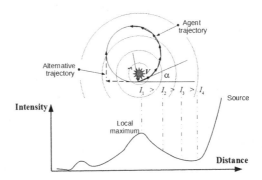

Fig. 4. An agent trajectory for escape from local maximum area, star – indicates the local maximum

Agent can leave the local maximum point, if inequality $I_i(t+2) < I_i(t+1)$ is true. However if agent follow the algorithm it would force him to commit a turn at the angle α. If the local maximum covers large area, the agent will make a circle and return to the point of departure (Fig. 4). Successful exit can be done, if the agent's trajectory will finish in the lower part of local maximum slope. Considering the geometry shown in Fig. 4 the following condition has to be met:

$$\left(2r \sim \frac{V}{\sin \alpha} \right) > \frac{D_{\mathrm{av}}}{2} \tag{4}$$

where: r is an agent trajectory radius, V is the vector of an agent speed, α is the angle at which an agent turns, D_{av} is an average diameter of filed intensity changes (fluctuations).

There are two ways to fulfil the Formula (4). The speed can be increased or the angle α can be reduced. In reality increasing the speed is not possible behind physical limits. Too small angle of turn is undesirable due to the bad algorithm convergence for small angles. It is worth noting, that the intensity of the field decreases with the square of the distance. Thus it leads to much simpler solution, i.e. if the trajectory line leads through the sides of square (described as alternative trajectory in Fig. 4), going to the next corner is the way to exit fluctuations.

Shown in Fig. 4 scheme can be also used as ability to find the source. If we assume that this scheme is not local but the maximum search, then it shows that at small angles agent requires a lot of steps to get closer to the source. The most acceptable for quick approximations are rotation angles close to 90°.

Programs for mobile agents are more complicated and their flowchart is shown in Fig. 5. During the experiment the agents were pointed in the same direction at the beginning, as shown in Fig. 1. It was necessary due to the fact that robots do not know their position in space. At the beginning the agents move forward and the initial movement vector and RSSI value are recorded. After this the XBee packet is sent to coordinator and the RSSI value is obtained. This value is compared with the previous one and if it is lower, the agent moves in the same direction. If the RSSI value is higher than the previous one then the robots make a turn left or right (alternately) and move forward. Due to the inertia of the platform and slight differences in the DC motors the turns are not 90° but 80°. If the RSSI value decreases the new movement vector is designated and the angle between initial and new vector is calculated and sent to the other agent together with the RSSI value and agent name.

An autonomous robot looks for information from the second one and if the received RSSI value is lower than the agent possesses, then it starts to move at an angle. If the RSSI value decreases, the agent keeps the movement, otherwise it turns until it starts to move in the direction mirrored by the vector along the line of the initial movement.

To obtain the RSSI value in Arduino environment the getAtCommandResponse function was used:

```
xbee.getResponse().getAtCommandResponse(atResp);
resp = atResp.getValue();
dBm = resp[0];
```

To reduce the fluctuations, the RSSI value was averaged. Additionally too big and too little values were discarded. To prevent processing such random bursts of RSSI the filter described beneath was implemented. If the current value does not exceed the bounds of plus or minus 20 % of the average value for a previous period, then the agent should respond to it. Otherwise it pauses for a moment and restarts the calculation. The signal value that is passed through the filter contributes to the calculation of the average signal over a long period of time. To calculate the average signal value the following formula was used [21]:

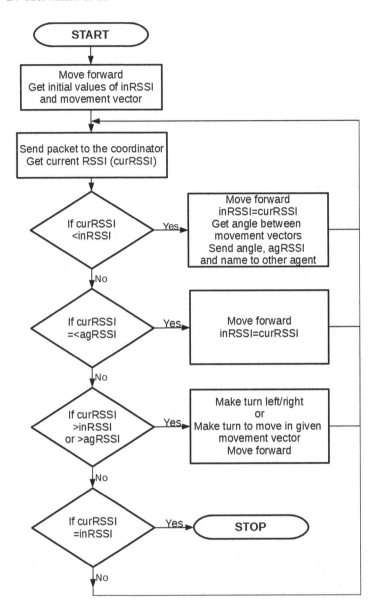

Fig. 5. Flowchart for agent node operation

$$I_{av}(t+1) = \frac{I_{av}(t) \cdot t + I(t)}{t+1} \tag{5}$$

where: t – subsequent time step number, $I_{av}(t)$ – average value for t time steps, $I(t)$ – current RSSI value.

3   Experimental Results

The idea of the experiment was to check the programmed algorithms by recording the movement of the mobile agents. The experiment relayed on three scenarios is shown in Fig. 6.

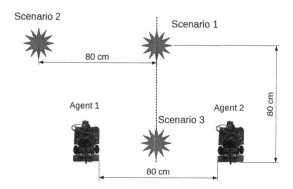

Fig. 6. Scenarios of the experiment

The radio wave source was placed in different positions:

- ahead of the agents on symmetry line,
- ahead of the agents, 80 cm to the left of symmetry line,
- between the agents on the symmetry line.

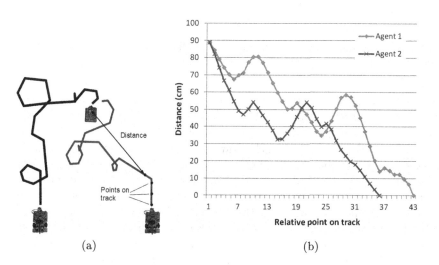

(a)   (b)

Fig. 7. Results of scenario 1: (a) recorded tracks, (b) distance from source versus point on track

The experiments were conducted in a room about the size of 6×3 m. The movements of the mobile agents were recorded with an Web camera (Logitech c270 HD Webcam), placed 2.3 m above the floor surface. Each mobile agent has the LED diode on it, which helps in the object tracking process. The camera was connected to a computer, on which the dedicated software for coloured object tracking was running. The application uses the Open Source Computer Vision Library (OpenCV). The algorithms placed in the library allow for: face detection and recognition, object identification, human actions in video classification, camera movement tracking, moving object tracking, extracting 3D models of objects, finding similar images from an image database, removing red eyes from images taken with a flash, following eye movements, recognising scenery and establishing markers to overlay it with augmented reality, etc. [22–26].

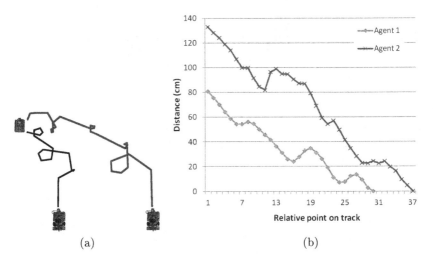

(a)   (b)

Fig. 8. Results of scenario 2: (a) recorded tracks, (b) distance from source versus point on track

The results of experiments were shown in Figs. 7, 8 and 9. The dedicated software, which uses the OpenCV library, was written in Java and allowed for detection and tracking objects in video based on object colour. This made it possible to record mobile agents' tracks drawing, during their move. The software detects the position of the blue LED on mobile robots (see Fig. 10). Each object with the same colour is marked as rectangle and the software records the position of the rectangle centre. Software records the rectangle ID and centre (X, Y) coordinates in the text file and after this further processing is possible.

It can be noticed that autonomous robots can find the source of radio waves using RSSI value in all scenarios. To compare the distance between source and agents the authors made an assumption that agent speed is constant. After this on recorded paths starting from the beginning at every 8 cm the points were added. Next the distance between the centre of the Xbee module and added

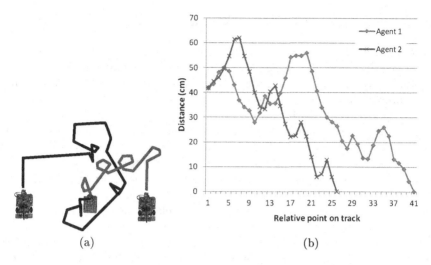

(a)   (b)

Fig. 9. Results of scenario 3: (a) recorded tracks, (b) distance from source versus point on track

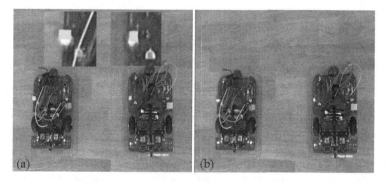

Fig. 10. LED position detection: (a) Java software marked rectangles – inset shows zoomed areas of LEDs, (b) recorded frame

relative points was measured as shown in Fig. 7(a). The results were presented in Figs. 7(b), 8(b) and 9(b). It can be noticed that agents exchange the knowledge on the RSSI value, however their decision of making right or left turn is not always correct. For example in Fig. 8(b) at point number 10 the Agent 1 turns correctly (RSSI value increases, distance from source decreases) but the Agent 2 turns badly. The Agent 2 recovered from this situation basing on the signal from Agent 1 and turning correctly, what can be noticed at point 13.

It can be also observed that the agents sometimes made small loops. It looks like agent position is in the local maximum area and an agent tries to escape from this as described in Sect. 2.2.

4   Conclusions and Future Work

The article presents a system of localisation of a radio wave source by means of mobile agents. The system works with Digi XBee S2 modules and its main search procedure is based on the RSSI value. The authors focused on the implementation of proper communication between robots using the IEEE 802.15.4 protocol stack and movement algorithms.

In all experiment scenarios autonomous agents can find the source of radio waves. However, it can be noticed that for the developed algorithm in the case of source localisation on the symmetry line the agents need more moves to reach the target. The developed algorithms and software enable to run them on the Arduino embedded platform, which is a good starting point for further expansion of the system to more complex structures of wireless sensory networks.

During the implementation of the system and the measurements, the authors noted that a mobile robot platform is a reliable source of errors due to its inertia and the small difference in the power of DC motors. In future work the use of mobile robots odometers is planned.

To improve the shape of the trajectory of robots the authors want to implement the Kalman filter. There is also ongoing research on the influence of the other radio wave sources on the robots' localisation capabilities.

References

1. Han, G., Xu, H., Duong, T.Q., Jiang, J., Hara, T.: Localization algorithms of wireless sensor networks: a survey. Telecommun. Syst. **52**(4), 2419–2436 (2013)
2. Xueqing, W., Shuxue, Z.: Comparison of several sensor deployments in wireless sensor networks. In: 2010 International Conference on E-Health Networking, Digital Ecosystems and Technologies (EDT), vol. 1, pp. 236–239, April 2010
3. Mistry, H.P., Mistry, N.H.: RSSI based localization scheme in wireless sensor networks: a survey. In: Fifth International Conference on Advanced Computing and Communication Technologies, pp. 647–652, February 2015
4. Kwiecień, A., Maćkowski, M., Kojder, M., Manczyk, M.: Reliability of bluetooth smart technology for indoor localization system. In: Gaj, P., Kwiecień, A., Stera, P. (eds.) CN 2015. CCIS, vol. 522, pp. 444–454. Springer, Heidelberg (2015)
5. Capriglione, D., Ferrigno, L., D'Orazio, E., Paciello, V., Pietrosanto, A.: Reliability analysis of RSSI for localization in small scale WSNs. In: IEEE International Instrumentation and Measurement Technology Conference, pp. 935–940 (2012)
6. Bernas, M., Płaczek, B.: Energy aware object localization in wireless sensor network based on Wi-Fi fingerprinting. In: Gaj, P., Kwiecień, A., Stera, P. (eds.) CN 2015. CCIS, vol. 522, pp. 33–42. Springer, Heidelberg (2015)
7. Adewumi, O.G., Djouani, K., Kurien, A.: RSSI based indoor and outdoor distance estimation for localization in WSN. In: IEEE International Conference on Industrial Technology (ICIT), pp. 1534–1539 (2013)
8. Larranaga, J., Muguira, L., Lopez-Garde, J., Vazquez, J.: An environment adaptive ZigBee-based indoor positioning algorithm. In: International Conference on Indoor Positioning and Indoor Navigation (IPIN), pp. 1–8 (2010)
9. Oliveira, L., Hongbin, H., Abrudan, T., Almeida, L.: RSSI-based relative localisation for mobile robots. Ad-Hoc Netw. **13**(B), 321–335 (2014)

10. Kouril, J.: Using RSSI parameter in tracking methods - Practical test. In: 2011 34th International Conference on Telecommunications and Signal Processing (TSP), pp. 248–251, August 2011
11. Chaochen, W., Yongxin, Z.: An improved localization framework based on maximum likelihood for blind WSN nodes. In: 2015 IEEE 17th International Conference on High Performance Computing and Communications (HPCC), pp. 1567–1572, August 2015
12. IEEE Standards Association: IEEE 802.15: Wireless Personal Area Networks (PANs). https://standards.ieee.org/about/get/802/802.15.html
13. Bedford, M.D., Kennedy, G.A.: Evaluation of ZigBee (IEEE 802.15.4) time-of-flight-based distance measurement for application in emergency underground navigation. IEEE Trans. Antennas Propag. **60**(5), 2502–2510 (2012)
14. Elango, S., Mathivanan, N., Gupta, P.K.: RSSI based indoor position monitoring using WSN in a home automation application. Acta Electrotechnica et Informatica **11**(4), 14–19 (2011)
15. Arexx: Arexx AAR-04 robot specification, December 2015. http://www.conrad.com/ce/en/product/191694/Arexx-AAR-04-Programmable-Arduino-Robot
16. Digi International: XBee and XBee-PRO ZB Modules – Datasheet (2011). https://cdn.sparkfun.com/datasheets/Wireless/Zigbee/ds_xbeezbmodules.pdf
17. Arduino: Arduino Uno board, September 2014. https://www.arduino.cc/en/Main/ArduinoBoardUno
18. Arduino: Arduino Software (IDE), September 2014. https://www.arduino.cc/en/Guide/Environment
19. Andrew Wrapp: Arduino Xbee Library, November 2015. https://github.com/andrewrapp/xbee-arduino
20. MacDonald, D.K.C.: Noise and Fluctuations: An Introduction. National Research Council, Ottawa (1962)
21. Sutton, S., Barto, G.: Reinforcement Learning: An Introduction, 2nd edn. The MIT Press, London (2012)
22. Itseez: OpenCV Library, January 2016. http://opencv.org/about.html
23. Kopniak, P.: The use of multiple cameras for motion capture. Przeglad Elektrotechniczny **90**(4), 173–176 (2014)
24. Ozturk, T., Albayrak, Y., Polat, O.: Object tracking by PI control and image processing on embedded systems. In: 23th Signal Processing and Communications Applications Conference (SIU), pp. 2178–2181 (2015)
25. Shopa, P., Sumitha, N., Patra, P.S.K.: Traffic sign detection and recognition using OpenCV. In: International Conference on Information Communication and Embedded Systems (ICICES), pp. 1–6 (2014)
26. Kuehlkamp, A., Franco, C.R., Comunello, E.: An evaluation of iris detection methods for real-time video processing with low-cost equipment. In: Czachorski, T., Gelenbe, E., Lent, R. (eds.) Information Sciences and Systems, pp. 105–113. Springer, Heidelberg (2014)

Evolutionary Scanner of Web Application Vulnerabilities

Dariusz Pałka[1]([✉]), Marek Zachara[1], and Krzysztof Wójcik[2]

[1] AGH University of Science and Technology,
30 Mickiewicza Av., 30-059 Krakow, Poland
{dpalka,mzachara}@agh.edu.pl
[2] Cracow University of Technology, Jana Pawła II 37 Av., 31-864 Krakow, Poland
krzysztof.wojcik@mech.pk.edu.pl

Abstract. With every passing year, there are more and more websites, which often process sensitive and/or valuable information. Due to models like Continuous Development, manual testing and code review are reduced to minimum, with new features implemented and deployed even on the same day. This calls for development of new automated testing methods, especially the ones that will allow for identification of potential security issues. In this article such a new method, which is based on automated web pages comparisons, clustering and grammatical evolution is proposed. This method allows for automated testing of a website and can identify outstanding (unusual) web pages. Such pages can then be further investigated by checking if they are legitimate, contain some unused modules or potential threats to application security. The proposed method can identify such anomalous pages within the set of interlinked web pages, but can also find web pages that are not linked to any other web page on the server by utilizing genetic-based generation of URLs.

Keywords: Web security · Vulnerability detection · Evolutionary algorithms

1 Introduction

Maintaining a secure and quality website is an important task for the majority of organizations. However, many of them fail to do so, and some spectacular failures in this area sometimes make the headlines of daily news. Symantec claim that, while running 1 400 vulnerability scans per day, they have found approximately 76 % of the scanned websites with at least one unpatched vulnerability, and 20 % of the servers having critical vulnerabilities [1]. In one of their previous reports they also stated that in the single month of May in 2012 the LizaMoon toolkit was responsible for at least a million successful SQL Injections attacks, and that approximately 63 % of websites used to distribute malware were actually legitimate websites compromised by attackers. In another report [2], WhiteHat Security stated that 86 % of the web applications they tested had at least one

© Springer International Publishing Switzerland 2016
P. Gaj et al. (Eds.): CN 2016, CCIS 608, pp. 384–396, 2016.
DOI: 10.1007/978-3-319-39207-3_33

serious vulnerability, with an average of 56 vulnerabilities per web application. A thorough overview of recent vulnerabilities can also be found in [3].

In this paper, a method is proposed to identify unusual outlier pages of a website. Such web pages are often left-overs from installation of some software or experiments with site development. Since they are generally unmaintained, they can serve as a back door to the website or provide means for staging a successful attack against it; e.g. by providing the attacker with valuable information.

The important feature of such outlier pages is that they are often not linked to the main content of the website, therefore, they cannot be accessed by typical methods of following the links in the pages. This requires the attacker to either know or guess the URL, or to provide certain parameters in the requests sent to the server. Such parameters can be supplied either as part of the URL:

protocol://website:port/path?p_1_name=p_1_val&p_2_name=p_2_val

where:

p_n_name – the name of n-th parameter,

p_n_val – the value of n-th parameter,

or as part of the request's body if POST method is used instead of GET.

Modification of the parameters sent to a website is a widely used attack technique known as *Parameter Tampering*, and it may lead to various exploits including privilege elevation or circumventing access control; there are a number of ready-made tools serving this purpose; e.g. the "Tamper Data" browser plugin.

Finding stray web pages at unknown URLs is a very difficult task, since checking all possible URLs is virtually impossible due to the size of the search space. However, evolutionary (genetic) algorithms are known for being able to cope reasonably well with such tasks, and they have been employed for vulnerability scanning, e.g. identifying buffer overflows [4], identifying possible XSS vulnerabilities [5] or other general vulnerability assessments [6,7].

Nowadays, however, so called "Semantic (or Friendly) URLs" are often adopted as a replacement for typical parameter-driven website navigation due to the following factors:

- they are favoured by Search Engines, so they are an important element of SEO (search engine optimization),
- they are more explicit for users and thus increase web service usability,
- they usually do not include any deployment details (such as server-side file types: .php, .asp, and physical paths inside the web application structure), they can remain unchanged even when the underlying web application is upgraded or replaced, ensuring that individual web resources remain consistently at the same URL,
- they do not directly expose parameters in a request (usually there is no visible connection between the name and the value of the parameter), which reduces the temptation of some users to manipulate these parameters.

The comparison of a traditional (non-semantic) form of an URL address and a friendly (semantic) form of this address is presented below:

http://example.com/products?category=kitchen&id=72 (non-semantic),
http://example.com/products/kitchen/72 (semantic URL).

As can be seen from this example, using the semantic URL makes the parameters indistinguishable from the request path. Scanning and analysing such application requires a method that manipulates the whole URL, like the Vulnerability Scanner described in this article below.

2 A Vulnerability Scanner

In a nutshell, the idea of an automatic vulnerability scanner is as follows: a scanner generates requests to a tested web application and checks if the response of a web application is normal or anomalous in comparison to typical application responses. If the response is anomalous, i.e. the web application does not behave in a typical manner, it can indicate some vulnerability in the application. The proposed method relies on evolutionary approach to generate new URLs and its main contribution is the transformation of the responses into a metric space that represents relative distance between the web pages retrieved from the server.

This study focuses only on single GET requests to a web application, but the same idea can also be used in case of POST requests. Due to the nature of the evolutionary scanning, the requests are generated semi-randomly and do not follow any fixed path. Some support for stateful scanning can be added by handling and forwarding a *Session ID* provided by the server, however, this is a simple programming issue, and it does not affect the proposed method.

2.1 Typical Application Responses

In order to detect anomalous web application behaviour, which can be determined by receiving anomalous responses, it is first necessary to identify the kind of responses that are typical. In this article it is assumed that only responses reachable by typical navigation over the website (i.e. by following existing links) are considered typical. So, the first step in detecting abnormal web application responses is collecting responses obtained while traversing web pages indicated by available hyperlinks. During this stage, a preliminary Standard Response Landscape (SRL) is created. The scanner uses a web crawler mechanism to collect all available links and responses sent by a web application obtained while visiting these links. For the purpose of this article, only responses with text content (i.e. html or xhtml) are processed, because binary file comparisons are not feasible at the moment with the employed techniques.

SRL represents the similarity between particular responses (web pages); it is created by computing the distance between each pair of web pages. As a result, the set of points in the metric space is obtained.

To compute a distance (i.e. similarity) between pairs of web pages, the algorithm relies on comparing their HTML code using typical text-related methods, such as Levenshtein distance [8]. Computing Levenshtein distance for a raw HTML code is computationally expensive – computation of edit distance for two strings of length n and m, where $n >= m$, using a classical algorithm has a complexity of $O(n \cdot m)$, and the best known algorithm for constant-size alphabet is only slightly better $O(n \cdot \max(1, \frac{m}{\log(n)}))$ [9]. Therefore, a hash representation of HTML code (webpage) instead of a raw (full) HTML code seems a better choice. As shown in the authors' previous work [10], this is a valid approach and the algorithms evaluated there are used in the presented method.

2.2 Projection from a Metric Space to the 2D Cartesian Coordinate System

Because the SRL is created in a metric space, it needs to be projected into the 2D or 3D Cartesian coordinate system in order to provide human-understandable visualisation. This transformation should preserve distances between objects; i.e. the distance in the Cartesian coordinate system should be equal or possibly very close to the distance between the same objects in a metric space. This approach is widely used in a technique of Multidimensional scaling (MDS) [11,12].

In the presented method, a spring model, similar to the one presented in [13], is used for projecting the points from a metric space into the 2D or 3D Cartesian coordinate system:

- Page hashes (the points in a metric space) are treated as mass points in the Cartesian coordinate system.
- Mass points are mutually connected by springs.
- The length of a spring between mass points p_k and p_n is equal to the distance between corresponding page hashes h_k and h_n in a metric space. The spring length between points p_k and p_n is denoted $l_{k,n}$.
- Restoring force value between a pair of points p_k and p_n is $F = -k \cdot (\|p_k - p_n\| - l_{k,n})$, where: k is the spring constant and $\|p_k - p_n\|$ is the current Euclidean distance between points p_k and p_n. The spring constant influences only the scale, so its value can be assumed as 1.
- The solution is obtained using a simulation approach (with discrete time) for a dynamic system described above, and the positions of mass points are found through an iterative method.
- Damping is introduced in order to find a static solution (a system which minimizes energy) without oscillations. The energy E of a dynamic system is:

$$E = \sum_{i=1}^{n-1} \sum_{j=i+1}^{n} \frac{1}{2} \cdot k \cdot (\|p_k - p_n\| - l_{k,n})^2. \tag{1}$$

As a result, the projection which minimizes the total difference between the distance between points in the Cartesian coordinate system and the distance in a metric space is obtained.

Standard Response Landscapes for two websites are presented below. Each contains 5000 unique points (responses) obtained while crawling these websites starting from their home page. Site A is one of the largest Polish news portal, Onet (home page: http://www.onet.pl), with about 14 million web pages indexed by Google, and the average page size of a few hundred kilobytes. Site B is the website of AGH University (home page: http://www.agh.edu.pl), consisting of

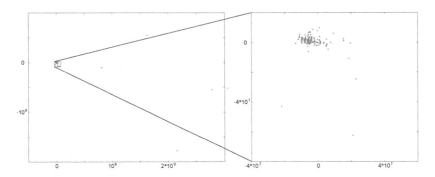

Fig. 1. 2D projection of SRL for Site A (Onet). The image on the left shows the full landscape, with the marked rectangular fragment zoomed in the image on the right.

Fig. 2. Examples of typical web pages for Site A from the primary agglomeration area

about 350 thousand indexed pages, and smaller page sizes – typically around 50 KB.

Figure 1 presents the projection of SRL for Site A into the 2D Cartesian coordinate system.

As can be seen from the above figures, the landscape in full scale for Site A contains a number of points (representing standard responses) relatively far from the main agglomeration (concentration). Two examples of web pages from the agglomeration area are presented in Fig. 2.

Some examples of web pages that are far from primary agglomeration are presented in Fig. 3. As can be noticed, these web pages differ considerably from typical pages presented in Fig. 2. The fact that they are linked and accessible from other web pages of the service is probably a webmaster's mistake.

For comparison, Fig. 4 shows a 2D projection of SRL for the other site (Site B). Similarily, typical web pages of this site are presented in Fig. 5 and examples of pages far from the primary agglomeration area are presented in Fig. 6.

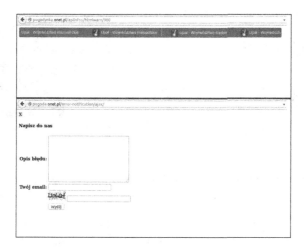

Fig. 3. Example of two web pages for Site A far from the agglomeration area

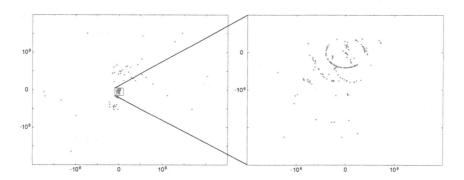

Fig. 4. 2D projection of SRL for Site B (AGH). The image on the left shows the full landscape, with the marked rectangular fragment zoomed in the image on the right.

Fig. 5. Examples of typical web pages for Site B from the agglomeration area

Fig. 6. Examples of web pages for Site B far from the agglomeration area

The analysis of Standard Response Landscapes can provide valuable information about the web service tested. It allows for finding unusual or abnormal web pages (the ones which are far from the agglomeration areas). The existence of direct links to such anomalous web pages can be a sign of mistakes or errors in the web service design.

This is analogous to one of the applications of MDS as a method of making these data accessible to visual inspection and exploration, which was presented in [11].

3   Identification of Anomalous Web Application Responses

One of the methods used in an attempt of finding website's vulnerabilities is to look for requests that generate anomalous application responses. As discussed above, responses are classified as typical or unusual on the basis of their deviation from the SRL. The assumption is as follows: if the response obtained in the process of searching for vulnerabilities differs from any response obtained during the process of web crawling, it is treated as anomalous. Such responses indicate a potential deviation from standard behaviour of the web application and may indicate a vulnerability of the application. Therefore, the process of searching for vulnerabilities of an application is based on looking for requests (in this study GET requests), which results in obtaining a web page far from the SRL for this application.

3.1   The Measure of Deviation of the Response from Standard Response Landscape

The measure of deviation of a point p (which represents the web application response) from SRL may be defined as a minimal distance between the point and SRL set

$$dist(p, SRL) = \min_{k \in SRL} dist(p, k). \tag{2}$$

Calculating the deviation from SRL requires calculating the distance between a given point and all the points of SRL. Because the number of distinct points (responses) in SRL is typically large and calculating the dissimilarity (distance) of pages (for example, Levenshtein distance) is a computationally time-consuming process, some optimizations need to be considered.

One of such optimizations can be based on assuming an arbitrary threshold for the distance if the distance between a given point and any point in SRL is lower that the assumed threshold, this given point is considered to belong to the SRL. This approach allows for limiting the number of distances calculated between a given point and points in the SRL. It has, however, certain disadvantages and one of them is related to the choice of a suitable threshold.

The threshold has to be dependent on particular SRL and can be assumed, for example, as an average distance between points in the SRL or a minimal distance between points in the Standard Response Landscape (although it may not be suitable in every case). Generally speaking, too small a threshold will result in the need to calculate a greater number of distances between the given point and points in SRL, whereas too big a threshold can lead to situations in which some anomalous responses (possibly vulnerabilities) will be treated as typical ones (i.e. belonging to the SLR).

Another disadvantage stems from the fact that even if a suitable threshold is chosen, it may be necessary to calculate all distances between a given point and all points in the SLR. It might happen if a given point does not belong to the SLR. However, this may not be a significant problem, as points which do not belong to the SLR are not often found.

Another optimization can be based on clustering the SRL set using the k-medoids algorithm and then calculating the distance between a given point and the medoid of each cluster. If the distance between this given point and the medoid is smaller than the radius of the cluster, the point is considered to belong to the SRL. Otherwise, the distance between this point and each member (point) of the nearest cluster should be calculated. The nearest cluster is the one for which the distance calculated as

$$d(p, C) = d(p, m_c) - r_c \tag{3}$$

is minimal, where:

$d(p, C)$ – the distance between a given point p and cluster C,
$d(p, m_c)$ – the distance between a given point p and the medoid of cluster C,
r_c – the radius of cluster C.

The minimal distance from any given point to the point from the nearest cluster can be considered the distance between this given point and the SRL.

However, this approach has several disadvantages, as the efficiency of clustering is very sensitive to a proper choice of the number of clusters (k parameter). A wrong selection of k leads to unnatural clustering and the created medoids do not represent the actual structure of the SRL. There is no effective automatic way of obtaining the k parameters (a desired number of clusters).

Another problem is associated with the situation when the point is not classi-fied as belonging to the SRL: a necessary number of calculations is decreased by k times (where k is the number of clusters). A proper number of clusters depends on the analysed web application, but is relatively small (between about a dozen and several dozens). One more problem occurs if there are several clusters in a similar distance as the closest one (or the distance to several closest clusters is comparable), and in such case the distance to all points from these clusters must be calculated. Another drawback results from the fact that the minimal distance to a point from the nearest cluster calculated this way does not guarantee to be the minimal distance to the points from the SRL.

3.2   The Process of Searching for Requests Leading to Anomalous Responses

The key part of a scanner searching for potential vulnerabilities of web appli-cations is the process generating GET requests. In the proposed method, this process is based on a Grammatical Evolution (GE) technique [14,15], which stems from the idea of genetic programming (GP) [16]. Two main differences between GE and standard GP are:

- In GP the genotype is represented as a tree structure, and genetic operators are applied directly to this tree structure. However, in case of GE the genotype is a linear string and genetic operators are applied to this linear representation (like in classical genetic algorithms).
- In GE the grammar is used to control the evolutionary process, for example, the grammar provides declarative search space restrictions and it allows for defining homologous operations.

The vulnerability scanner generates (using an evolutionary approach based on a GE technique) strings representing requests to the examined web applica-tion. The context free grammar describing valid GET requests is used to restrict the search space. The advantages of using a grammar are: the simplicity and flexibility in defining valid individuals created by an evolutionary process, in this case the URLs representing GET requests. Moreover, the same grammar allows for incorporation of the existing URLs into the pool of individuals used in the evolutionary process, as it is described below.

The evolutionary algorithm used by the vulnerability scanner can be described as follows:

1. The initial generation of individuals (URLs representing GET requests) is created.
2. Created requests are sent to the analysed web application and the responses of the web application for each request are registered.
3. The distance between each application response and the Standard Response Landscape is calculated.
4. On the basis of the calculated distance, the value of the fitness function for each request (individual) is calculated.

5. On the basis of the fitness values, using tournament selection, candidates (i.e. GET requests) for the next generation are selected.
6. The next generation of URLs is created from selected candidates (from Step 5) using genetic operations: crossover and mutation. The genetic operations are driven by the URL grammar (i.e. created URLs assume the form dictated by the grammar).
7. The process loops to Step 2.

As can be seen from the above considerations, this algorithm has no termination criteria and runs until a user requests its termination explicitly. This is due to the fact that the searching space (the space of all GET requests) is infinite, and there can always exist additional requests, which could lead to anomalous application responses, which in turn can indicate additional vulnerabilities.

3.3 Initial Generations

In the GP or GE the first generation of individuals is usually randomly created, but this approach is quite ineffective for the purpose of the presented method due to the fact that most randomly generated URLs (GET requests) will result in the web server returning the HTTP error 404 "not found".

In order to avoid this difficulty, the first generation is created on the basis of the URLs acquired in the initial phase, i.e. while creating the Standard Response Landscape. This guarantees that (for virtually all individuals of the first generation) the web server will return a correct response "200 OK".

The process of finding potential web application vulnerabilities starts, therefore, with legitimacy requests, which yield responses belonging to the SRL. In order to introduce new URLs to the pool of individuals, existing URL strings are parsed using the URL grammar, and, as a result, the sequence of grammar production numbers is obtained. This production number sequence is the chromosome for an individual representing a given URL. The grammar of this evolutionary approach has, therefore, two objectives:

- the first one is typical for the GE processes: it drives the evolutionary process and defines restrictions for the search space, and,
- the second one allows for introducing existing URLs (for example, the ones obtained while creating the SLR) into the pool of individuals. It is a part of the parsing mechanism.

3.4 Fitness Function

The fitness function for the evaluation of individuals is defined as follows:

$$fitness = \begin{cases} dist(p, SRL) & \text{if response} = 200 \text{ (OK)} \\ -1 & \text{otherwise} \end{cases} \tag{4}$$

where $dist(p, SLR)$ is the distance between the web application response p and the nearest point from the SRL.

The objective of the evolutionary algorithm is to find points in the search space (URLs) that maximise the values of the fitness function; i.e. to find the requests for which the responses of the web application are most distant from the SLR. Such responses denote anomalous pages, which can indicate a possible vulnerability in the examined application.

3.5   The Tabu List

One of the problems occurring while generating solutions in the search space using GE technique (and, generally, techniques based on genetic algorithms) is the creation of possible solutions in the next generation which are identical to the ones already checked. This may considerably decrease the efficiency of the search process.

It has even more significant impact on the discussed method, as evaluation of each newly constructed URL requires sending a request to the server and waiting for the response. Since this is the real bottleneck of the whole process, reducing the number of duplicate requests becomes a priority.

To prevent checking the same URL several times, a tabu list is used. The tabu list is a key concept used in the Tabu search heuristic method [17]. In order to limit the memory usage of the tabu list, a circular buffer is used. The tabu list complements the fitness function and together they govern the direction of the exploration of the search space.

3.6   Creating a Follow-Up Generation

After calculating the values of the fitness function for all individuals in a given generation, a new generation is created in the following way:

- $k1$ individuals for the new generation are created on the basis of the present generation. Tournament selection is used to select the best fitting individuals, which then undergo a crossover or a mutation (both with certain probability). Newly created individuals are compared against the tabu list, and – if they are not on the list – are added to the new generation and to the tabu list. However, if a given individual is already on the tabu list, it is rejected and the process of tournament selection is repeated.
- $k2$ individuals are created on the basis of individuals included in the SRL. Individuals included in the SRL are randomly selected, and then they also undergo a possible crossover or a mutation. Newly created individuals are also compared against the tabu list with the possible outcome described already in the previous point.

3.7   Genetic Operators

Two genetic operators are used in the evolution process: a crossover operator and a mutation operator; both are performed typically, the same way as in GE. The crossover operator operates on two individuals and is one-pointed, i.e. within each of the two individuals a random crossover point is selected and particular sections of the chromosome for the individuals are swapped.

The mutation operator operates on one individual. The number of mutations applied to a given individual is random within a predefined range. Next, a number of genes in the chromosome are selected, whose values are replaced by other randomly selected values.

4 Results and Future Work

Figure 7 illustrates several anomalous pages identified within one of the investigated domains (AGH). As can be seen in this figure, these web pages are not meant for public access and some contain information that could assist a malicious person in staging an attack, i.e. the output of *phpinfo()*. These are just a few examples of pages that were found even though there was no direct link leading to them from any of other web page within the domain.

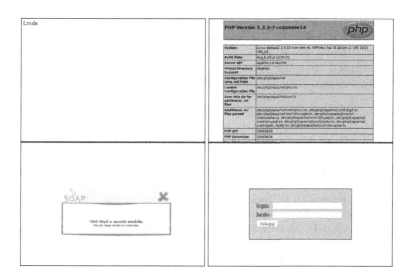

Fig. 7. Examples of anomalous web pages identified by the discussed method

This might serve as a proof that the proposed method has certain important advantages that are not available in standard vulnerability scanning tools. Unfortunately, as it was mentioned earlier, there is no clear indication referring to when the search of the possibly infinite space should be stopped, so it is up to the operator to decide when to terminate it. Further research is planned on assessing the probability of finding new anomalous pages depending on the current state of search, but this will depend on the availability of large enough statistical corpus. Another area that certainly may benefit from further research is the process of creating new generations, which could lead to more effective coverage of the search space.

References

1. Symantec: Internet Security Threat Report (2015). http://www.symantec.com/security_response/publications/threatreport.jsp
2. WhiteHat: Website Security Statistics Report (2013). http://info.whitehatsec.com/2013-website-security-report.html
3. van Goethem, T., Chen, P., Nikiforakis, N., Desmet, L., Joosen, W.: Large-Scale Security Analysis of the Web: Challenges and Findings. In: Holz, T., Ioannidis, S. (eds.) Trust 2014. LNCS, vol. 8564, pp. 110–126. Springer, Heidelberg (2014)
4. Rawat, S., Mounier, L.: An evolutionary computing approach for hunting buffer overflow vulnerabilities: a case of aiming in dim light. In: 2011 Seventh European Conference on Computer Network Defense, pp. 37–45 (2010)
5. Duchene, F., Rawat, S., Richier, J.L., Groz, R.: Kameleonfuzz: evolutionary fuzzing for black-box xss detection. In: Proceedings of the 4th ACM Conference on Data and Application Security and Privacy, CODASPY 2014, NY, USA, pp. 37–48. ACM, New York (2014)
6. Budynek, J., Bonabeau, E., Shargel, B.: Evolving computer intrusion scripts for vulnerability assessment and log analysis. In: Proceedings of the 7th Annual Conference on Genetic and Evolutionary Computation, GECCO 2005, NY, USA, pp. 1905–1912 (2005).http://doi.acm.org/10.1145/1068009.1068331
7. Dozier, G., Brown, D., Hou, H., Hurley, J.: Vulnerability analysis of immunity-based intrusion detection systems using genetic and evolutionary hackers. Appl. Soft Comput. **7**(2), 547–553 (2007). http://www.sciencedirect.com/science/article/pii/S1568494606000512
8. Levenshtein, V.: Binary Codes Capable of Correcting Deletions and Insertions and Reversals. Soviet Physics Doklady **10**(8), 707–710 (1966)
9. Andoni, A., Onak, K.: Approximating edit distance in near-linear time. SIAM J. Comput. **41**(6), 1635–1648 (2012)
10. Zachara, M., Pałka, D.: Comparison of text-similarity metrics for the purpose of identifying identical web pages during automated web application testing. In: Grzech, A., Borzemski, L., Świątek, J., Wilimowska, Z. (eds.) ISAT 2015, Part II. AISC, vol. 430, pp. 25–35. Springer, Heidelberg (2016)
11. Borg, I., Groenen, P.: Modern Multidimensional Scaling: Theory and Applications. Springer, New York (2005)
12. Torgerson, W.S.: Multidimensional scaling: I. theory and method. Psychometrika **17**(4), 401–419 (1952)
13. Kamada, T., Kawai, S.: An algorithm for drawing general undirected graphs. Inf. Process. Lett. **31**(1), 7–15 (1989)
14. O'Neill, M., Ryan, C.: Grammatical evolution. IEEE Trans. Evol. Comput. **5**(4), 349–358 (2001)
15. O'Neill, M., Ryan, C.: Grammatical Evolution: Evolutionary Automatic Programming in a Arbitrary Language. Genetic programming, vol. 4. Kluwer Academic Publishers (2003)
16. Koza, J.R.: Genetic Programming: On the Programming of Computers by Means of Natural Selection. MIT Press, Cambridge (1992)
17. Glover, F.: Future paths for integer programming and links to artificial intelligence. Comput. Oper. Res. **13**(5), 533–549 (1986)

Data Transformation Using Custom Class Generator as Part of Systems Integration in Manufacturing Company

Jacek Pękala[(⊠)]

Institute of Production Engineering, Cracow University of Technology,
Warszawska 24, 31-155 Kraków, Poland
pekala@m6.mech.pk.edu.pl
http://www.pk.edu.pl

Abstract. This paper attempts to bring closer one of the issues of data exchange between different subjects of IT structure in production enterprises which is the essence of their integration. The paper focuses primarily on transformation of data swapped between systems in open B2MML format. The article presents a design concept of solution for data transformation using custom class generator. Its operability was tested on sample data presented in B2MML format.

Keywords: Transformation · B2MML · Class generator · XSD · Schema definition

1 Introduction

Computer systems have become an integral part of any kind of human activity – from the entertainment through learning and processing information to assist in making important decisions. It is no different in the case of manufacturing operations and its various aspects. The development of information technologies allow to create new solutions in many different areas or improve existing ones [1]. In contemporary information structures, whose architecture is often created based on distributed solutions, using appropriate communication tools is essential [2]. This applies to manufacturing companies, which production activities take place often on large manufacturing plants and are subjected to the processes of automation and computerization. This results in a large amount of demanding top-down management or control information entities which are generating greater amounts of (often redundant) data [3]. Due to the specific operation of enterprise information systems, data exchange between them must be free of randomness and the method of communication must be determined in advance. In the mechanisms for exchange of information, in addition to the transmission, equally important is the role of transformation, which is necessary because of differences even in the way they are stored [4]. It is particularly important for the smooth flow of information between the two classes of

© Springer International Publishing Switzerland 2016
P. Gaj et al. (Eds.): CN 2016, CCIS 608, pp. 397–409, 2016.
DOI: 10.1007/978-3-319-39207-3_34

systems occurring in the data structure of production enterprise: Manufacturing Execution Systems (MES) and Enterprise Resource Planning (ERP). In the flow of data between the systems, Business-To-Manufacturing Markup Language (B2MML) is used. This language specification has been built on the basis of the ISA-95 standard in compliance with the specification of Extensible Markup Language (XML) [5]. B2MML has been described broadly in Sect. 2.2. In papers [6,7] an Extensible Stylesheet Language (XSL) was used as a tool for data conversion. XSL seemed to be natural candidate for this task since it's also based on XML specification [8]. Despite having the right tools, data conversion is not flawless. This process causes the loss of information due to incorrectly defined rules of transition, or their lack. The purpose of this paper is to present the concept of a solution that deals with the problem by using custom class generator.

2   Data Flow Between Production Enterprise Systems

2.1   Information Structure in Production Enterprise

In all manufacturing companies, data acquisition and their management on production layer should serve to raise the efficiency and reliability of production. The acquisition is a key element in the decision making process and at every level of management in the company – from the operational services and maintenance, through the departments of engineering, to the administrative units [9].

These levels also have their reference in the hierarchical information structure of the company [10]. At the lowest level there are sensors, actuators and a variety of industrial automation devices having a direct connection with the company's shop floor layer. Level above is domain of the Supervisory Control And Data Acquisition (SCADA), Computer Numerical Control (CNC), Programmable Logic Controller (PLC) and other industrial control systems. These systems operate in real time and despite of being responsible for data collection they have also control of machines and production lines components in their agendas. In execution of production and high-level management areas in an industrial enterprise, there are two classes of systems used – Manufacturing Execution Systems (MES) and Enterprise Resource Planning (ERP), respectively. MES systems are responsible for the effective realization of the production process on the basis of accurate and up-to-date production data from lower-level systems [10]. ERP domain is the management of the company which includes supply chain management, human resources, finance, etc. The above-mentioned systems are mutually complementary solutions and their possible interoperability adds value to the company. Their co-operation and the associated mutual communication is as important as any functionality that different systems provide, regardless of the presence of any other systems instances in the company's information structure. An important element of a modern MES system is the ability to seamlessly integrate with automation systems. To do that, common and open communication standards such as ISA-88 and OPC are used [11]. In the case of exchange of information from MES system with the parent ERP

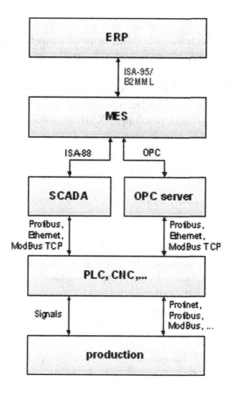

Fig. 1. IT infrastructure model in an industrial enterprise

system, ISA-95 standard and based on the XML-functional implementation – Business To Manufacturing Markup Language (B2MML) is used. The exchange of information between those two class of systems (MES and ERP) is as important to the company as the data flow between the other levels. It is the subject of discussion of this paper, in particular, the analysis of the loss arising from the information. Figure 1 shows the hierarchical model of information systems in the industrial structure including communication standards.

2.2 ISA-95 and Business-To-Manufacturing Markup Language (B2MML)

ANSI/ISA-95 Enterprise-Control System Integration is an international standard approved by a group of manufacturers, systems' providers and their contributors. It is described in several documents as broadly understood systems' integration methodology, which consists of five parts. If we assume that the ISA-95 standard presents a theory on the integration of management (ERP) and production systems (MES), the B2MML language should be recognized as the executive arm of that standard. In [12] the author gives a simple definition B2MML language as XML-based implementation of the ISA-95 standard.

B2MML contains a set of XML schemas defined in the Extensible Schema Definition (XSD) language, which include definitions of object-oriented models drawn from the content of the ISA-95 standard. The main goal for the language is to mediate in the process of integration by conversion of data and the structure of messages exchanged between the systems. The combination of XML and ISA-95 provides many tangible benefits in the process of information transfer. Besides its openness, simplicity and independence, XML schemas are easily adaptable to the needs of data exchange, which requires preserving the uniformity and consistency of the data structure. A significant advantage of XML and B2MML is the legibility of the information resulting from a clear structure [13]. However, it's important to notice that the B2MML is not standard but an interpretation of standard and that the small details can be understood differently by different system vendors and users.

B2MML language is used for the transmission and transformation of messages between integrated systems and middleware. When the message is sent by the one system it goes first to one of the layer of middleware that translates metadata and structure of messages into B2MML document (schema conversion), and then transforms it repeatedly into the language appropriate for the target system. Not only schema is transformed but also semantic part, if necessary (data conversion). Only after this actions and obtaining a guarantee that it will be received by a proper system, the document can be sent to its receiver (intelligent routing) [14]. Figure 2 shows a simplified diagram of the flow of information with inclusion of message transformation to B2MML intermediate format.

3   Style-Sheet-Based Data Transformation

Extensible Stylesheet Language (XSL) in an XML-based transformation language of XML documents. It allows to translate documents from one format such as XML to any other format compatible with XML syntax, including the above-mentioned B2MML. The input in the transformation process is the source XML document and XSL stylesheet that specifies an XML document transformation. A stylesheet is made up of templates. Each template describes how to

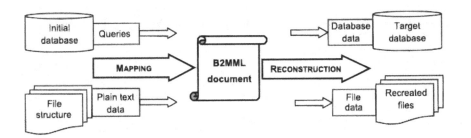

Fig. 2. Schema of the information flow between IT systems including the transformation process

convert a part of the input document to the output document fragment. These data are processed by an XSLT processor – an application that can interpret XSLT stylesheet and – basing on input – generate the output document. Execution of transformation is based on evoking a template matching a specific input element. In addition to the above description, it should be added that the document B2MML (like any XML document) is not a "flat" file. It has a tree structure, and the data stored in it are hierarchical. XSLT uses templates to the tree elements that match selected patterns, and therefore XSLT provides a set of rules describing the transformation of one XML tree to new one. The processor in the transformation process creates a new tree. Figure 3 shows a schematic representation of the XSL transformation process.

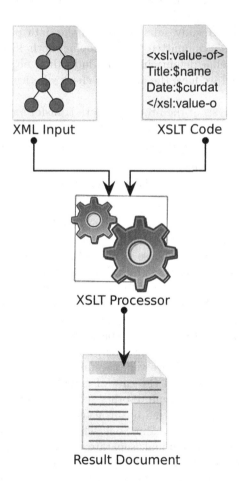

Fig. 3. Diagram of the basic elements and process flow of Extensible Stylesheet Language Transformations [8]

This concept for data transformation was used in author's previous work [7]. Although XSL seemed to be the best possible solution to the needs of converting data to and from the format B2MML, however, it causes some problems described in Sect. 4.

4   Problems with Data Exchange

4.1   Data Schemas

As a result of transformation, output package of information should be ready to transfer to the target system. However, it needs to be checked beforehand whether the structure of the data is correct. For this purpose an XML Schema Definition (XSD) can be used [15]. It comes from the same family of languages as XML, and therefore ideally suits to control the data packets stored in the B2MML format. SP-95 committee by developing both ISA-95 standard and B2MML provides schema definitions of the individual object models. Having a message transformed we have the ability to check whether the created file has the correct definition.

Basing on the analysis of the process of data transformation, content of files involved in conversion, as well as files with the schema definitions, three fundamental problems associated with data validation are defined by the author:

– multi-level nested structure,
– frequent references to external schema definitions,
– recursive construction.

Virtually every object containing a range of information consists of other objects, which have their own attributes. Subjection of the items is often presented in the form of the tree. These trees, in the case of objects developed in the B2MML language are very powerful. There are elements that have their own data structure at each level of the information in the diagram. The more elements of a data tree, the harder it is to ensure its correctness as a whole, because even a slightest mistake can bring unwanted consequences. Such multi-level, nested data structures, despite the transparency of writing, require very complex schema definitions. Cascade structure increases the risk of making a mistake in it, so complex structures require very careful schema definition preparation.

Another frequent problem in the case of complex data structures, is reference in the definition of the schema to other schema definition. For example, an object can be defined in the schema but it also contains a subordinate element, which has its own definition file. Thus, one schema definition can contain multiple elements defined in individual, separate diagrams. This fact complicates the validation of data, especially in the cascade structure, because the subordinate nodes with their own schema definitions, also have their own subordinate elements with their own schema definitions, etc. Importing them dynamically during the process of validation is merely possible.

The final problem identified by the author are recursive references contained in the schema definitions. One object may contain element, which has the same definition as its parental node. Therefore, schema definitions must comprise records that allow for the formulation of data with reference to itself. Recursive schemas can in extreme cases cause infinite nesting structures, highly undesirable in the context of performing data validation [7].

4.2   Defects of Data Processing

The biggest drawback of transformations performed by XSL stylesheets is formation of incomplete output files stripped of the part of data from the input document. Some of the data from the source file is lost during the conversion. This is primarily the result of incorrectly defined XSL transformation files, which are devoid of template definition responsible for the conversion of specific information. In the absence of such a template, information is completely ignored by the XSLT processor and overlooked when creating a new data structure in the output file.

Another issue is the common problem of lack of certain information in the source file, which is expected by the target system. This is the consequence of specificity of certain system operation, which does not provide the information needed by other co-operating systems because it does not need it to work for itself. Stylesheet alone does not create non-existing data even if they have templates prepared for their conversion. Therefore, it is impossible to supplement an output document with a missing data during the transformation process. The question arises what lacks after conversion and what are the structural differences of the data between the source and output files. To answer this, a look at the structure of documents as well as particular transaction operations carried out in the conversion process is needed. When sending messages by input system to another dedicated system, they pass double transformation process. They go to the middleware layer that translates metadata and structure of messages into B2MML document (schema conversion), and then translates data from the B2MML format to the target figure. Transformation of metadata and data into appropriate for target system requires not only a transformation of metadata but also its semantic parts, if necessary (data conversion). This situation can occur for example, when you change the format of a date, when one of the systems determines the order by *day-month-year* and another by *month-day-year*. Only after this treatment and obtaining a guarantee that a proper system will receive it, the document can be forwarded.

Issues mentioned in Sect. 4 are the cause of action related to the search for better solution.

5   Custom Class Generator

Custom class generator (CCG) is a program that generates source code based on the content recorded using a specific formal language. Generator, basing on the language description, usually in the form of grammar and syntax, creates classes from elements that occur in formal record. In this case, the generator receives

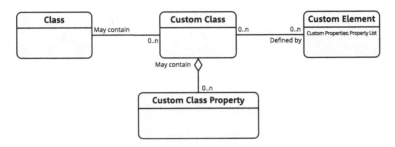

Fig. 4. Custom class pattern

an XSD file, which contains a description of the XML schema in the form of its syntax and provides the classes corresponding to all possible elements of the XML tree. Classes are created according to a certain pattern shown on Fig. 4. In the process of parsing XSD file and code generating, the name of XSD file becomes namespace. The root node in the XSD file is recognized as the base class. Each element subordinate to root node is treated as property (field of value of a simple type), or instance of custom class (if schema element is of a complex type). All custom classes may also consist of fields or objects of other custom classes. In addition, generator creates for each field of value (property) a set of accessors methods (getters and setters).

6   Differences Between XSLT and CCG

The main difference between the concepts of Extensible Stylesheet Language Transformation (XSLT) and custom class generator (CCG) concerns belonging to different programming paradigms. XSL is a declarative language, CCG allows you to operate on code written in an imperative language. XSL instead of defining how to deal with solution, that is a sequence of steps leading to a result, describes the solution itself. In other words, XSL focuses on describing the outcome and not how to reach it (no description of flow control). Source code written in imperative language describes the process of performing a sequence of instructions changing the status of the program. As imperative in linguistics expresses the demands of some activities to do, imperative programs consist of a sequence of commands to be executed by the computer. The advantage of using XSL is greater autonomy of the transformation process. Expected outcome remains independent of the program but only interpretation of the XSLT processor. XSL provides greater consistency of the conversion process, however, it has limitations on data processing. This language is dedicated exclusively to such solutions and applying it comes down to creating and altering transformation rules so that the result (output file) was in expected shape. Operating on the data only in the form of files is the weakness of XSLT. Custom class generator allows to use much more tools to shape the transformation process beyond the scope of defining the rules of transformation. It is possible to perform more complex calculations, implement sophisticated algorithms, use of specific libraries or resources (eg. by established database connection). Another advantage is popularity of imperative programming techniques. A huge base of pre-existing imperative code guarantees finding proper solution. Imperative code has an advantage because its already out there, in great volume. And as long languages like Java will be developed, we'll be writing more of it.

7 Application

7.1 Concept

An application has been developed by the author to pursue the purpose of this paper. The primary purpose and essence of the program is to create XML tree from the schema definition. Assumed functionality dealing with problems described in Sect. 4 determine the construction and operation of the program. Therefore, after the application is started, the user can load the XSD file and parse it for creation of XML file. Simultaneously, program process the XSD file in order to create new classes. Each and every autogenerated class in application refers to specific node in XML tree. This way, every piece of information can be handled separately. The value of nodes can be changed. Thanks to such solution we can also change the XML file structure by adding or removing nodes from the tree before saving it. More importantly, this way the rules of transformation can be implemented as methods operating on class instances instead of the elements in XSL file.

The user interface in this solution is negligible (Fig. 5). The approach to the problem is to reduce the role of the human. In the visual layer, program consists of text box where the XML tree is presented and few functional buttons.

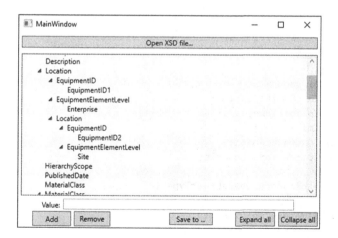

Fig. 5. Application window

The program was developed in C# with .NET Framework 4.5 involved.

7.2   Application Description

Main window class contains the methods of functional buttons. Method *cmd-LoadXml_Click()* is responsible for loading a file with .xsd extension, a method call, which will convert the XSD file to a XML file and then loading the file to the *TreeView* control. After pressing the button opens *OpenDialogBox* thanks to whom file selection can be conducted. Selected file name without the extension is passed to field *_selectedFileName.* Then, using the *XmlTextWriter* class object a new file is created, which by *XmlSampleGenerator* generate XML structure. At a later stage using the *XmlDocument* class object file is loaded into the *TreeView* control.

Method *cmdExpandAll_Click()* allows you to display all the elements (nodes) of the tree. The method *cmdCollapse_Click()* acts contrary to the method *cmd-ExpandAll_Click(),* allows you to display the main node (parent) and node one level down (children). *Button_Click* method allows you to add a node (parent) to an existing tree. Value field is contained in *textBox. Remove_Click* method is responsible for removing the selected nodes from the displayed tree structure. *Save_Click()* method allows you to save the tree structure to XML file. After pressing *SaveDialogBox* is opened. It is possible to indicate the recording folder and rename the file.

With XSD file loaded program also starts generation of classes and designing them accordingly to schema definition. In this case the library *XSD2Code* was used [16]. Listing presents fragment of generated code.

Fragment of generated code

```
//   <auto-generated>
//     Generated by Xsd2Code.
Version 3.4.0.23984 Microsoft Reciprocal License (Ms-RL)
//   [...]
//   </auto-generated>
// -------------------------------------------------------
namespace test123 {
using System;
using System.Diagnostics;
using System.Xml.Serialization;
using System.Collections;
using System.Xml.Schema;
using System.ComponentModel;
using System.Collections.Generic;

public partial class PurchaseOrderType {

    private USAddress shipToField;

    private USAddress billToField;

    private string commentField;

    private List<ItemsItem> itemsField;
```

```
private System.DateTime orderDateField;

private bool orderDateFieldSpecified;

public PurchaseOrderType() {
    this.itemsField = new List<ItemsItem>();
    this.billToField = new USAddress();
    this.shipToField = new USAddress();
}

public USAddress shipTo {
    get {
        return this.shipToField;
    }
    set {
        this.shipToField = value;
    }
}

public USAddress billTo {
    get {
        return this.billToField;
    }
    set {
        this.billToField = value;
    }
}

public string comment {
    get {
        return this.commentField;
    }
    set {
        this.commentField = value;
    }
}

[System.Xml.Serialization.XmlArrayItemAttribute("item",
    IsNullable=false)]
public List<ItemsItem> items {
    get {
        return this.itemsField;
    }
    set {
        this.itemsField = value;
    }
}
```

[...]

Problems encountered when creating a project processing XSD file for XML:

- Setting appropriate indexes of elements of a tree when adding or removing nodes. Each node in the tree has its own index (number). Updating the array of indexes that when a new node was inserted between two existing and indexed nodes, this new node had to take over the index after subsequent node. All subsequent nodes need to have previous value of index to be changed. The same problem applies to node removal.
- Adequate representation of the schema tree of the XML file. After generating XML tree from the XSD file, XML file structure posed problem. The algorithm which generates an XML file was not able to interpret some attributes that occur in the XSD file. They were treated as a comment. Comments were regarded as plain text or the values what made that file structure was not compatible with XML specification, what could prevent from its further processing.

Having above problems solved, data transformation can be based on using class generator. It allows to translate documents from one format such as XML to any other format compatible with XML syntax, including the above-mentioned B2MML. Operating on schema definitions is more accurate than on the data.

8 Conclusions

Elaborated solution responsible for converting the data into desired format can be a useful element in the exchange of information between computer systems. Standard ISA-95, along with the implementation in the form of language B2MML gives a lot of opportunities. It allows to manage data on production, scheduling, maintenance, materials, personnel, etc. It seems that contemporary systems in manufacturing company are so diverse that without middleware solutions they won't be able to communicate.

Tangible benefit of using custom class generator instead of Extensible Stylesheet Language is greater ability to shape the transformation process. XSLT processor is in fact a black box on which operation we have no influence. Code compiler allows for greater influence and control of the process.

Elaborated software application has been tested on sample XSD documents provided by SP-95 committee. The results of the program are at the present stage of its development in line with expectations. This solution is part of the theory associated with the data exchange between ERP and MES systems as a middleware element which resides in environment of those systems. It allows to transform and validate data from different backgrounds together. It can also be used as analysis tool for data mapping processes, completeness or data correctness. Each of the feature is an added value to the solution, which ultimate goal is to provide interoperability to the systems at the management level in a production enterprise.

Further work on application will be focused on development of data processing algorithms, presentation layer (e.g. Gantt charts for production schedule), use of Internet technology, in particular Web services.

References

1. Hasselbring, W.: Information system integration. Commun. ACM **43**(6), 33–38 (2000)
2. Stonebraker, M.: Integrating Islands of Information. eAI J. **1**(10), 1–5 (1999)
3. Chwajoł, G.: The Evolution of middleware used in distributed manufacturing control systems. In: III Ukrainian-Polish Conference of Young Scientists "Engineering and Computer Science", pp. 18–20 (2005)
4. Chen, D., Vernadat, F.: Standards on enterprise integration and engineering - a state of the art. Int. J. Comput. Integr. Manufact. **17**(3), 235–253 (2004)
5. Extensible Markup Language (XML), W3C Recommendation, (2015). http://www.w3.org/XML
6. Pękala, J.: Data completeness verification in the process of providing the interoperability of production enterprise systems. In: Knosala, R. (ed.) Innovations in Management and Production Engineering, pp. 214–223. Oficyna Wydawnicza Polskiego Towarzystwa Zarządzania Produkcją, Opole (2013)
7. Pękala, J.: Some aspects of interoperability of enterprise and manufacturing systems. In: Hajduk, M., Koukolova, L. (eds.) Industrial and Service Robotics. Applied Mechanics and Materials, vol. 613, pp. 368–378. Trans Tech Publications, Switzerland (2014)
8. XSL Transformations (XSLT) Version 2.0, W3C Recommendation, (2007). http://www.w3.org/TR/xslt20/
9. Kwiecień, B.: Data integration in computer distributed systems. In: Kwiecień, A., Gaj, P., Stera, P. (eds.) CN 2010. CCIS, vol. 79, pp. 183–188. Springer, Heidelberg (2010)
10. Kletti, J.: MES - Manufacturing Execution System. Springer, Heidelberg (2007)
11. van der Linden, D., Mannaert, H., Kastner, W., Vanderputten, V., Peremans, H., Verelst, J.: An OPC UA interface for an evolvable ISA88 control module. In: 2011 IEEE 16th Conference on Emerging Technologies and Factory Automation (ETFA), pp. 1–9 (2011)
12. Gould, L.: B2MML Explained. Automotive Design & Production, (2007). www.autofieldguide.com/articles/b2mml-explained
13. Brandl, D.: Business to Manufacturing Integration Technologies that Provide for Manufacturing Independence. WBF white paper
14. Scholten, B.: The Road to Integration: A Guide to Applying the ISA-95 Standards in Manufacturing. ISA, USA (2007)
15. XML Schema (XSD), W3C Recommendation, (2010). https://www.w3.org/XML/Schema
16. xsd2Code.net class generator from XSD Schema, (2013). https://xsd2code.codeplex.com/

Signal Recognition Based on Multidimensional Optimization of Distance Function in Medical Applications

Krzysztof Wójcik[1](✉), Bogdan Wziętek[2], Piotr Wziętek[2],
and Marcin Piekarczyk[3]

[1] Production Engineering Institute, Cracow University of Technology,
Al. Jana Pawla II 37, 31-864 Cracow, Poland
krzysztof.wojcik@mech.pk.edu.pl
[2] Medical Center Artroskop, Wyszyńskiego 17a, 32-503 Chrzanów, Poland
dr@wzietek.pl
[3] Institute of Computer Science, Pedagogical University of Cracow, Podchorążych 2,
30-084 Cracow, Poland
marp@up.krakow.pl

Abstract. The paper presents an idea of the method of creating the signal classifier which is based on the optimization of the metric (distance) function. The authors suggest that the proper choice of metric function parameters allows to adapt the whole classification operation to solve certain problems of the time-varied signal recognition, especially in medical applications. The main advantage of the described approach is a possibility to interpret the obtained solutions. This may enable to progress the doctor's skills, as well as improve the automatic classification method. The paper presents a brief example of the method usage in a practical application. It deals with the classification of the signals obtained from MEMS (3-axis accelerometer) sensors during the Lachman knee test. The authors point to main conditions which determine an increase in the efficiency of the described approach. Particularly, they are involved in developing efficient optimization methods of discontinuous criterion functions and algorithms for detection the cohesive group of points that define the relevant signal regions.

Keywords: Signal processing · Human-computer interaction · Pattern recognition

1 Introduction

An automatic interpretation, classification or understanding [1] of time-varying signals may be regarded as an extension of well-known methods of signal processing. Performing the operations, the automatic systems frequently utilize the knowledge of the analyzed phenomena. The knowledge is obtained during the learning process by the help of the domain expert (teacher) [2]. This idea involves

© Springer International Publishing Switzerland 2016
P. Gaj et al. (Eds.): CN 2016, CCIS 608, pp. 410–420, 2016.
DOI: 10.1007/978-3-319-39207-3_35

a cooperation between the expert and the system developer. The expert's support may vary from simple pointing to some examples of learning sequence [2] (as belonging or not to known classes), to supporting the process of method creation. In the second case the expert's knowledge about technical aspect of classification methods is needed. The main goal of the paper is to describe the proposal of such classification method construction, which makes easier this expert's support.

Obviously, for the sake of limited space, the article cannot cover many important matters of the described approach. The authors decided to omit also a regular bibliography survey as a separated section. Instead, the references to the bibliography were put into the main text (Sect. 2). Before the main idea, we will discuss a few possible methods of automatic signal classification.

2   Automatic Classification of the Signals

For simplicity's sake, we assume that we will deal with the problem of classification of discretized time-varied signals, or more generally, signals represented by discretized, real functions having only one argument.

2.1   Artificial Neural Networks

The classification of the signals is commonly solved by the usage of the artificial Neural Networks (NN) [3]. First, we should make some assumptions about architecture of the used network. Assuming that rather a short signal is analyzed, it may be represented by (at most) several dozen samples. An input layer of the network consists of nodes which receive the signals related to the exact samples. Output layer neurons should indicate the class to which the input signal belongs. The classification can be achieved by three or four-layer feed forward NN. As an activation function a sigmoidal function may be applied. In such case the back-propagation learning algorithm may be effectively used [3].

This method, like the others considered in the paper, utilizes a so-called "learning sequence" [2]. This is a string of pairs containing the signals (generally – objects) and indices of classes to which the objects should belong. The suitable choice of the indices is carried out by the expert (teacher). Thus, the learning sequence contains the knowledge about the objects.

The process of the Neural Network learning, despite of usage of sophisticated algorithms, remains very time-consuming. However, the network in the above form can solve any task of signal classification. Unfortunately, the knowledge which is hidden in the node parameters (synaptic weights), is very difficult to interpret and modify. For this reason, the role of the expert is limited to judging the elements of the learning sequence.

2.2   Pattern Recognition Methodology Based on a Distance Function

We will briefly describe an idea of many classification methods which utilize distance (or metric) function. The initial operation consists in establishing the properties of the classified objects. In our case the concept "object" refers to a set of discrete samples of signal. We may utilize many kinds of signal properties, e.g. the value and position of maximum (or minimum), the amplitude, the width of the signal peak, its energy, and so on. The selected signal properties constitute a feature vector. In the space of these vectors a metric function may be defined. The classification can be performed by computing the values of metric (distance) function between examined object (signal) and some objects which affiliations to the particular classes are known. In the simple method called Nearest Neighbour Algorithm [4], the object for which the distance value is minimal, points to the class of recognized, unknown signal. There are many variants of this approach. Frequently, some sets of the known objects are utilized to choose (or produce) special, significant objects called patterns.

We omit the discussion about the other known methods of the signal recognition, let us only note some of them:

1. Syntactic Pattern Recognition, especially using a string or graph grammars [1,5–7], in this approach the signal is represented by the list or graph of symbols which relate to some signal artifacts or properties.
2. Methods that apply conceptual clustering and classification approaches [8–10].
3. Dynamic Time Warping (DTW) [11,12].
4. Wavelet Transform.

At the end of this section let us also note the problem of evaluation of the generated classifier. One of the possible approaches consists in building the classifier on the base of only a part of learning sequence (e.g. on the base of a half of randomly selected elements). The second part can be applied to examine the classification accuracy. For instance, a number of misclassification errors may be used (the obtained performance function is discontinuous). The other measures of classification quality may utilize statistical parameters of the distance function values between objects belonging to some particular classes, or belonging to different ones.

3   Searching of the Proper Form of the Distance Function

Let us assume, that some minimal distance method is used. A vital question is how to develop the form of the distance function that allows to properly classify unknown signals. Let us start with the simple distance function defined as a sum of squares:

$$d_0(x, y) = \sum_{i=1}^{n} (x_i - y_i)^2 \tag{1}$$

where: $\mathbf{x} = (x_1, x_2, \ldots, x_n)$, $\mathbf{y} = (y_1, y_2, \ldots, y_n)$ – sample vectors of signals x and y, n – number of samples.

This simple form enables to compare two signals, however cannot be apply in real, specific cases in which some ranges of analyzed signals are more valid than others. The way out may be the usage of distance function based on generalized Euclidean metric, i.e.:

$$d_1(x, y) = \sum_{i=1}^{n} w_i(x_i - y_i)^2 \tag{2}$$

where: $\mathbf{w} = (w_1, w_2, \ldots, w_n)$ – weight vector (we omitted the operation of calculating the square root of the total sum). By the proper choice of the weight vector $\mathbf{w}$ we can control the behavior of the distance. So the task of generating the suitable classifier may be regarded as the optimization of the performance function in the multidimensional space. It seems to be properly defined, however let us point to some serious difficulties. The performance function that assesses the classifier quality is generally discontinuous (see previous section), moreover its computation is very time-consuming (multiple execution of the distance function between signals belonging to certain classes). We should be also fully aware that the number of parameters to optimize (dimension of the search space) may extend to a few hundred.

For these reasons the task of proper choice of $\mathbf{w}$ vector is extremely difficult. We must radically limit the number of parameter (length of vector) or/and propose an adequate method. The length of $\mathbf{w}$ vector corresponds to number of samples that represent discretized signal. The obvious solution is its limitation. This leads to decrease in the accuracy of the signal representation and in consequence may lead to misclassification errors. In the previous section we flag up the idea of signal representation by using the string of symbols (syntactic pattern recognition). Each symbol may relate to a range of signal that corresponds to certain signal artifacts (local extremum, "slope" of signal value, etc.). The validity of each "piece" (artifact) may be expressed by some parameter. Then, the task of finding the suitable distance function leads to finding a suitable set of the parameters. Assuming that cardinality of this set is significantly lower than the number of samples, the optimization may be much simplified. Nevertheless, there arise some new problems. They are connected with building the distance function between two strings of symbols. We may use different ways, however they commonly lead to problems of the definition of the similarity (or distance) function between two single symbols. Particularly, for each pair of symbols we may establish the concrete distance value. These values may be treated as new parameters to optimize. As we can see, this may undo the whole effect of the parameter number reduction.

Let us propose another approach. Let us assume, like before, that artifacts (they will be also called sub-objects) correspond to certain ranges of the signal domain. We also assume that each artifact relates to some weight parameter (which assesses its validity). The distance function between two signals we may calculate as follows:

$$d_2(x, y) = \sum_{k=1}^{N} w_k \sum_{i \in I^k} (x_i - y_i)^2 \tag{3}$$

where:

N – number of artifacts in signal $\mathbf{x}$,
$\mathbf{w} = (w_1, w_2, \ldots, w_N)$ – weight vector, each value refers to exact artifact,

I^k – set of indices of discrete points belonging to kth artifact, the establishing of I^k sets is carried out by the process of object detection (see details in Sect. 4), $\mathbf{x} = (x_1, x_2, \ldots, x_n)$, $\mathbf{y} = (y_1, y_2, \ldots, y_n)$ – sample vectors, like in (1).

The distance between signals is calculated according to distances between values of discrete points corresponding to artifacts. So, the only reason for their detection is establishing the set of points which relate to individual parameter value w_k. This weight value corresponds to the sub-object position instead of its type. This brings important advantages. The possible errors of artifact identification do not significantly influence the value of the distance function between signals. First, of course, there do not exist the errors of the bad identification of the artifact type (they have not types). The improper artifacts detection may be a result of unsuitable establishing the set of points used to its creation. Let us assume that a group of discrete points has been incorrectly omitted. Such a group may form another sub-object for which a proper weight parameter may be calculated.

The presented idea has just another important advantage. The value of the distance function depends directly on the values of discrete point of the considered signals, not on the values of some chosen features of the sub-objects. So, the impact of an improper choice of the kind of sub-object feature is limited. Thus, this key problem of the minimal distance methodology may be partially solved (see Sect. 2.2). The sub-objects relate to coherent (i.e. having some common properties) set of discrete point. In consequence, the impact of the inappropriate choice of objects properties may be visible in the task of sub-object detection only.

In our case the object may be constituted by points that:

- correspond to signal extrema,
- correspond to linear, quadratic, etc. growth of signal function; this behavior should be defined by additional parameters.

Regardless the possible sophisticated method of the sub-objects creation, the computation of the distance between signals is very simple (it refers to the distances between discrete points). This makes the cooperation between the domain expert and the system developer easier. Particularly, the expert (e.g. physician) may not be fully aware of the details of the object detection method, knowing that "this region" is valid, and the difference between the certain points must have been exposed. This agrees with our assumptions expressed in Sect. 1.

Let us return to main problems of described method. The discontinuity of evaluation functions of the classification process demands the usage some specific techniques of problem solving [13]. Let us notice here only a few of them.

1. Direct search method [13].
   The method is extremely time-consuming, nevertheless, in simple cases it may be used to generate and study all possible solutions. So, let us estimate the number of all solutions. We assume, that the length of the parameter vector is n. The i-th parameter may have one of p_i discrete values. Thus, the number of all possible solutions might be easily expressed by:

$$h_{\text{all}} = \prod_{i=1}^{n} p_i. \tag{4}$$

Assuming additionally that all numbers p_i roughly equal p, number of possible solutions approximately equals p^n. This shows its radical (exponential) growth with increasing in the length of the parameter vector.

2. Evolutionary algorithms (a particular parameter vector is considered as an individual) [13].
3. Monte Carlo methods.
   There may be used some variants of Monte Carlo methods that combine random searching and bundle methods [13].

4   Example of the Method Usage

The described method has been tested in the problem of automatic diagnostic of the person having a disease of the locomotor system (tearing a ligament in a knee, particularly). We will use medical time-varied signals acquired from the MEMS (Micro Electro-Mechanical Systems) sensors. These devices usually combine 3-axis accelerometer, 3-axis gyroscope as well as 3-axis magnetic sensor. The MEMS accelerometers can reach a height accuracy (in the order of 0.005–0.05 [m/s2]). This allow (by the use of discrete integration operation) to estimate the velocity, and (after the second integration) the location. The MEMS gyroscopes allow to detect a sensor rotation, in consequence, the calculation of the acceleration with respect to an inertial coordinate system (related to the ground) is possible. In our case however, we decide to use only the direct values of the acceleration.

The signals are obtained from two accelerometers. During a specific medical test of knees, so-called Lachman test [14] , the doctor try to shift (using his hands) a shank (near the knee) of the patient. In the analogous way a Pivot-shift test is executed [14]. The obtained signals of acceleration should help to identify the ligament tearing. However, its proper classification is difficult even for experienced physicians. The signals are short, and their number is very limited (the tests are painful). Nevertheless, on the other hand, this difficult problem may be used for testing the proposed methodology.

Let us briefly enumerate some details of the used method.

1. The signals (accelerations) are digitalized (sampling rate is 500 Hz) in the MEMS device and send to minicomputer via USB port.
2. The preprocessing of the signal consists in reducing the noise. We apply the low-pass gaussian filter [4].
3. The signal artifacts (sub-objects) correspond to the cohesive group of points related to linear growth of the signal function. Let us start from the point (t_0, y_0) belonging to $y(t)$ signal, i.e.: $y_0 = y(t_0)$, which is positioned next to the "end" of the previously detected object. We may found a gradient parameter a of two parallel line segments which border (over the time $t_{\min} \leq t \leq t_{\max}$) the signal $y(t)$, that is:

$$\forall_{t_{\min} \leq t \leq t_{\max}}\; a(t - t_0) + y_0 - d \leq y(t) \leq a(t - t_0) + y_0 + d \qquad (5)$$

where: $d, t_{\min}, t_{\max}$ are chosen parameters.

For digitized signal the computation of the gradient a and establishing the points that satisfy the condition (5) is very simple. These points create our sub-object.

4. The distance function between two signals is calculated according to (3), however the direct use of this metric causes an error connected with not precise signal matching. There is a necessity to shift and scale one of the signal in the time and value. In a very simplifield form we may define the modified function as:

$$d_s(x, y) = \min_{s_t \in S_t, d_t \in D_t, s_y \in S_y, d_y \in D_y} d_2\big(x, \mathrm{scale}(y)\big) \qquad (6)$$

where: $\mathrm{scale}(y)$ denotes an operation of scaling and shifting of y signal according to parameters: s_t, d_t, s_y, d_y which belong to finite sets: S_t, D_t, S_y, D_y respectively[1]. We will not discuss here the used, uncomplicated optimization process leading to the calculation of the (6) statement.

5. The classification method relates to checking the affiliation of the examined signal to one class only (the patient has, or has not the ligament tearing). This is done by the use of single pattern signal, which corresponds to typical signal "shape". This pattern is selected by the expert (doctor).

6. The classifier optimization refers to search of the $\mathbf{w}$ weight vector of the distance function that minimize the performance function of the classifier. The computation of the function is performed in two phases. In the first one we take into consideration the number of misclassification errors. We examine the distance to only one pattern, thus, in order to conduct the classification some threshold value is needed. It is an additional parameter to optimize. The number of misclassification errors (for the sake of very limited count of training sequence – 12 in our case) must be zero. If not, the function return a constant big value. If the error count is zero, the function value is defined as a ratio: $\frac{m_1}{m_0}$, where: m_1 is the average value of distance function between pattern object and all objects from learning sequence which relate to person having the ligament tearing (according the doctor decision), m_0 – like above for healthy person.

7. Because of the discontinuity of used evaluation functions the direct search method which generate all solutions has been used.

Figure 1 illustrates two signals obtained during the Lachman test (in Z-axis which is parallel to the movement direction).

The black colored signal corresponds to the pattern signal (ligament tearing), the blue colored relates to some ill patient. There were also shown the time periods corresponded to detected artifacts (regions of linear growing or decreasing the signal). They are symbolized by the small segments.

---

[1] More precisely, time-varied signal $y(t)$ is scaled and shifted according to following rule: $\mathrm{scale}(y(t)) = s_y\, y(s_t t + d_t) + d_y$.

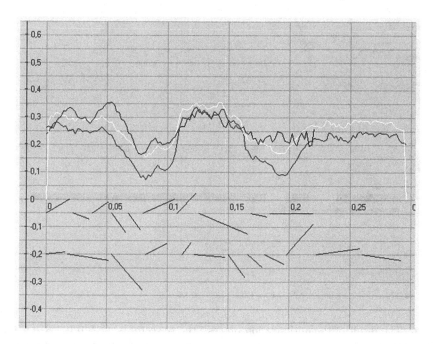

Fig. 1. Pattern signal and a selected signal corresponded to ill patient. The horizontal axis is a timeline (time is expressed in seconds), the vertical axis refers to the acceleration (in $[\mathbf{m/s^2}]$). (Color figure online)

Figure 2 depicts the result of optimization process. The values of the found weights are listed on the bottom of the picture. They are also illustrated by the color bars which are put near the segments symbolizing the artifact periods. The biggest values of weight vector indicate the regions of the signal that are significant for proper classification. Its interpretation is very clear and may be utilized by the doctor to detect the valid artifacts. This improves the method (choose of the significant pattern) as well as it allow to training the doctor skills. There is also possible an iteration route of the above processes.

Figure 3 shows the result of another run of the optimization process. For the test purpose one of the element of training sequence was incorrectly classified as relating to the ill person (before it had represented the healthy patient). There is also the possibility to produce the right operating classifier (having however a worse performance value – compare Figs. 2 and 3). The similar experiments show a big potential of the whole described method to generate "any" correct solution. However, at that point of method testing (there are not available a statistically significant number of training examples), we cannot judge, whether the good results corresponds to a random match only.

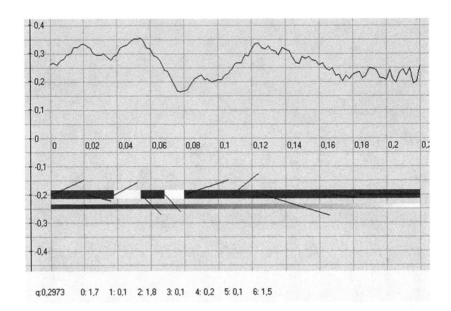

Fig. 2. The result of the optimization process; the **w** vector is: (1.7, 0.1, 1.8, 0.1, 0.2, 0.1), $n = 6$, first two segments are omitted; the last parameter (equals 1.5) represents the threshold value of distance function; the value of the classifier performance function is 0.2973 (left, bottom corner).

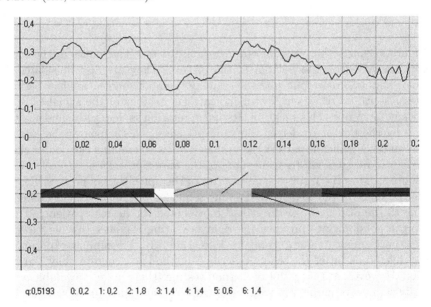

Fig. 3. The result of another optimization process; the **w** vector is: (0.2, 0.2, 1.8, 1.4, 1.4, 0.6); threshold value is 1.4; the value of the performance function is 0.5193, it is significantly less (worse) than in the previous case.

5   Conclusions

The article presents a proposition of the scheme of creating the classifier of the time signals which tries to overcome the problem of proper interaction between domain experts and the computer system. The proposed methodology consists in the optimization of the parameters of the distance function. We may draw the conclusion that the whole presented approach has, to some degree, properties halfway between the features of the "classical" pattern recognition methods and methods based on direct processing of the sampled signal. From the first methods it borrows the stages of sub-object detection and the object feature calculation. The usage of the very simple calculation method of the distance between discrete points relates to second kind of methods. This simplifies the interaction between the man (expert) and computer recognition system and can be regarded as a main advantage of the presented approach.

However, the main difficulty is the exponential growth of a search space with the number of samples of the examined signals. This limits the signal recognition to relatively short sample only.

In this connection we can point to the conditions of gainful usage of the proposed approach:

- Developing methods for reducing the length of the weight vector.
- Proposing the algorithms for detect the cohesive group of points that minimize the group cardinality.
- Developing efficient, adequate methods of optimization of the discontinuous criterion function (e.g. genetic algorithms).

We may consider the solution of listed above questions as the main challenge of a successful implementation of the proposed method.

References

1. Tadeusiewicz, R.: Automatic understanding of signals. In: Klopotek, A., Wierzchoń, S., Trojanowski, K. (eds.) Intelligent Information Processing and Web Mining. Advances in Soft Computing, vol. 25, pp. 577–590. Springer, Heidelberg (2004)
2. Mitchell, T.M.: Machine Learning. McGraw-Hill Science, New York (1997)
3. Bishop, C.M.: Neural Networks for Pattern Recognition. Oxford University Press, New York (1995)
4. Duda, R.O., Hart, P.E., Stork, D.G.: Pattern Classification. John Wiley and Sons, New York (2000)
5. Piekarczyk, M., Ogiela, M.R.: Hierarchical graph-grammar model for secure and efficient handwritten signatures classification. J. Univ. Comput. Sci. **17**(6), 926–943 (2011)
6. Tadeusiewicz, R., Ogiela, M.R.: Medical Image Understanding Technology. Studies in Fuzziness and Soft Computing, vol. 156. Springer, Heidelberg (2004)
7. Piekarczyk, M., Ogiela, M.R.: Matrix-based hierarchical graph matching in off-line handwritten signatures recognition. In: Proceedings of 2nd IAPR Asian Conference on Pattern Recognition, IEEE (2013)

8. Michalski, R.S., Steep, R.: Learning from observation: conceptual clustering. In: Michalski, R.S., Carbonell, J.G., Mitchell, T.M. (eds.) Machine Learning: An Artificial Intelligence Approach, vol. 2. Morgan Kaufmann, San Mateo (1986)

9. Wójcik, K.: Hierarchical knowledge structure applied to image analyzing system - possibilities of practical usage. In: Tjoa, A.M., Quirchmayr, G., You, I., Xu, L. (eds.) Availability, Reliability and Security for Business, Enterprise and Health Information Systems. LNCS, pp. 149–163. Springer, Heidelberg (2011)

10. Wójcik, K.: Knowledge transformations applied in image classification task. In: Choraś, R.S. (ed.) Image Processing and Communications Challenges 5. Advanced in Intelligent Systems and Computing, vol. 233, pp. 115–123. Springer, Heidelberg (2013)

11. Adistambha, K., Ritz, C.H., Burnett, I.S.: Motion classification using dynamic time warping. In: International Workshop on Multimedia Signal Processing, Wollongong (2008)

12. Wójcik, K., Piekarczyk, M., Golec, J., Szczygiel, E.: Simple system for supporting learning of human motion capabilities. Appl. Mech. Mater. **555**, 673–680 (2014)

13. Russell, S., Norvig, P.: Artificial Intelligence: A Modern Approach, 3rd edn. Prentice Hall, Englewood Cliffs (2010)

14. Labbé, D.R., Li, D., Grimard, G., de Guise, J.A., Hagemester, N.: Quantitative pivot shift assessment using combined inertial and magnetic sensing. Knee Surg. Sports Traumatol. Arthroscopy **23**(8), 2330–2338 (2015). Springer-Verlag, Berlin, Heidelberg

Determining the Popularity of Design Patterns Used by Programmers Based on the Analysis of Questions and Answers on Stackoverflow.com Social Network

Daniel Czyczyn-Egird$^{(\boxtimes)}$ and Rafal Wojszczyk

Koszalin University of Technology, Sniadeckich 2, 75-453 Koszalin, Poland
daniel.czyczyn-egird@cicomputer.pl, rafal.wojszczyk@tu.koszalin.pl

Abstract. User-generated content in social networks constitutes tremendous stores of knowledge to be analysed. The article presented results of research on the popularity of design patterns on the basis of data gathered in the specialised social networks. The conducted analyses concerned i.a. general popularity of questions about design patterns and indicating a group of patterns which cause possible problems during implementation. The research results were obtained thanks to using data mining techniques.

Keywords: Social networks · Data mining · Design patterns

1 Introduction

The communication via the internet has significantly changed since few last years. Currently e-mails, static web sites, are often substituted by more and more popular direction of development of using the Internet is the so-called Web 2.0 [1]. The popular social services are based on Web 2.0 assumptions and they influence the speed and range of distributed information considerably. Simultaneously, by engaging the basic assumptions of semantic networks (through tagging information with metadata), one can obtain combination that offers more than access to information – the knowledge resulting from data mining.

The modern information technologies would not exist if it weren't for programming craft. Programming is strictly connected with the use of different patterns, after [2]:

- Architectural patterns are high-level strategies that are concerned with large-scale components and global properties and mechanisms of a system.
- Design patterns are medium-level strategies that are concerned with the structure and behavior of entities, and their relationships.
- Idioms are paradigm specific and language-specific programming techniques that fill in low-level internal or external details of a component's structure or behavior.

P. Gaj et al. (Eds.): CN 2016, CCIS 608, pp. 421–433, 2016.
DOI: 10.1007/978-3-319-39207-3_36

The design patterns presented in [3] were created on the basis of the experience of the community which was then interested in the issues of object-oriented programming and good practices. The main purpose of the patterns is solving problems of designing and object-oriented designing. Design patterns are also considered as a certain type of a language of communication. Thanks to this language, the code of software which includes the patterns will be easier to get to know and understand than the one without the patterns. Therefore, people who know the patterns, spontaneously create the community which communicates using the language of patterns [4].

The objective of the article is to analyse popularity and trends of changes of using design patterns on the basis of specialised social networks. The research results will be used to indicate the direction of further research on the quality of design patterns implementation, while the methodology will contribute to the selection of the direction of further analyses in the domain of data mining.

The second section of the article presented a review of the selected issues connected with using community networks to support the work of software developers and describe information on design patterns. The third section included the description of the research environment and the grounds for the choices made. The fourth section introduced research results and relevant conclusions. The last, fifth section constitutes the summary.

2   Review of Community Network Solutions

2.1   General Support for Developers

Nowadays, developers are constantly being provided with new platforms with product and technical documentation which are characterised with an extensive base of information useful during developing new IT solutions. Many companies providing own information solutions and products knowingly and intentionally try to make well-documented libraries and programming interfaces accessible to all the interested. This kind of approach is in a way advantageous; more specifically, it provides relevant promotion of products and their wide distribution in the market, for instance, through providing their users with the possibility of creating personalised extras or new functions of the already existent computer systems.

In the 21st century, the amount of paper documentation has dramatically decreased to the benefit of online documentation, which, thanks to a common access to the global Internet network, is world-wide and open round-the-clock.

Developers equipped with the access to the global Internet network are able to find technical descriptions, construction calculations, plans, figures, schedules and cost calculations easily. Therefore, new solutions can, to a great extent, be created without any contact between a developer and products' owners.

An additional element supporting the work of developers are social networks, which give a possibility of analysing technical documentation together and offering new solutions. Ready intermediate products, packages and libraries are made accessible in network repositories which connect developers by exchanging confidential fragments of software. One of the specific tools, which is worth mentioning in the context of working with repositories of that type, is NuGet[1]. It is a so-called package manager or a package management system for the Microsoft development platform, including .NET platform. NuGet client tools provide an easy use of packages or a possibility of creating one's own, and the whole thing is easily integrated with Visual Studio development environment. The content of packages included in NuGet repository is constantly being updated by the authors of packages gathered in a social network. Similar solutions available on the market are, for instance: GetIt for development tools from Delphi family, or Maven for Java.

Such platforms as CodePlex[2], Github or SourceForge, which are gathering information on IT projects, can also be called social networks. These give access to free hosting for open source software and paid private repositories. This type of portals is also used by developers to exchange information and source codes, in this way, contributing to development of software products, and especially those licensed openly and freely.

The network community has also contributed to development and popularisation of file repositories available online as version control systems. Those tools are, for example, Git and SVN, which are software products tracking changes mostly in a source code, and assisting software developers with combining changes made in files by many developers in different moments. The version control systems can be divided into:

- local ones, enabling one to save data only to a local hard drive (e.g. SCCS and RCS),
- centralised ones, characterised with a client-server architecture (e.g. CVS, SVN),
- distributed ones, based on a peer-to-peer architecture (e.g. BitKeeper, Git).

The first type of the system only saves file versions to a local hard drive, and it is a very convenient, still, not very save solution (with regard to data loss prevention) which blocks a possibility of sharing one's codes with other developers. In case of centralised solutions, there is one central repository used by all the system users to synchronise their changes. The distributed solutions give the possibility of simultaneous maintenance of independent but equal branches which can be freely synchronised, e.g. through e-mail. The type of the used version control system depends on the assumptions of a project and should be adjusted to developers' needs.

A great way of using social networks are also Internet forums joining together users whose interests corresponded with topics of these portals. For instance, the biggest programming forum in Poland – 4programmers.net – attracts software developers, administrators, webmasters, in other words, people connected with

---

the IT branch. Thousands of topics and comments prove the power of these portals where users can post new topics and the interested can add new comments concerning a particular issue. Over time, the demand for even more technically specialised portals appeared, where sharing specialist knowledge is based on Q&A pattern, that is "question-answer". One of such portals is Polish service devpytania.pl which gives a possibility of asking a question or answering a question for free. Currently there are approximately 4000 questions. However, it is way better to take advantage of a worldwide portal as, for instance, stackoverflow.com service, where the question base is really impressive and it reaches nearly 11 million, and the answers to the questions are given by the specialists from all over the world. This kind of services offer assistance in solving problems that developers encounter each day; therefore, these services' growing popularity should not be surprising. This publication is also partly focused on analysing and mining of data from the above-mentioned portal and, more specifically, on investigating popularity of design patterns in relation to knowledge collected in the database of the service which was established in 2008. Since that time, the service has developed a huge social network which contributes to the portal's constant development – as of today, there are approximately 4.8 million users asking and answering questions.

2.2 Stackoverflow.com Service

The stackoverflow.com service is one of the biggest Q&A portals in the scope of broadly taken subjects connected with software engineering. It is the platform enabling its users to ask questions and find answers. The portal community has got a possibility of assessing asked questions and given answers "up" or "down". It is a type of natural selection where "bad" solutions are rejected and "good" ones are distinguished. The users vote according to their knowledge and preferences, therefore, one should bear in mind that their choices may not be objective. However, the voice of the community is a very good indicator of popularity, and this was used in further research. The users are gratified for their actions by receiving points and batches which create internal and, what's more important, positive relation between them. All the content generated by the users is based on Creative Commons license.

The selection of stackoverflow.com as a source of information is justified by the popularity measured by Google Trends and its use in other research studies [5]. Figure 1 shows the chart of stackoverflow.com popularity in relation to its direct competitor – experts-exchange.com.

2.3 Information on Design Patterns

The basic literature concerning design patterns is the one mentioned above [3], where have been collected and described the most usable patterns [6]. The most popular catalog of patterns [7] is literature thus other authors often base their work on the catalogue offering other variants or implementations adjusted to

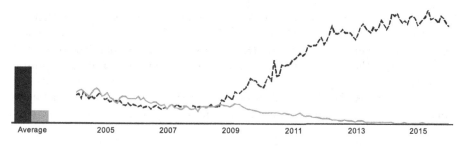

Fig. 1. The chart of popularity of the competing services; the doted line stands for the popularity of stackoverflow.com queries, the solid one means experts-exchange.com queries; status as of 6 January 2016

a specific language, e.g. [6,8]. The form of description of design patterns introduced in the above-mentioned literature is, by definition, intended for learning to use them, and it includes: verbal description in a natural language, class diagrams, and an exemplary implementation code based on simple examples. A similar objective is pursued by the repository of knowledge of design patterns introduced in [9]. The aim of this repository is disseminating knowledge of the patterns, learning and supporting their implementation. Thanks to the author's Internet application from [9], a dialogue between a user and the system is possible in a form of question-answer. Knowledge of the patterns and questions are predefined with a possibility of adding new resources.

Some of the formal ways of representation of design patterns are based on semantic networks; more specifically, they use ontologies [10–12]. In philosophy, ontology describes the nature of beings of the real world, while in computer science, it is used for formal representation of knowledge in a form of a set of notions of a particular field and relations between those notions. A very important advantage of ontology is a possibility of transforming data from and into different sources. In [11,12], one can find ontologies describing basic information on design patterns. This knowledge can be transformed into other approaches, for instance, the mentioned above [9] or [10] used to search for instances of design patterns in a source code.

3   Preparation for the Research

3.1   Design Patterns as the Research Subject

Design patterns should be treated as universal and verified in practice solutions of frequent design problems. They show the occurring relationships and class and object dependences; they also facilitate creation, modification and maintenance of source codes of IT systems. Using the patterns in small projects usually contributes to increasing labour intensity of a production process. On the other hand, in more extensive projects, it improves the general quality of produced systems and facilitates maintenance. Taking the advantages of using design patterns into consideration [13], one should assume that the popularity of the patterns should be increasing constantly. The subsequent part of the article presented the results of research on the popularity of design patterns promoted by the so-called the Gang of Four (Erich Gamma, Richard Helm, Ralph Johnson and John Vlissides) [3]. The research study was carried out through analysing the topics gathered from stakcoverflow.com, on the basis of keyword instances connected with design patterns, using data mining techniques.

3.2   The Course of Research

The research was conducted for four separate data sets which were created by means of engineering techniques for acquiring information. Subsequently, the set were adequately processed in order to obtain specific answers and represent them by means of specific measures. In order to perform data mining out of stackover-flow.com, StackExchange API programming interface in its newest 2.2[3] was used. By means of relevant interface methods, it was possible to create parametrized server queries and, subsequently, obtain specific answers meeting the preset criteria. All the feedback were obtained in JSON format [14], processed adequately by means of the author's software and placed in a database as records; Fig. 2 presents a schematic diagram of the research environment. The fact that API considerably facilitates the access to the information base in the portal is worth mentioning here. Unfortunately, it also means certain limitations for developers. The service uses IP address validation, that is, within 24 h, one can only call approximately 300 queries which can only return 100 back objects at one time. In case of willingness to obtain a set of questions concerning a selected keyword with a single query, one can only obtain up to 100 results. Preparing a query itself means appropriate parameter and filter management, where, for instance, one can filter data according to creation date, phrases in titles of questions or as keywords, as well as, sorting according to a number of answers or popularity. The most serious inconvenience concerning the current API version is connected with the lack of possibility of searching through the service database according to a queried phrase with regard to its occurrence in both question title and content. The problem is quite significant since, as initial identification study showed, the users not always give a specific phrase in the title of their messages or as a tag – a looked up phrase is only occurring in the question content which is currently unavailable from API level. The solution to this problem is manual use of the service search engine which takes question content into consideration. Unfortunately, this solution is unacceptable from the perspective of the research objective, therefore, only one analysis was conducted in this way. In other cases, the search was conducted only according to the occurrence of a selected phrase in a question title since it turned out that searching for keywords (tags) was ineffective – it ignored too many questions.

All the obtained results were processed and placed in relational data-base PostgreSQL 9.3[4], which enables easy data mining and classification.

---

[3] https://api.stackexchange.com/docs/search.

[4] http://www.postgresql.org/.

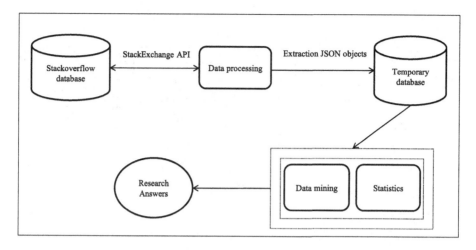

Fig. 2. The research environment scheme

Figure 3 presents a data model which was designed in a way that it provides an easy access to data and trouble-free data comprehension. The entities correspond with the following objects: questions, questions' authors and keywords, with a relevant field describing certain dependences.

4   Research Results

4.1   General Questions Popularity of Design Patterns

The first data set was fed by manual querying of the search engine of the service, using in this aim 23 names of desined patterns as key words. The obtained result, shown in a form of a histogram, is presented in Fig. 4. When analysing the obtained results, one can notice a dominant group of four patterns: Command, State, Singleton and Factory. The analysis of the general popularity does not provide additional information which would stem from the context of the questions asked on stackoverflow.com. High popularity of questions means that the software developers attempt to implement particular design patterns; however, they encounter obstacles or difficulties in understanding hereby patterns. Low popularity shows less considerable need of using certain patterns or easy implementation. Still, this is not confirmed by high popularity of Singleton pattern, which is considered as one of the easiest. In order to provide more specific conclusions, further research was carried out.

4.2   Relative Popularity of Design Patterns

Design patterns are the so-called good practices which are often created on the basis of the experience of a community. Therefore, comparison of current

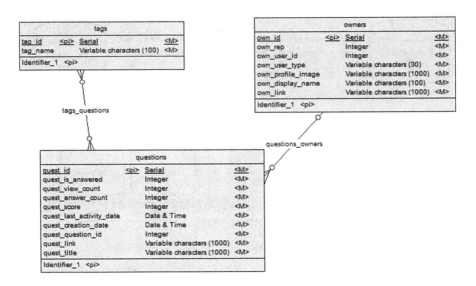

Fig. 3. Entity-relation model for the database for objects from stackoverflow.com

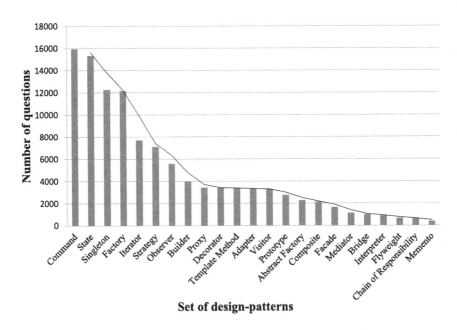

Fig. 4. Histogram showing a number of questions for the particular design patterns [3] on stackoverflow.com, together with a moving average trendline, status as of December 2015

popularity of design patterns [3] in relation to similar good practices, in this case, architectural patterns [9] as MVC, MVP and MVVM, is justified.

The second data set was composed of 13 000 questions which appeared in December 2015 and acquired from the service by means of API. The database was fed exclusive of duplicates which resulted in reducing the set to 12 095 unique questions. Figure 5 presents the obtained results. The obtained result shows considerably higher popularity in relation to similar good practices. The reason could be the fact that design patterns require understanding and contribution from a developer with every implementation. The popular programming environments do not offer an automatic implementation of patterns. The opposite situation is found in case of architectural patterns where automated tools integrate with programming environments and it often goes unnoticed to a developer. Therefore, the lack of tools automating implementation of the analysed patterns increases interest in this subject.

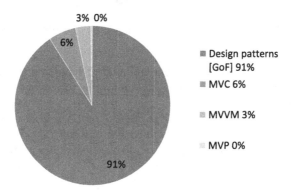

Fig. 5. Pie chart showing the distribution of questions for design [3] and architectural patterns

4.3   Popularity of Design Patterns in a Time Function

The aim of the last analysis based on popularity measures was showing trends of changes for the selected patterns in the preset time function.

The data set prepared for this analysis was composed of the questions acquired for 3 design patterns which were most popular with regard to a number of questions in the service, that is, for: Command, State, Singleton. There were questions acquired for each of the patterns from the beginning of 2011 to the end of 2015. Therefore, an initial number was nearly 25 000 questions; still, after exclusion of duplicates there were 21 640 questions in the database. The search was again conducted according to the criterion of the selected phrase's presence in the question title. The obtained results are presented in Fig. 6. The analysis of the obtained results enabled one to draw the following conclusions:

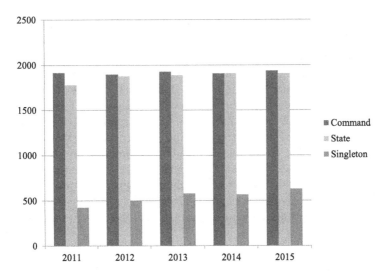

Fig. 6. Popularity of the selected patterns in a time function

- the patterns selected for the research maintain constant popularity trend, slight fluctuations result from different random factors which are numerous in active communities,
- popularity proportion between the particular patterns is also constant which was confirmed by the result of analysis concerning general popularity measure,
- it can be predicted that the popularity of the selected design patterns will not decrease in 2016.

4.4   Discussion Measure of Design Patterns

The aim of the last analysis was determining discussion measure in the implementation of design patterns on the basis of a number of visits, number of answers to questions, and the ratio of a summary number of question to the answers. The discussion measure is an indirect determinant of a number of problems occurring in the implementation of the patterns. When there are difficulties in using a particular design pattern, there are different implementation variants discussed and there are more diverse opinions. The last data set was composed of questions acquired for all 23 design patterns promoted by [3]. For each pattern, there were 100 newest questions acquired; the initial total number was 2300 questions; however, after excluding duplicates, there were 2175 questions in the database. All data was presented in Table 1. The analysis of the obtained calculations enabled one to propose the following theses:

- the patterns which questions are characterised with high percentage of answers ($>75\%$), can be defined as patterns relatively easily understandable and implementable. One can notice a certain correlation between understanding the complexity of question and a level of complexity of a particular pattern,

Table 1. Research results for the discussion measure

Name of design pattern	Number of questions with answers [%]	Number of questions with no answer [%]	Number of questions to a number of answers ratio	Classification
Abstract factory	69.44	30.56	0.67	1
Builder	43.14	56.86	1.11	2
Factory	65.57	34.43	0.71	1
Prototype	75.61	24.39	0.61	1
Singleton	69.81	30.19	0.76	1
Adapter	33.82	66.18	1.1	3
Bridge	30.36	69.64	1.56	5
Composite	46.38	53.62	1.17	3
Decorator	73.81	26.19	0.72	1
Façade	56.25	43.75	1.09	2
Flyweight	81.82	18.18	0.48	1
Proxy	14.29	85.71	2.59	5
Chain of responsibility	88.24	11.76	0.74	1
Command	40.00	60.00	1.36	4
Interpreter	51.22	48.78	1.08	3
Iterator	64.86	35.14	0.81	2
Mediator	48.21	51.79	1.1	3
Memento	75.00	25.00	0.57	1
Observer	38.98	61.02	1.05	3
State	45.00	55.00	1.02	3
Strategy	28.79	71.21	1.89	5
Template method	84.51	15.49	0.71	1
Visitor	46.05	53.95	0.94	3
Summary	54.00	46.00	1.04	2

- the patterns for which percentage for questions with no answers is high (>75 %), can be treated as difficult and complicated patterns; questions regarding more difficult patterns can be more difficult to understand. A number of users who can give useful answers is commensurately lower,
- the patterns characterised with high answers to asked questions ratio (>1.0) can be described as the ones encouraging to discussion on solutions to a particular problem or different ways of implementation of a particular problem.

Aiming at indicating the patterns which can possibly show problems in the implementation, the classification expressed by the following formula was proposed:

classification = % of quest. with no ans. × a num. of quest. to ans. ratio.   (1)

The high value of both coefficients in Formula (1) will stigmatise the patterns which obtained unfavourable results, while the opposite situation, that is, low values of both coefficients, emphasizes a problematic character of patterns. Classification was divided into 5 levels, evenly distributed in relation to the result (1):

1. 0–25,
2. 26–50,
3. 51–75,
4. 76–100,
5. 101 and more.

The lower the result of classification, the implementation of the pattern is potentially: more intuitive, unequivocal, easily understandable in relation to assumptions and intention of use. The design patterns classified as 3 or lower are characterised with the opposite properties, namely, the implementation is: hardly intuitive, there are many variants, extensive assumptions and intention of use is difficult to understand. Therefore, the patterns classified as 3, 4 and 5 can be called in into question in the community of software developers; especially, when a beginner makes an attempt of implementation.

5 Summary

The article presented the results of the analysis of the popularity of design patterns. The results were obtained thanks to data mining techniques supported by the author's software. The research data was acquired from stackoverflow.com, which is one of the examples of social networks. An important advantage of using this type of networks is the fact that the community of software developers is obliged to verify the content occurring in the service.

The conducted analyses concerned: general popularity of design patterns, popularity of patterns in relation to similar good practices, an attempt of predicting the popularity in 2016 on the basis of previous years and indicating a group of patterns which cause possible problems during implementation. The obtained results showed that the most popular design patterns are: Command, State, Singleton and Factory. The mentioned patterns were used as examples to show the constant popularity over the 5 last years, which enables one to predict that the popularity will not change. What is more, it was shown that in relation to popularity of architectural patterns, design patterns [3] are more popular. Finally, a group of patterns, which can be called in into question by the software developers, and which is connected with potential obstacles in implementation, was distinguished.

Further research on data mining techniques will be connected to the development of the method and use of Business Intelligence tools. Subsequently, a change of research subject into other issues of software engineering is projected. The results obtained in the above-mentioned analyses will contribute to the selection of the direction of further works in the scope of quality assessment of implementation of design patterns.

References

1. Heath, T., Motta, E.: Ease of interaction plus ease of integration: combining Web2.0 and the semantic web in a reviewing site. Web Semant. Sci. Serv. Agents World Wide Web **6**(1), 76–83 (2008). Elsevier, Amsterdam
2. Tesanovic, A.: What is a pattern?. Linkoping University, Department of Computer and Information Science, Linkoping (2008)
3. Gamma, E., et al.: Design Patterns: Elements of Reusable Object-Oriented Software. Addison-Wesley Professional, Boston (1994)
4. Alexander, C.: A Pattern Language: Towns, Buildings, Construction. Oxford University Press, Oxford (1977)
5. Wang, S., Lo, D., Jiang, L.: An Empirical Study on Developer Interactions in Stack-Overflow. Research Collection School Of Information Systems, Singapore (2013)
6. Metsker, S.K.: Design Patterns in C#. Addison-Wesley Professional, Boston (2004)
7. Tsantalis, N., et al.: Design pattern detection using similarity scoring. IEEE Trans. Softw. Eng. **32**(11), 896–908 (2006)
8. Bruegge, B., Dutoit, A.: Object-Oriented Software Engineering Using UML, Patterns, and Java, 3rd edn. Pearson Education, New York (2009)
9. Pavlic, L., et al.: Improving design pattern adoption with ontology-based design pattern repository. Informatica Int. J. Comput. Inform. **33**, 189–197 (2009). Ljubljana, Slovenia
10. Kirasić, D., Basch, D.: Ontology-based design pattern recognition. In: Lovrek, I., Howlett, R.J., Jain, L.C. (eds.) KES 2008, Part I. LNCS (LNAI), vol. 5177, pp. 384–393. Springer, Heidelberg (2008)
11. Alnusair, A., et al.: Rule-based detection of design patterns in program code. Int. J. Softw. Tools Technol. Transfer **16**, 315–334 (2013). Springer-Verlag, Berlin, Heidelberg
12. Dietrich, J., Elgar, C.: A formal description of design patterns using OWL. In: Software Engineering Conference, Australian (2005)
13. Wojszczyk, R.: The model and function of quality assessment of implementation of design patterns. Appl. Comput. Sci. 11(3), (2015). Institute of Technological Systems of Information, Lublin University of Technology, Lublin
14. Kasprowski, P.: Choosing a persistent storage for data mining task. Studia Informatica **33**(2B), 509–520 (2012). Wydawnictwo Politechniki Śląskiej, Gliwice

Author Index

Printed in the United States
By Bookmasters